# Mechanical Engineering Series

**Series Editor**

Vish Prasad, Department of Mechanical and Energy Engineering, University of North Texas, Denton, USA

The Mechanical Engineering Series presents advanced level treatment of topics on the cutting edge of mechanical engineering. Designed for use by students, researchers and practicing engineers, the series presents modern developments in mechanical engineering and its innovative applications in applied mechanics, bioengineering, dynamic systems and control, energy, energy conversion and energy systems, fluid mechanics and fluid machinery, heat and mass transfer, manufacturing science and technology, mechanical design, mechanics of materials, micro- and nano-science technology, thermal physics, tribology, and vibration and acoustics. The series features graduate-level texts, professional books, and research monographs in key engineering science concentrations.

Oveepsa Chakraborty · Biplab Das

# Parabolic Solar Trough Collector

## Design, Development and Usage

Springer

Oveepsa Chakraborty
Department Mechanical Engineering
Girijananda Chowdhury University
Guwahati, Assam, India

Biplab Das
Department of Mechanical Engineering
National Institute of Technology Silchar
Silchar, Assam, India

ISSN 0941-5122 ISSN 2192-063X (electronic)
Mechanical Engineering Series
ISBN 978-3-032-11228-6 ISBN 978-3-032-11229-3 (eBook)
https://doi.org/10.1007/978-3-032-11229-3



This Springer imprint is published by the registered company Springer Nature Switzerland AG
The registered company address is: Gewerbestrasse 11, 6330 Cham, Switzerland

# Preface

As the world shifts toward cleaner and more sustainable sources of energy, solar power stands out as one of the most promising solutions. It is abundant, renewable, and environmentally friendly. Among various solar technologies, parabolic solar trough collectors (PTCs) offer a reliable method to convert solar radiation into thermal energy, which can be used for electricity generation, industrial processes, and water heating. PTCs focus sunlight onto a central receiver tube using a parabolic-shaped reflector. A heat transfer fluid flows through this tube, absorbing concentrated solar energy, which is then used for practical applications. This simple yet powerful mechanism plays a critical role in the global transition away from fossil fuels.

Despite their effectiveness, PTC systems face technical challenges that limit their performance. One key issue is the non-uniform distribution of solar heat on the surface of the receiver tube. This uneven heating can cause thermal stress, reduce energy efficiency, and lead to long-term material degradation. Existing literature often underrepresents this issue, creating a gap in research and innovation. This book aims to fill that gap by exploring solutions to improve the thermal performance and reliability of PTCs through advanced design, simulation, and experimentation.

The book begins with foundational knowledge on solar energy, including the types of solar radiation, solar tracking, and comparisons with fossil fuel systems. It then introduces various types of concentrating solar collectors before focusing on the structure, operation, and evolution of PTC technology. Readers gain a complete understanding of how these systems function and the parameters that influence their thermal and optical performance.

A major portion of the book investigates receiver tube geometry and the use of internal inserts to enhance heat transfer. Traditional cylindrical tubes are compared with elliptical and trapezoidal shapes, which can increase energy absorption and improve fluid dynamics. The book also examines turbulence-inducing inserts such as twisted tapes and spinning devices that enhance convective heat transfer. These methods aim to maximize energy capture without significantly increasing pressure losses.

The use of secondary reflectors is explored as a solution to the problem of non-uniform solar flux. These additional reflectors help redirect sunlight to less-illuminated areas of the receiver tube, creating a more uniform temperature distribution and boosting efficiency. Another key topic is the application of nanofluids, where nanoparticles are added to conventional heat transfer fluids to improve their thermal conductivity. The book analyzes mono, binary, and ternary hybrid nanofluids, as well as the influence of different base fluids on thermal performance.

This book is intended for researchers, students, and professionals working in solar energy and thermal system design. It offers detailed insights into design innovations and performance enhancement techniques backed by experimental and computational studies. By addressing current challenges and proposing practical solutions, this book contributes to the ongoing development of efficient, clean, and sustainable solar thermal technologies.

Guwahati, India — Oveepsa Chakraborty
Silchar, India — Biplab Das

# Contents

# Chapter 1
# Introduction to Solar Energy Sources

This foundational chapter introduces the reader to the core concept of solar energy, positioning it as a sustainable and clean alternative to fossil fuels in the fight against climate change and energy scarcity. It explores the Sun as an energy source by explaining key concepts such as the solar constant, types of solar radiation, and solar time. Techniques for solar ray tracking and surface orientation are discussed in detail, laying the groundwork for technical understanding in later chapters. The chapter also reviews the technologies used to convert solar radiation into usable energy, alongside their advantages and limitations. It compares solar energy to other renewables, emphasizing its potential in reshaping the global energy landscape. As solar power becomes increasingly viable through technological advances and carbon reduction efforts, this chapter sets the stage for a deeper dive into the practical and innovative aspects of solar energy utilization.

## 1.1 The Sun's Energy

As a G-type main-sequence star, the Sun is classified within this category (surface temperatures between 5300 and 6000 Kelvin, characterized by their yellowish color and moderate mass), a stable-phase star, often referred to as a yellow dwarf, and it lies at the core of the solar system's composition. It is the primary energy source for life on Earth, influencing weather and climate. The Sun's significance extends beyond merely providing light and warmth; it is a fundamental force that influences the orbits of planets and the stability of the entire solar system. Approximately 4.6 billion years ago, the Sun emerged from the collapse of a molecular cloud within a large molecular cloud, possibly triggered by shock waves from a nearby supernova. This process led to the creation of a protostar encircled by a protoplanetary disk, from which the planets eventually formed. Figure 1.1 shows a near photograph of the Sun taken by NASA.

 
O. Chakraborty and B. Das, *Parabolic Solar Trough Collector*, Mechanical Engineering Series, https://doi.org/10.1007/978-3-032-11229-3_1

**Fig. 1.1** Sun's near view. *Source* https://education.nationalgeographic.org/resource/sun/

Regarding physical characteristics, the Sun is an enormous body, with a mass of around $1.989 \times 10^{30}$ km. This makes it 333,000 times more massive than Earth, holding ninety-nine point eighty-six percent of the solar system's total mass, which is composed of the Sun. Its vast mass creates an intense gravitational field that keeps the planets, asteroids, comets, and other bodies orbiting around it. With a diameter of ~1.39 million kilometers, the Solar Giant is around 109 times larger than Earth. This colossal size gives the Sun a surface area of about ($6.09 \times 10^{12}$ km$^2$) and a volume of roughly ($1.41 \times 10^{18}$ km$^3$). Despite its large size, the Sun's average density is relatively low at 1.408 g per cubic centimeter, which is about a quarter of Earth's average density [1].

Hydrogen, making up about 74% of the Sun's mass, and helium, which accounts for around 24%, are its primary components. The rest, 2%, consist of elements like oxygen, carbon, neon, and iron, which are heavier. These elements, though present in small quantities, play a vital part in the Sun's makeup and the processes that occur within it [2]. The sunlight energy is produced at its core, where the temperatures exceed an astonishing 15 million degrees Celsius. At these extreme temperatures, hydrogen atoms undergo nuclear fusion [3], transforming into helium and emitting vast amounts of energy as light and heat. This process not only powers the Sun but also generates the flow of charged particles, known as solar wind, impacting the entire solar system, including Earth's magnetosphere.

At the center of the solar star lies the nucleus, where stellar fusion happens, followed by the radiation layer, which transfers energy outward via radiation. The convective layer sits above the radiative zone, transporting energy to the surface through convection currents. Far into space extends the Sun's outermost layer, the corona, which is visible during solar eclipses. The corona's temperature is surprisingly higher than the surface, reaching up to 2 million degrees Celsius, although the reason for this remains a topic of active research [4].

The Sun's life cycle is a critical aspect of understanding its future and the future of the solar system. Currently in the middle of its life for roughly the next 5 billion years, the Sun will continue to burn hydrogen. In the distant future, the Sun will deplete its hydrogen reserves, undergo a vast expansion into a red giant, and expel its outer layers, culminating in the formation of a compact, dense core known as a white dwarf. The solar system will experience significant consequences from this transition, with the inner planets, including Earth, potentially being engulfed.

In summary, the solar powerhouse is the heart of our stellar system and the essential energy source for all life on Earth. Its immense size, mass, and energy production define the characteristics of our solar system, shaping the conditions directly that make life possible. Understanding the Sun is crucial not only for astronomy but also for appreciating the delicate balance that sustains life on our planet [5].

## 1.2 The Basics of Solar Energy

At its core, the Sun's nuclear fusion reactions generate solar energy, which produces an immense amount of energy. This energy reaches Earth in the form of sunlight that can be employed through various technologies. Solar energy is classified into two principal forms: photovoltaic (PV) systems, which convert sunlight into electricity, and solar thermal systems, which harness solar heat for various applications. Each category employs distinct methods to convert sunlight into usable energy. Figure 1.2 represents the classifications of the solar system.

(i) **Photovoltaic Systems**

Photovoltaic systems, often recognized by the solar panels seen on rooftops, capture sunlight and transform sunlight into electricity using the photovoltaic phenomenon. Semiconductor materials, usually silicon-based, are used in this process to produce electrical current when sunlight-illuminated panels. Key components of a PV system include solar panels, an inverter that converts DC to AC, and, occasionally, batteries for energy storage.

A photovoltaic (PV) panel works based on the principles of the p–n junction within semiconductor materials, typically silicon. The panel consists of many photovoltaic modules; each containing a p–n junction is created where n-type (negative) and p-type (positive) semiconductors converge. When sunlight strikes these cells, photons from the light energize the semiconductor material, electrons acquire sufficient energy to release electrons from their atomic bonds, generating pairs of electron–hole. At the p–n junction, the electric field pushes electrons into the n-type region and holes into the p-type region, forming the depletion zone.

Charge carrier separation creates an electric field, resulting in a potential gradient across the junction. When an electric field that drives current is connected, electrons

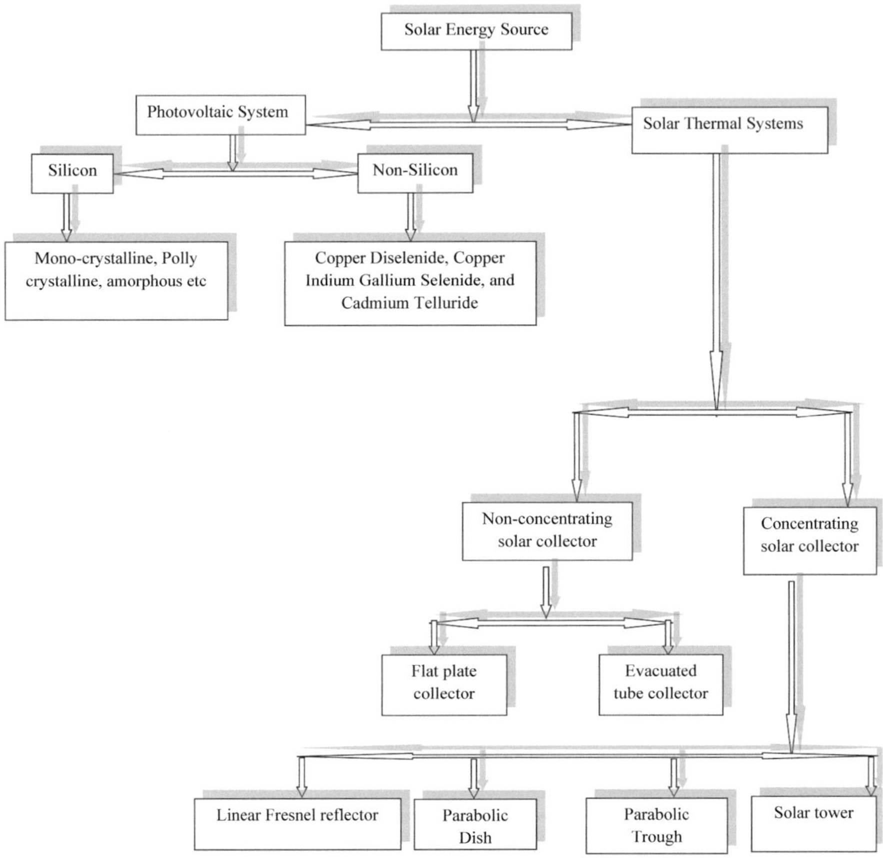


**Fig. 1.2** Classifications of solar system

move from the n-type zone through the circuit to the p-type zone, producing an electric current. This DC can power devices, be stored in batteries, or be converted into AC for use in homes and businesses through an inverter. The efficiency of this process depends on the features of the materials, the design of the cell, and ecological factors such as temperature and sunlight intensity. Figure 1.3 shows the photograph taken for the PV panel, and Fig. 1.4 is an illustration of the working of the PV panel.

The efficiency of photovoltaic systems has improved significantly over the years. Early solar panels had conversion efficiencies of around 6–10%, but modern panels can achieve efficiencies of 15–20% or higher. Advancements in materials and design, like thin-film cells, are lightweight; multi-junction cells are efficient, and are continually expanding the possibilities. Solar panels can be installed in a variety of settings, including residential rooftops, commercial buildings, and solar farms.

**Fig. 1.3** PV panel

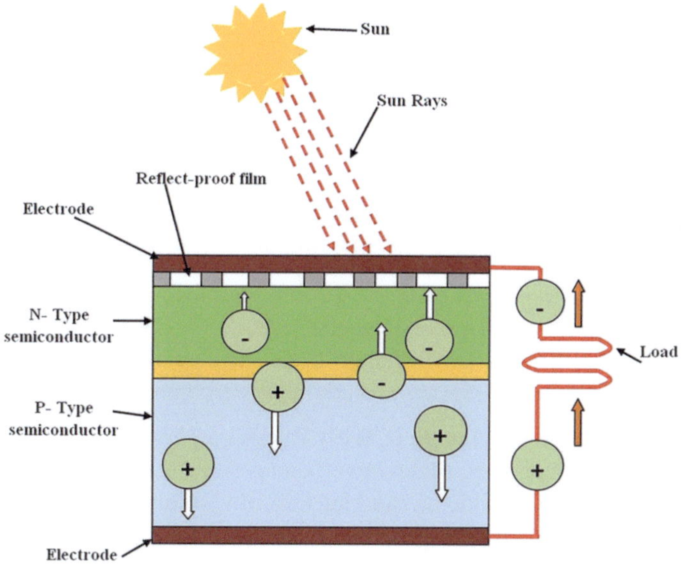


**Fig. 1.4** Working of a PV panel

## Silicon-Based Photovoltaic (PV)

Silicon-based photovoltaic (PV) panels are widely used for solar energy applications and come in several types, each with unique attributes.

**Monocrystalline Silicon Panels** are fabricated from a monocrystalline lattice, exhibiting superior efficiency due to the uninterrupted crystalline structure, usually ranging from 15 to 22%, because of their pure silicon composition. Distinguished by their uniform black hue and contoured, rounded edges, they exhibit a distinctive appearance. Although they perform well in low-light and high-temperature conditions, their higher cost and production complexity are notable drawbacks.

**Polycrystalline (or Multi-crystalline) Silicon Panels**, on the other hand, are created from silicon melted and poured into molds, resulting in wafers with multiple crystal structures. Generally, these panels demonstrate lower efficiency, ranging from 13 to 16%, and are recognizable by their speckled blue color. Due to simpler manufacturing processes, these panels are more cost-effective than monocrystalline panels, though they require more space for equivalent power output and experience slight performance degradation at high temperatures.

**Amorphous Silicon Panels**, a type of thin-film technology, utilize a non-crystalline form of silicon deposited onto substrates like glass or plastic. These panels are the least efficient, with ranges from 6 to 10%, but offer benefits such as flexibility, lightweight design, and better effectiveness in both low-light and high-temperature situations. Their lower production costs and adaptability make them suitable for specific applications, although they typically have a shorter lifespan and less power density compared to crystalline panels. Different types of silicon-based PV panels come with their benefits and drawbacks, and the best option is determined by efficiency, cost, and installation demands.

**Non-silicon Photovoltaic (PV)**

Non-silicon photovoltaic (PV) technologies encompass a range of alternative materials and methods for converting sunlight into electricity.

**Cadmium Telluride (CdTe) Panels** use a compound semiconductor to create thin-film solar cells, offering efficiencies between 9 and 18%. They are known for their cost-effectiveness and good performance under high temperatures and low-light conditions, although they involve toxic cadmium, which requires careful handling.

**Copper Indium Gallium Selenide (CIGS) Panels** employ a compound of indium, copper, gallium, and selenium, achieving efficiencies from 10 to 20%. These panels are noted for their high efficiency among thin-film technologies, flexibility, and lightweight nature, though their production costs are higher compared to other thin-film options.

**Organic Photovoltaics (OPV)** utilize carbon-based organic materials, providing flexible, lightweight, and potentially lower cost solutions, but they generally offer lower efficiency, ranging from 2 to 11%, and have a shorter lifespan.

**Perovskite Solar Cells** use perovskite-structured compounds and have shown remarkable efficiency improvements, reaching up to 25% in lab settings. They promise low production costs and flexible applications, but face difficulties regarding long-term stability and environmental durability concerns due to the use of lead in some materials. Each non-silicon technology possesses a unique balance of benefits and limitations, which makes it suitable for specific usage based on factors like cost, flexibility, and efficiency.

(ii) **Solar Thermal Systems**

Instead of electricity, solar thermal units use the Sun's energy to create heat. The captured heat can be applied to various usages, including Industrial processes, water heating, and space heating. The two primary types of solar thermal systems are concentrating solar collectors as well as non-concentrating solar collector systems.

**Non-concentrating Solar Collectors**

Non-concentrating solar collectors are intended to collect and utilize solar energy into heat without focusing sunlight through mirrors or lenses. Known as flat-plate collectors, these solar thermal systems are designed to absorb solar energy and convert solar energy into heat. They consist of a flat, horizontal panel typically made of metal or a similar material. The panel includes a collector plate, usually coated with a selective surface or painted black, which enhances absorption to enhance heat absorption, and a network of pipes or tubes through which a heat-carrying liquid (commonly water or antifreeze) moves in a continuous loop.

When sunlight strikes, the collector plate absorbs the Sun's energy, causing it to heat up and transfer thermal energy to the fluid circulating through the pipes. This heated fluid is then directed to a storage tank or employed in space heating, capturing solar energy efficiently, water heating, or manufacturing operations.

Flat-plate collectors are widely utilized for their simplicity, dependability, and affordability. They are most efficient in regions with ample sunlight and are ideal for residential and commercial use. Their performance is contingent upon variables such as the installation angle, collector integrity, and prevailing climatic conditions. They are a popular choice in solar thermal systems due to their proven track record and relatively low maintenance requirements. Figure 1.5 features the structure of a flat-plate collector.

**Evacuated Tube Collectors** are cutting-edge solar thermal apparatuses meticulously engineered to capture and transmute solar radiation into thermal energy with unparalleled efficiency. These collectors are composed of several glass tubes, each featuring an inner receiver tube and an outer glass casing. A vacuum is formed between the two tubes, effectively reducing heat loss via conduction and convection.

The inner receiver tube features a selective surface that maximizes solar energy absorption and converts it into heat. As sunlight strikes the absorber, it heats the fluid inside the tube. Typically, water or the heat-conveying fluid is circulated through a network of pipes, transferring the heat to a storage tank or for immediate use in applications such as water thermal heating, space temperature control, and industrial operations.

These collectors, known as evacuated tube collectors, are exceptionally effective, especially in colder regions or when solar radiation is diffused, as the vacuum insulation reduces thermal loss. They can achieve better temperatures than flat-plate

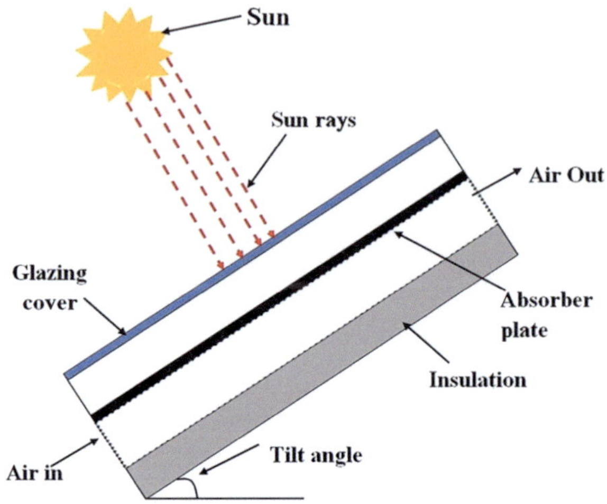


**Fig. 1.5** Flat-plate solar collectors

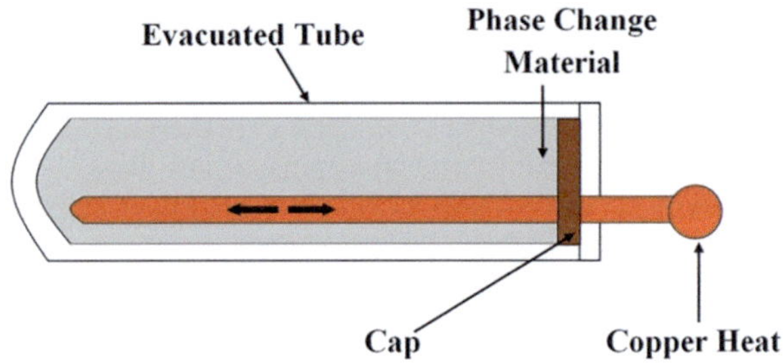


**Fig. 1.6** Evacuated tube solar collectors

collectors and are often used in applications requiring high performance and reliability. Their effectiveness in a variety of conditions makes them a widely chosen solution for solar thermal systems in residential and commercial applications. Figure 1.6 showcases the schematic structure for evacuated tube solar collectors.

**Integral Collector-Storage Systems (ICS)** integrate the collector and storage tank into one unit, simplifying installation and reducing costs, though they may offer lower efficiency and less flexibility compared to systems with separate components. **Air collectors** utilize air as the working medium and are used primarily for space heating and ventilation. They are simpler and less expensive but typically have lower efficiency, ranging from 30 to 50%, compared to liquid-based collectors. Each type of non-concentrating solar collector is suited to different applications and environments, providing a range of options based on factors like efficiency, cost, and installation complexity.

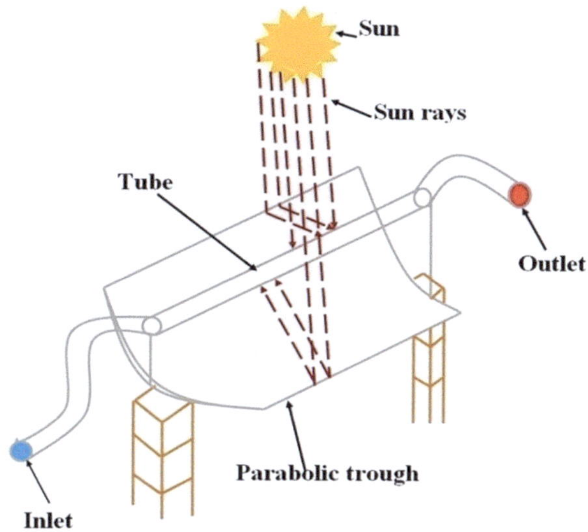


**Fig. 1.7** Parabolic solar trough collector

## Concentrated Solar Power (CSP)

**Concentrated Solar Power (CSP)** technologies harness solar energy by utilizing lenses or mirrors to focus solar radiation onto a compact area, producing concentrated heat that is then converted into electricity. **Parabolic trough systems** employ curved mirrors; sunlight is focused onto a receiver tube, heating fluid to produce steam that powers a turbine. This well-established technology is efficient and can incorporate thermal storage, though it requires significant space and direct sunlight. Figure 1.7 shows the structure of a solar parabolic trough unit.

**Solar Power Towers**, or central receiver systems, utilize a collection of mirrors, known as heliostats, that focus sunlight onto a central receiver at the tower's peak, resulting in higher temperatures and efficiencies, typically ranging from 15 to 20%. They are suitable for large-scale plants and can also use thermal storage, but they involve high costs and complex infrastructure. Figure 1.8 elaborates on solar power towers.

**Parabolic Dish Systems** feature dish-shaped mirrors focusing the sunlight onto a receiver located at their focal point, which can directly power a Stirling engine or raise the temperature of a fluid, offering high efficiency up to 30% but requiring precise Sun tracking and complex mechanics. Linear Fresnel reflector systems utilize long, flat, or mildly curved mirrors to concentrate sunlight onto a collector tube, offering a more affordable and simpler solution compared to parabolic troughs and power towers, though with generally lower efficiency. CSP technologies work best in regions with abundant direct sunlight and are primarily used for large-scale power generation, with each system offering unique benefits and trade-offs in terms of efficiency, cost, and complexity. Figure 1.9 features a parabolic solar dish collector.

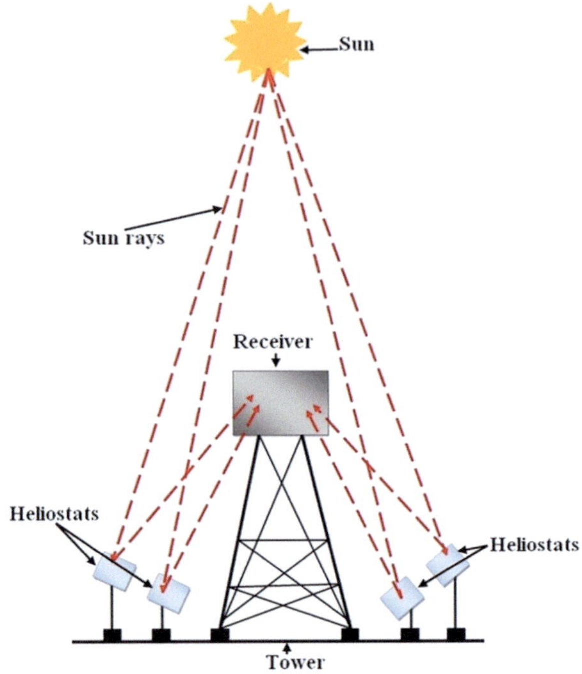


**Fig. 1.8** Solar power towers

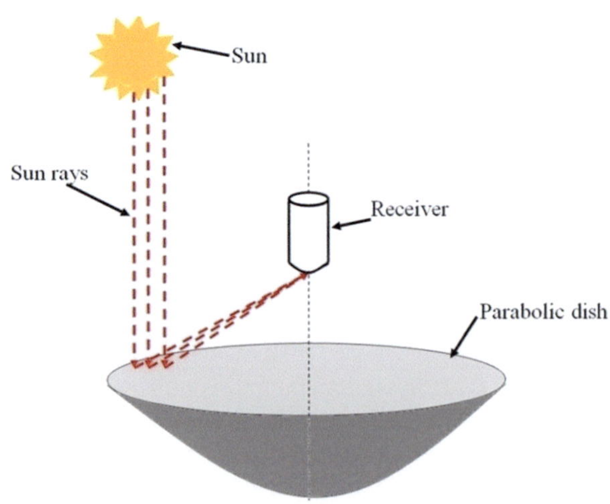


**Fig. 1.9** Parabolic solar dish collector

## 1.3 Environmental Impacts

The growing global demand for energy, driven by rapidly expanding major economies such as India and China, has significantly increased fossil fuel consumption in the last decade [6–8]. As shown in Table 1.1, the top 10 countries are listed according to their energy consumption per capita and overall energy usage. Currently, developing nations primarily rely on fossil fuels to satisfy their energy demands, contributing to global warming. Carbon is the primary contributor to environmental pollution, as depicted in Fig. 1.10, which shows the cumulative carbon emissions over the past century.

In response to worldwide issues about climate change and the depletion of fossil fuel stocks, the manufacturing industry is increasingly turning to clean energy sources as alternatives [9, 10]. Many countries have set a target to realize a 40% reduction in greenhouse gas emissions by 2030 and an 80–95% contraction by 2050 [11, 12]. Prioritization of investments in sustainable energy technologies—tidal, wind, solar,

**Table 1.1** Leading for energy consumption of 10 countries [world bank data]

| Rank | Country | Energy use per capita (units) | Total energy use (units) |
|---|---|---|---|
| 1 | Qatar | 17,922.73 | – |
| 2 | Iceland | 17,916.12 | – |
| 3 | United States | 6962.55 | 2,216,186,625,000 |
| 4 | China | – | 3,051,503,511,000 |
| 5 | India | – | 824,742,957,000 |
| 6 | Russia | – | 710,882,706,000 |
| 7 | Japan | – | 441,744,848,000 |
| 8 | Germany | – | 306,070,275,000 |
| 9 | Canada | 7897.86 | 279,879,745,000 |
| 10 | World* (average) | 1922.71 | 13,949,013,151,887 |

https://worldpopulationreview.com/country-rankings/energy-consumption-by-country
*The data is given in kilograms of oil equivalent

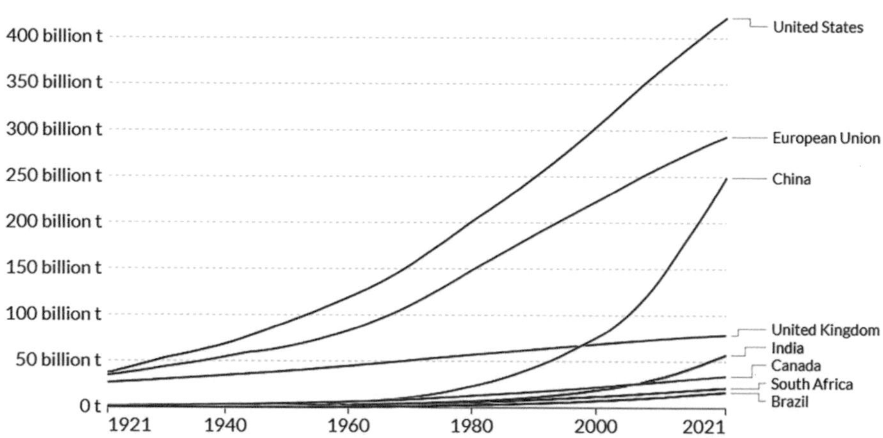


**Fig. 1.10** Total $CO_2$ Emissions from the Past Century. https://ourworldindata.org/contributed-most-global-$CO_2$

hydropower, biomass, geothermal, and wave: resilient, sustainable, transformative energy solutions—is underway in leading nations, including India, the United States, China, Japan, Germany, and the Netherlands: global leaders. Figure 1.11 delineates the trajectory of sustainable energy investment across these geopolitical entities [13].

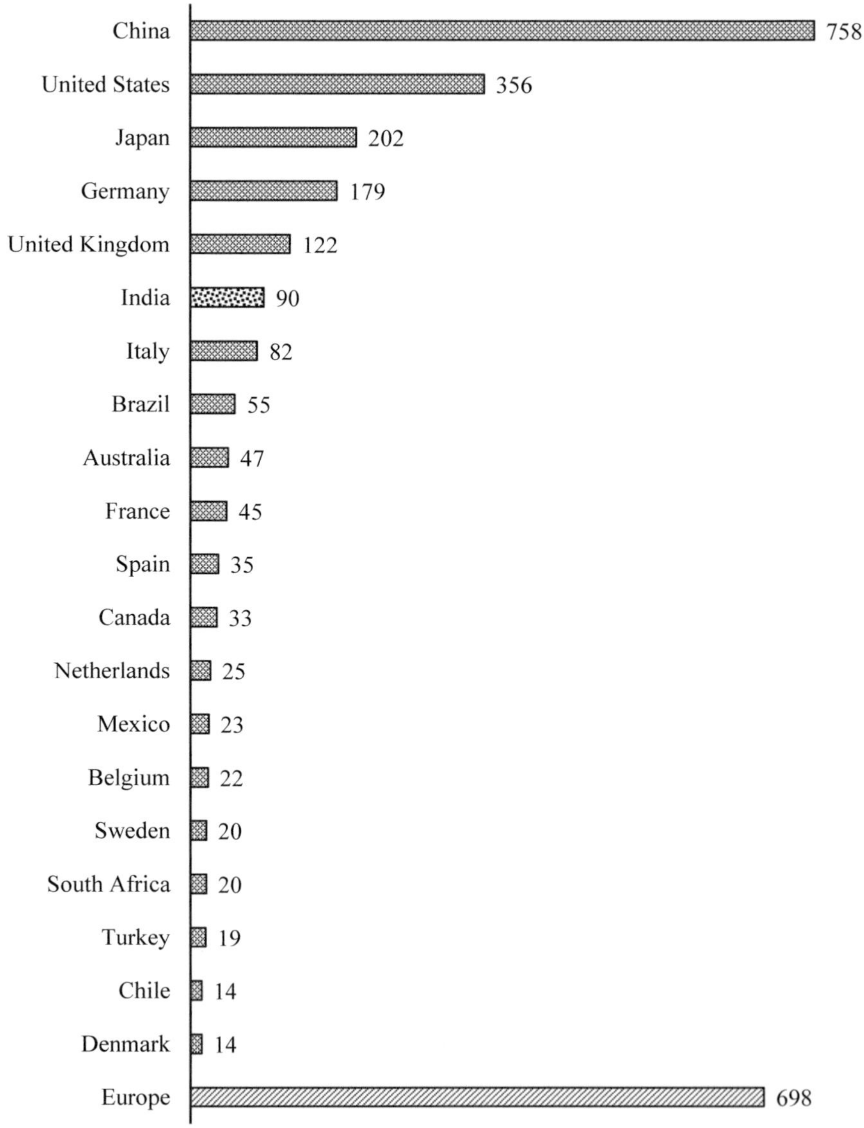


**Fig. 1.11** Investment in sustainable energy capacity by several countries, from 2010 to the first half of 2019, in U.S. dollars ($) (billions) [13, 22]

Recent developments and future objectives concerning carbon emissions and sustainable energy in Europe, as well as Asia, include.

In Europe:

1. Targeting 32% green energy by 2030, the European Union commits [14].
2. According to a Eurostat report, renewable energy contributed 17.5% of the EU's overall energy use in 2020 [15].
3. Approximately two-thirds of the cumulative discharge reduction from 1990 to 2020 occurred within the energy generation units, emphasizing its central importance in the EU's decarbonization strategy [16].
4. The EU aims to reduce its greenhouse gas emissions by ~55% by 2030, relative to levels of 1990 [17].

In Asia:

1. By 2030, China aims to generate 20% of its total energy from renewable sources, positioning itself as an international frontrunner in sustainable energy production [18].
2. China has set ambitious goals to reach carbon neutrality by 2060 and cap its $CO_2$ emissions by 2030 [19].
3. India aims to derive 40% of its power capacity from renewable energy by 2030 [20].
4. India ranks as the third-leading factor in global warming, following China and the United States [21].

Again, the energy production utilizing solar sources is not free from emissions. The technical, economic, and environmental factors are listed in Table 1.2. Some components of solar energy sources may emit cadmium as well as mercury [23]. From Table 1.1, it is observed that emissions of $SO_2$ and $NO_x$ from the solar energy sources are a little higher relative to other eco-friendly energy sources. Again, the energy conversion ratio for energy is observed to be superior to wind. The primary benefit of solar panels is their easy installation, although the cost to produce 1 kWh of electricity generation is marginally higher compared to other renewable sources.

## 1.4 Technological Advancements and Innovations

The field of solar energy is characterized by rapid technological developments that continually improve capability, reduce costs, and expand applications. Key innovations include.

1. **Bifacial Solar Panels**: These panels can harness sunlight from both sides, boosting energy output. They perform especially well in areas with high albedo, like snowy or reflective surfaces.
2. **Floating Solar Farms**: By placing solar panels on water surfaces, floating solar farms can reduce land use and increase efficiency through the cooling effect of water.

**Table 1.2** Analysis of the environmental and techno-economic aspects of renewable energy sources [24]

| | Criteria | Wind | Solar | Hydro | Biomass | Geothermal |
|---|---|---|---|---|---|---|
| Environmental effect | Air pollution | * | * | * | * | * |
| | Water pollution | – | – | * | * | * |
| | Sound pollution | + | * | * | – | + |
| | $CO_2$ emission (mg/Kwh) | 17,652 | 49,174 | 22,696 | 58,000 | 18,913 |
| | $SO_2$ emission (mg/Kwh) | 54 | 257 | 33 | 76 | 280 |
| | $NO_x$ emission (mg/Kwh) | 32 | 178 | 23 | 1325 | 280 |
| Techno-economic impacts | Land requirement ($Km^2$/1000 MW) | 100 | 35 | 750 | 5000 | 18 |
| | Water requirement (kg/kWh) | 1 | 10 | 1.25 | 156 | 36 |
| | Energy input output ratio | 0.20 | 0.14 | 0.04 | 0.14 | 0.02 |

*Means negligible; + contributes; – does not effect

3. **Building-Integrated Photovoltaics (BIPV)**: This technology seamlessly incorporates solar panels into architectural components, including windows, facades, and roofing, facilitating energy generation without necessitating supplementary spatial allocation.
4. **Advanced Energy Storage**: Advancements in energy storage technologies, notably lithium-ion and solid-state batteries, represent significant strides, and enhance the storage of solar energy for use during low sunlight periods, improving solar power reliability.
5. **Solar Tracking Systems**: These units modify the angle of solar panels to track the solar movement, optimizing exposure and energy capture throughout the day.

Sun's energy is gaining popularity as an inexhaustible energy source because of its numerous advantages; however, it also presents a unique set of obstacles. Here's a breakdown of the advantages and limitations.

## 1.5 Advantages and Limitations

### Advantages

1. Solar energy is abundant and inexhaustible on a human timescale. As long as the Sun shines, solar energy can be harnessed.
2. Electricity bills can be lowered or even eliminated by generating electricity, depending on the size of the solar system.
3. Electricity is generated by solar power without the emission of greenhouse gases, mitigating pollution and climate change.

4. Once installed, solar panels require very little upkeep, with a typical lifespan often exceeding 25 years.
5. By diversifying the energy portfolio, solar energy diminishes dependence on fossil fuels and bolsters energy security.
6. Through its contributions to manufacturing, installation, and maintenance, the solar industry acts as a catalyst for economic growth and job creation.
7. Ranging from small residential installations to large utility-scale solar farms, solar systems can be deployed at diverse scales.

**Limitations**

1. Solar energy production relies on weather context and the part of the day. It's less reliable compared to continuous energy sources like coal or nuclear.
2. The upfront cost of buying and the setup of solar panels, along with the necessary equipment, can be substantial, though prices have been falling and financial incentives are available in many regions.
3. Solar panels need considerable space to produce significant amounts of electricity, which could pose a challenge in densely populated or small-property areas.
4. Energy storage solutions, such as batteries, are required to use solar energy at night or during cloudy periods, but these can be costly and have environmental impacts of their own.
5. Solar panels have a variable efficiency rate, typically around 15–20%, meaning not all sunlight is converted into electricity. Efficiency can also be affected by geographic location and panel orientation.
6. Manufacturing solar panels involves energy and raw materials, which can result in environmental impacts, but efforts are underway to enhance the sustainability of their production processes.
7. Solar panels have a finite lifespan and can create waste generated at the end of their lifespan, though recycling technologies and also programs are improving.

## 1.6 Recent Trends in Solar Energy Research and Development Compared to Other Forms of Renewable Energy

Recent trends in solar energy research highlight significant advancements, setting it apart from other renewable sources. In photovoltaics (PV), innovations like Bifacial panels, perovskite solar cells, and transparent PV cells are expanding the limits of efficiency and integration. Perovskite cells are particularly promising owing to their strong efficiency and prospects for low-cost manufacturing, making solar energy more affordable. The ability of bifacial panels to capture sunlight from both sides leads to increased electricity generation, while transparent PV cells enable integration into windows, expanding their applications in urban settings.

Concentrated solar power (CSP) technologies, including technologies that include parabolic troughs, solar power towers, and parabolic dishes, have also seen considerable improvements. Recent developments in CSP focus on enhancing efficiency through better heat storage systems and reducing costs, making it attain greater competitiveness relative to conventional energy sources.

In comparison, advancements in wind energy have centered on the development of larger, more efficient turbines and enhanced grid integration techniques. These improvements help capture more wind energy and stabilize the electricity supply. Hydropower, while a mature technology, is evolving with small modular systems and innovations in turbine design aimed at minimizing environmental impact.

Geothermal energy research is advancing with enhanced geothermal systems (EGS) and new drilling techniques that access deeper and hotter reservoirs, expanding its geographic reach. Meanwhile, bio-energy is progressing with better conversion technologies and diverse feedstocks, such as agricultural residues and algae, improving the sustainability of bio-fuels and bioelectricity production. Figures 1.12 and 1.13 showcase the annual financial commitments and energy generation in various renewable energy sources.

Overall, solar energy leads the way with rapid technological advancements and declining costs, positioning it as an essential element in the worldwide transition to renewable energy. As research continues, solar power is likely to play an increasingly crucial role in attaining a sustainable, low-carbon future. Table 1.3 represents the recent developments in different renewable energies.

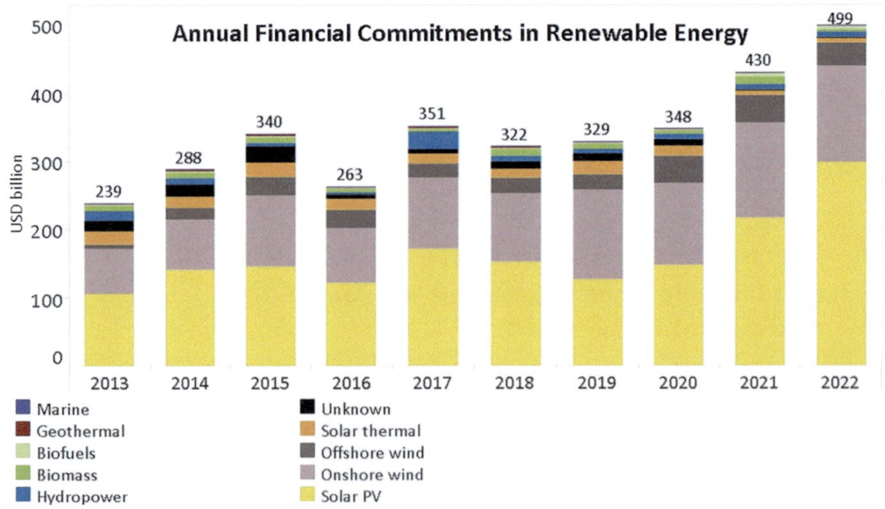


**Fig. 1.12** Annual financial commitments in renewable energy. *Source* https://www.irena.org/Energy-Transition/Technology/Solar-energy

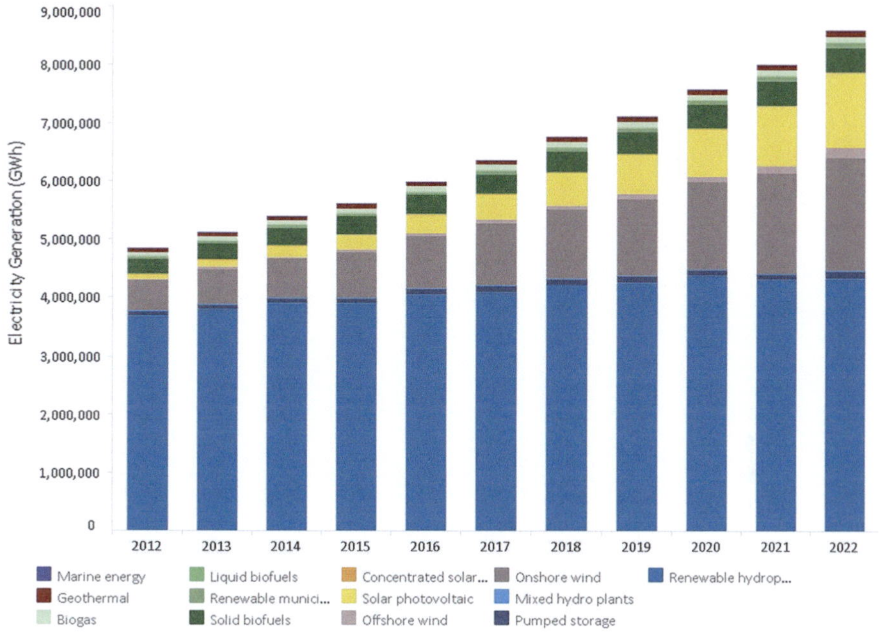


**Fig. 1.13** Annual energy generation for different renewable energy sources. *Source* https://www.irena.org/Energy-Transition/Technology/Solar-energy

## 1.7 Tracking Algorithm for Calculating Solar Position

The solar position tracking involves a deep understanding of various astronomical and geographical terms that describe the Sun's location relative to the Earth. This section provides an overview of key concepts such as Earth–Sun geometry, coordinate systems, and their relevance in calculating solar positions.

### Earth–Sun Geometry

The ecliptic is a fundamental concept in understanding the Sun's visible trajectory in the atmosphere. It represents Earth's elliptical trajectory in its heliocentric orbit and serves as a reference for determining the locations of celestial bodies. The celestial equator, on the other hand, is the depiction of Earth's equatorial plane in the celestial sphere. It separates the sky into the Northern and Southern Hemispheres, playing a crucial role in determining the positions of stars and the Sun.

**Equinoxes** are specific points where the celestial equator intersects the ecliptic. These points of intersection denote pivotal junctures in Earth's orbital trajectory. The vernal equinox, typically occurring around March 21, heralds the onset of the vernal season in the Northern Hemisphere, while simultaneously ushering in autumn in the Southern Hemisphere. In contrast, the autumnal equinox, occurring near September 23, marks the commencement of autumn in the Northern Hemisphere and spring in

**Table 1.3** Recent developments in different renewable energy [25–37]

| Aspect | Solar energy | Wind energy | Hydropower | Geothermal energy | Bio-energy |
|---|---|---|---|---|---|
| Technological advancements | • **Photovoltaic technologies:** Perovskite cells, multi-junction cells<br>• **Parabolic solar trough collectors:** improved mirror coatings, enhanced heat transfer fluids, modified tubes, and inserts<br>• **Parabolic dish:** high-efficiency concentrators<br>• **Solar power tower:** advanced heliostat designs, high-temperature storage | • Larger turbines<br>• Offshore wind turbines<br>• Floating turbines | • Small modular systems<br>• Pumped storage innovations | • Enhanced geothermal systems (EGS)<br>• Binary cycle plants | • Advanced bio-fuels<br>• Biogas systems |
| Efficiency improvements | • **Photovoltaic:** up to 24% for silicon cells, Perovskite tandem cells >30%<br>• **Parabolic trough:** 10–20% thermal efficiency<br>• **Parabolic dish:** 20–30% thermal efficiency<br>• **Solar power tower:** 15–25% thermal efficiency, enhanced with advanced materials | • Larger capacity turbines (up to 15 MW)<br>• Floating turbines | • High-capacity factors (above 40%) | • Improved drilling<br>• Enhanced systems for lower temps | • Improved conversion technologies<br>• Next-gen bio-fuels |

(continued)

**Table 1.3** (continued)

| Aspect | Solar energy | Wind energy | Hydropower | Geothermal energy | Bio-energy |
|---|---|---|---|---|---|
| Energy storage | • **Photovoltaic:** battery<br>• **Parabolic trough:** thermal storage (e.g., molten salts)<br>• **Parabolic dish:** limited thermal storage, often paired with other technologies<br>• **Solar power tower:** advanced thermal storage (e.g., molten salts) | • Storage and grid integration challenges<br>• Enhanced forecasting | • Pumped storage for grid stability<br>• High initial costs | • Base-load power<br>• Minimal storage needs | • Variable feedstock availability<br>• Integration challenges |
| Integration challenges | • **Photovoltaic:** improved with smart grids and microgrids<br>• **Parabolic trough**: requires high direct sunlight, land, and space requirements<br>• **Parabolic dish**: requires precise tracking, land space<br>• **Solar power tower**: land-intensive, requires direct sunlight | • Intermittency requires grid upgrades<br>• Storage solutions | • Base-load stability<br>• Environmental and social impacts | • Stable base-load power<br>• High upfront costs | • Feedstock variability<br>• Complex grid integration |

the Southern Hemisphere. The angular relationship between the ecliptic plane and the celestial equator, termed the obliquity of the ecliptic, is ~23.5°, though it experiences minor variations due to Earth's axial precession. This axial tilt is a fundamental factor in determining the Sun's apparent trajectory across the sky throughout the annual cycle.

**Latitude and Longitude**

The reference framework used for locating positions on Earth is defined by **latitude** and **longitude**. **Longitude** is measured in degrees of longitude, east or west of the Greenwich Meridian, which serves as the prime meridian. It helps in determining the east–west location of a location on Earth. **Latitude**, in contrast, denotes the latitude relative to the equator, with values starting at 0° at the equator and extending to 90° at the poles. This system allows for the precise identification of any position on Earth and is crucial for determining solar position relative to specific geographic points.

**Right Ascension and Declination**

For celestial objects, a diverse coordinate system is used: right ascension and declination. This system is analogous to longitude and latitude but is applied to the celestial sphere. The angular measurement of Right Ascension (RA) is defined eastward from the vernal equinox along the celestial equator, a critical element in celestial navigation and coordinate systems. It is akin to longitude but for celestial objects, providing a way to specify an object's position relative to a fixed reference point in the sky. Right ascension (RA) is measured in hours, minutes, and seconds, with 24 h corresponding to a full 360° circle.

**Declination (DECL)** is similar to latitude but measures the angle south or north from the celestial equator serving as the reference point for declination. In the celestial coordinate system, positive declination values signify positions to the north of the celestial equator, whereas negative values indicate southern positions. Declination is measured in degrees, minutes, and seconds. Together with right ascension, these parameters establish an accurate framework for pinpointing celestial bodies, including the Sun.

In the context of solar position calculations, these concepts help ascertain the Sun's location in the sky at a particular moment. The Sun's right ascension and declination change daily due to Earth's elliptical path, which introduces alterations not seen in the positions of distant stars. Understanding these changes is essential for accurate solar position tracking, influencing applications ranging from astronomy to solar energy optimization. Figure 1.14 shows the Earth–Sun geometry.

**The Solar Constant**

The solar constant, around 1361 W/m$^2$, epitomizes the equilibrium radiative flux impinging per unit surface area at the exobase of Earth's air envelope, contingent upon Earth's aphelion, situated at an average heliocentric distance of 149.6 million kilometers (93 million miles), or one astronomical unit (AU). It is a critical parameter

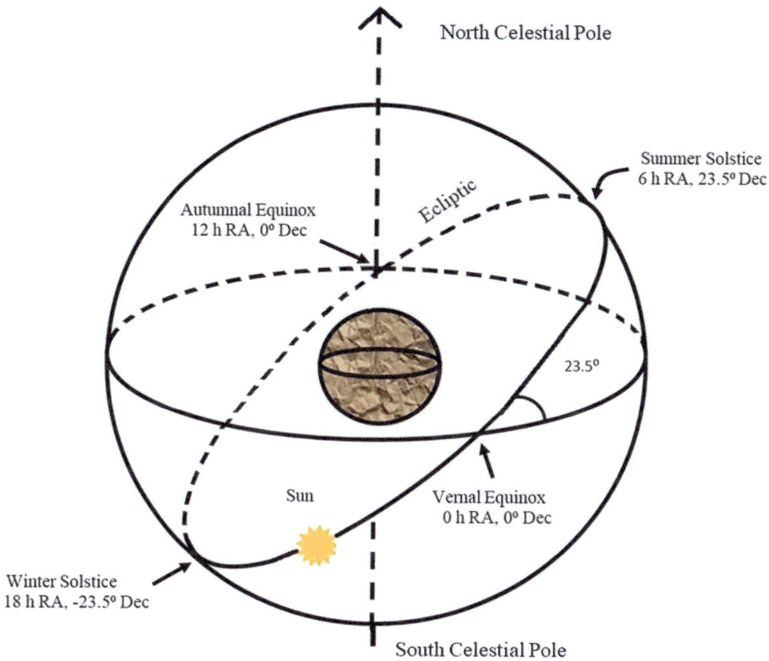


**Fig. 1.14** Earth–sun geometry

for interpreting Earth's energy balance, climate dynamics, and solar energy potential. The solar constant is determined through satellite-based measurements, such as those derived from NASA's Solar Radiation, the European Space Agency's Solar and Heliospheric Observatory (SOHO), and Climate Experiment (SORCE), which meticulously quantify and analyze the solar irradiance with high precision and temporal resolution providing accurate and continuous data essential for climate modeling and solar energy applications. Even though the solar constant is relatively stable, it exhibits slight variations because of Earth's oval-shaped orbit, which causes changes in solar radiation obtained at varying times throughout the year. Furthermore, the Sun is subject to an 11-year solar cycle, wherein periodic oscillations in heliospheric activity, exemplified by the modulation of sunspots and the eruption of solar flares, engender substantial perturbations in the Sun's radiative flux and overall output. These variations influence global temperatures, weather patterns, and the performance of solar energy units. Understanding the solar constant is vital for predicting climate changes and optimizing solar panel designs, making it an essential factor in both climate science and renewable energy technology (Fig. 1.15).

**Types of Solar Radiation**

Solar flux is an essential aspect of Earth's energy unit, affecting everything from weather patterns to technological applications. The study of solar radiation encompasses various types, including direct, diffuse, and reflected radiation, each with

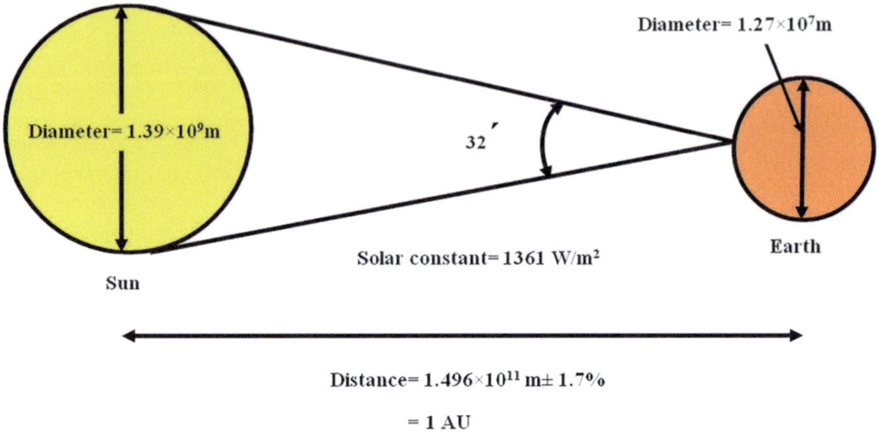


**Fig. 1.15** Solar constant

its unique characteristics and implications. Understanding these types is crucial for harnessing solar energy effectively and for comprehending the broader impacts on climate and environmental systems.

### Direct Radiation (Beam Radiation)

Beam radiation or direct solar radiation constitutes the fraction of solar energy that propagates along a rectilinear trajectory from the Sun to the surface of Earth, unimpeded by atmospheric scattering or absorption, commonly referred to as beam radiation or direct solar radiation. This radiation is most intense when the Sun is at its zenith, creating well-defined shadows. Direct radiation is measured using pyrheliometers, which are specialized instruments designed to capture sunlight directly from the Sun's disk. The intensity of direct radiation is crucial for solar energy applications, particularly concentrated solar power systems, which rely on direct sunlight, are designed to harness solar energy to generate electricity. These systems include technologies like solar power towers, parabolic dishes, and parabolic troughs, all of which focus direct sunlight to produce heat.

### Solar Declination ($\delta$)

Solar declination represents the angular separation between the incident Sun's rays and Earth's equatorial plane. This value undergoes periodic variation over the year as a consequence of Earth's rotational obliquity and its heliocentric orbital motion. Solar declination affects the length of daylight and the angle at which solar energy strikes the Earth's surface. During solstices, solar declination reaches its maximum and minimum values, resulting in the most extended and shortest days of the year, respectively. Understanding solar declination is essential for forecasting solar angles and maximizing the efficiency of solar panel placement. Figure 1.16 explains the solar declination:

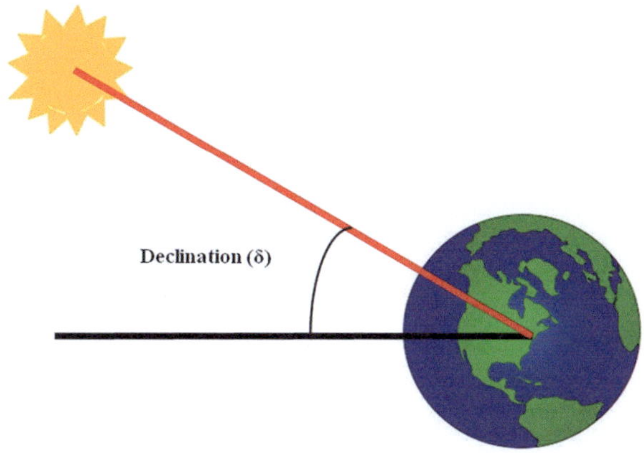


**Fig. 1.16** Solar declination

$$\delta = 23.45^{\circ} \times \sin\left(\frac{360^{\circ}}{365} \times (n - 81)\right) \tag{1.1}$$

"$\delta$" is the Sun's angular declination.

"$n$" is date of the year (1 for New Year's Day, 365 for the final day of the year).

23.45° represents the maximum obliquity of Earth's axis with the plane of its orbit (ecliptic).

$\frac{360^{\circ}}{365}$ converts the day number $n$ (where $n = 1$ is January 1) into an angle that corresponds to the Sun's position along its annual path relative to the Earth.

"$(n - 81)$" shifts the calculation so that $n = 81$ corresponds roughly to the spring equinox (around March 21), converging with the Sun's trajectory across the celestial equator from the southern to the northern celestial hemisphere; this formula is indispensable in fields such as astronomy, solar energy analysis, and celestial navigation. It facilitates the precise determination of the Sun's spatial coordinates on any specific day, thereby elucidating the dynamic interplay of diurnal durations, seasonal oscillations, and the Sun's zenithal altitude within the celestial dome.

In summary, the equation is derived from the geometry of Earth's orbit, its axial tilt, and its orientation relative to the Sun, providing a means to compute the Sun's declination angle throughout the annual cycle.

**Understanding Declination Angle ($\delta$)**

- The declination angle ($\delta$) denotes the angular displacement between the incident Solar irradiance and the plane of Earth's equatorial axis, undergoing periodic variation throughout the annual orbital trajectory of the Earth around the Sun.
- The apex of the declination angle, 23.45°, is attained during the solstices (approximately June 21 and December 21).

**Earth's Orbit and the Ecliptic**

- The Earth's axis is tilted at an angle $\varepsilon = 23.45°$ to the plane of its revolution around the Sun (the ecliptic).

- As the Earth orbits the Sun, the declination angle $\delta$ varies sinusoidally over the year.

**Sinusoidal Nature of Declination**

- The Earth's orbit can be approximated as circular with a span of 365 days. The Sun's obvious motion due to the axial tilt, the path along the ecliptic, can be represented by a sine function.
- The general form of the sinusoidal function describing $\delta$ is

$$\delta = \varepsilon \times \sin(\text{angular frequency} \times \text{time} + \text{phase shift}) \tag{1.2}$$

**Angular Frequency**

The angular frequency $\omega$

$$\omega = \frac{2\pi \text{ radians}}{365 \text{ days}} \tag{1.3}$$

However, since degrees are being used instead of radians, this becomes

$$\omega = \frac{360^{\circ}}{365 \text{ days}} \tag{1.4}$$

**Time Factor (*n*)**

- ***n*** represents the day of the year, beginning with January 1 (n = 1), also ending on December 31 (n = 365).

**Phase Shift (*n* − 81)**

- The phase shift accounts for the fact that the spring equinox, when $\delta = 0°$, occurs around March 21 ($n = 81$). Thus, the time factor is adjusted to $n - 81$.

  Combining the above and putting them in Eq. 1.2, the equation for declination δ

$$\delta = \varepsilon \times \sin(\omega \times (n - 81))$$

Substituting $\varepsilon = 23.45°$ and $\omega = \frac{360°}{365 \text{ days}}$

$$\delta = 23.45^{\circ} \times \sin\left(\frac{360^{\circ}}{365} \times (n - 81)\right) \tag{1.5}$$

**Hour Angle (*H*)**

The hour angle is an indicator of the angular distance used in solar energy and astronomy to determine the Sun's position in the sky at a specific location that is defined by variables such as the time of day, the date, and the geographic coordinates (latitude and longitude) of the location. It represents the time elapsed since solar noon; in angular terms, it is known as the hour angle, measured in degrees. The hour angle represents the angular separation between the Sun and the local meridian (a longitudinal line extending from north to south). It increases by 15° per hour as temporal displacement from solar noon occurs. The hour angle is 0° at solar noon. For each hour preceding or succeeding solar noon, the hour angle deviates by ±15°.

In the realm of solar energy applications, the hour angle is indispensable for precisely calculating the solar zenith and azimuth, thereby optimizing the orientation of photovoltaic systems to ensure maximal insolation capture throughout diurnal cycles:

$$H = 15^{\circ} \times (T - 12) \tag{1.6}$$

"$H$" is Hour angle expressed in degrees.
"$T$" is Local solar time in hours.

#### Relationship Between Time and Hour Angle

The Earth rotates 360° in 24 h, so each hour corresponds to a rotation of 15°:

$$360^{\circ}/24 = 15^{\circ}$$

#### Formulating the Equation

- At noon, the Sun is on the local meridian, so $H = 0^{\circ}$.
- For any time $T$ after noon, the Sun has moved westward from the meridian.
- The equation expresses this angular displacement.
- $T - 12$ represents the time difference from noon in hours.
- Multiplying by 15° converts the time difference directly into angular degrees of displacement.

#### Derivation Process

- Start from the fact that 1 h corresponds to 15°:
- $H = 15^{\circ} \times T$

Since $H = 0^{\circ}$ at $T = 12$ (noon), it is adjusted for this zero point.

This equation is fundamental in celestial navigation and astronomy for calculating the position of celestial objects relative to the observer's local meridian at any given time $T$.

#### Solar Zenith Angle ($\theta_z$)

The solar zenith angle is the angular displacement between the Sun's position and the zenith, the point directly overhead at a specific geographical location, quantifying the deviation of the Sun from its nadir. It determines how directly solar radiation strikes the Earth's surface. A smaller zenith angle signifies that the Sun is positioned higher in the sky, resulting in more direct solar radiation and an increase in intensity.

Conversely, a larger zenith angle means the Sun is lower in the sky, resulting in less direct radiation and increased scattering. The solar zenith angle varies throughout the day and year, affecting solar energy collection and climate conditions. Accurate measurement and prediction of the solar zenith angle are essential for optimizing solar panel orientation and maximizing energy generation.

Note: Based on the spherical law of cosines for a triangle on the celestial sphere

$$\cos(\theta_z) = \sin(\emptyset) \times \sin(\delta) + \cos(\emptyset) \times \cos(\delta) \times \cos(H) \tag{1.7}$$

"$\theta_z$" is the angular distance from the Sun to the zenith in degrees.
"$\phi$" is the latitude coordinate of the location in degrees.
"$\delta$" is Sun's declination angle in degrees.
"$H$" is angular hour displacement in degrees.

**Solar Azimuth Angle**

The solar azimuth angle quantifies the directional bearing, relative to true north, from which solar radiation is incident at any given instant, delineating the Sun's horizontal location on the celestial sphere. It is measured from the north and ranges from 0° to 360°. The azimuth angle fluctuates throughout the day as the Sun traverses its diurnal path across the sky. For example, at sunrise, the azimuth angle is close to the east, while at sunset, it approaches the west. Understanding the solar azimuth angle is integral to the engineering of solar energy systems, as it governs the precise alignment of photovoltaic modules, ensuring optimal solar insolation by determining the most advantageous orientation for maximal energy capture throughout the diurnal cycle. Figure 1.17 features the zenith and azimuth angles.

**Solar Elevation Angle**

The **solar altitude** represents the angular distance between the Sun and the horizon, reflecting the Sun's height above the Earth's surface at a specific moment. It denotes the **solar elevation** above the horizon at any moment. During sunrise and sunset, the solar elevation angle is 0°, when the Sun is at the horizon, while at solar noon, it attains its highest value for a specific location. The solar elevation angle influences the incident radiative flux density of solar irradiation at the Earth's surface. It is a vital element in energy dynamics and radiative flux measurement, playing an crucial role in the design of solar energy systems, as it affects the amount of energy that solar panels capture throughout the day:

$$\sin(\alpha) = \sin(\emptyset) \times \sin(\delta) + \cos(\emptyset) \times \cos(\delta) \times \cos(H) \tag{1.8}$$

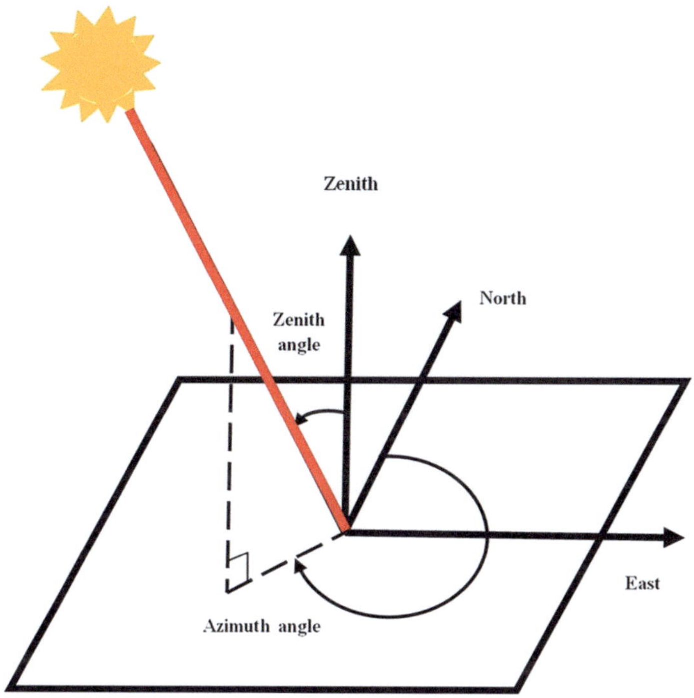


**Fig. 1.17** Zenith and Azimuth angles

## Extraterrestrial Radiation ($I_{b0}$)

Extraterrestrial radiation $I_{b0}$ is the quantity of the Sun's radiation received at the uppermost layer of Earth's surrounding. It denotes the aggregate solar irradiance incident per unit area, orthogonal to the Sun's rays, at the exobase of Earth's atmosphere. The intensity of this radiation is influenced by the solar constant, which is nearly 1361 W/m$^2$, as well as the Sun–Earth distance, which fluctuates throughout the year because of the Earth's elliptical orbit around the Sun. To calculate $I_{b0}$, the solar constant has to be adjusted based on the Earth–Sun distance factor, which is given by

$$I_{b0} = G_{sc} \times \frac{1 + 0.034 \times \cos\left(\frac{360^\circ}{365} \times n\right)}{d^2} \tag{1.9}$$

"$I_{b0}$" is extraterrestrial radiation in W/m$^2$.
"$G_{sc}$" is the solar constant (~1361 W/m$^2$).
"$d$" is the Earth–Sun distance factor, ~1.
$n$: The day number of the year (1 for January 1, 365 for December 31).
Note:

**Inverse-Square Law**

Following the inverse-square law, the solar irradiance I at a distance d from the Sun is expressed as

$$I = \frac{G_{sc}}{d^2} \tag{1.10}$$

where $d$ represents the Earth–Sun distance in astronomical units (AU).

**Variation in Earth–Sun Distance**

An approximation based on the Earth's oval-shaped path is used to explain the fluctuations in the varying distance d over the entire year. The Earth's orbit is an oval shape, so the Earth and Sun distance d can be expressed as

$$d = 1 + 0.034 \times \cos\left(\frac{360^\circ}{365} \times n\right) \tag{1.11}$$

**Derivation and Application**

1. **Ellipse Approximation**

To derive this formula, consider the Earth's orbit as an oval orbit with the Sun situated at one focus, wherein the Earth–Sun distance $d$ can be approximated through a simple harmonic function, reflecting the periodic variation in radial separation as Earth traverses its elliptical trajectory:

$$d = a(1 + e \times \cos(\theta)) \tag{1.12}$$

where

$a$ is the principal axis of the ellipse (average distance, ~1 AU).

$e$ is the eccentricity of the oval shape.

$\theta$ is the true anomaly, or the angular position of Earth in its orbit.

Since $\theta$ varies with time, the cosine function captures this variation, but for simplicity, an approximation with a fixed amplitude (0.034) is used to model the distance variation over the year, where $n$ represents the ordinal date; this equation delineates the oscillatory variations in the Earth–Sun distance induced by the elliptical orbital shape deviation, with 0.034 encapsulating the maximum deviation from the mean distance (1 AU).

**Substitute Distance into the Inverse-Square Law**

Substitute the expression for $d$ into the inverse-square law:

$$I = \frac{G_{sc}}{\left(1 + 0.034 \times \cos\left(\frac{360^\circ}{365} \times n\right)\right)^2} \tag{1.13}$$

This gives the solar energy flux at the atmosphere's upper boundary, adjusted for the varying distance from the Sun.

**Simplify**

The formula for the extraterrestrial solar irradiance $I_{b0}$ can be simplified by combining terms:

$$I_{b0} = G_{sc} \times \frac{1 + 0.034 \times \cos\left(\frac{360^\circ}{365} \times n\right)}{d^2} \tag{1.14}$$

**Direct Radiation ($I_b$)**

Strength of direct sunlight reaching the Earth's surface:

$$I_b = I_{b0} \times \cos(\theta_z) \tag{1.15}$$

"$I_b$" is direct radiation in W/m$^2$.
"$I_{b0}$" is extraterrestrial radiation in W/m$^2$.
"$\cos(\theta_z)$" is the cosine of the Sun's zenith angle.
Note:

**$\cos(\theta_z)$**: This factor accounts for the variation in the effective intensity of solar flux reaching a surface owing to the incident angle. As the solar zenith angle increases, the solar beams strike the surface at a more oblique angle, and the effective irradiance decreases.

**Diffuse Radiation**

Diffuse radiation occurs when solar energy is scattered by atmospheric particles, molecules, and clouds before reaching the ground. Unlike direct radiation, diffuse radiation does not create sharp shadows and contributes to the overall illumination of the sky. The scattering of sunlight is affected by factors such as the size and density of surrounding particles and the angle of solar incidence.

Diffuse radiation is measured with pyranometers, which can capture radiation from all directions. This type of radiation plays a significant role in areas where direct sunlight is limited due to weather conditions or geographical location. For instance, diffuse radiation is a crucial consideration in the layout of photovoltaic units, particularly in areas with persistent cloud coverage.

**Reflected Radiation**

Reflected solar radiation is the segment of solar energy that is reflected off the Earth's surface or clouds back into the surrounding. The magnitude of reflected radiation is influenced by the surface's reflectivity, albedo, and emissivity characteristics, which are measures of its reflective properties. Surfaces characterized by high albedo, including ice and snow, exhibit significant reflection of incident solar radiation, while darker surfaces, such as forests and oceans, preferentially absorb a larger fraction of this energy. Measuring reflected radiation helps in understanding the Earth's energy balance and assessing the impact of different surfaces on global temperatures. It also has implications for climate modeling and weather forecasting, as changes in albedo can affect local and global climate patterns.

**Albedo ($\alpha$)**

Albedo refers to the proportion of solar radiation that a surface reflects. It varies depending on the type of surface (e.g., snow, water, forest). The albedo is a dimensionless number between 0 and 1.

$$\text{Reflected Radiation}(I_d) = I_{\text{in}} \times \alpha \tag{1.16}$$

where
"$I_{\text{in}}$" is the approaching solar radiation (diffuse direct).
"$\alpha$" is the albedo of the surface.

**Total Solar Radiation**

Total solar radiation is the aggregate of direct, diffuse, as well as reflected solar energy, that reaches the Earth's surrounding. It provides a comprehensive quantification of the solar energy accessible at a given site and is essential for evaluating the potential for solar energy generation. Total solar radiation is measured using instruments like pyranometers, which capture both direct and diffuse components. Precise measurement of total solar radiation is vital for optimizing solar energy system efficiency and understanding the Earth's energy balance:

$$I_{\text{in}} = I_b + I_d \tag{1.17}$$

**Solar Time**

Solar time is a temporal framework predicated on the diurnal motion of the Sun, with time intervals defined by the Sun's apparent location relative to a fixed meridian defined by longitude. It contrasts with standard clock time, which is based on mean solar time and standardized through time zones. Solar time is ascertained through local solar noon, the instant at which the Sun attains its apex in the celestial sphere for a given locale. This concept is important for applications such as solar energy systems, where understanding the precise timing of solar events can optimize energy capture.

Solar time is a temporal metric derived from the Sun's position in the sky relative to an observer's location a specific location, and it can be different from standard clock time owing to the Earth's rotation and its elliptical orbit around the Sun influences the Sun's position in the sky. Here are the key steps and equations to calculate solar time.

**Solar Noon**

Solar noon is the instant when the Sun attains its maximum altitude in the celestial sphere, marking the zenith for a specific geodetic location, and it occurs when the solar hour angle ($H$) is zero.

**Equation for Solar Noon (in Local Solar Time)**

$$\text{Solar Noon(LST)} = 12:00 - \frac{\text{Equation of Time}}{60} \quad (1.18)$$

where the Equation of Time (EoT), expressed in temporal increments, quantifies discrepancies arising from the Earth's oval-shaped orbit as well as the tilt of the axis, which impact the precise alignment of solar time with clock time.

**Equation of Time (EoT)**

The temporal variance between solar time and mean time is rectified by the Equation of Time, which accounts for the Earth's orbital eccentricity and axial tilt obliquity.

**Equation for EoT**

$$\text{EoT} = 7.5\sin(0.9856 \times (n-3) + 1.914 \times \sin(0.9856 \times (n-2))) \quad (1.19)$$

where

"$n$" is the day of the calendar year.

Note:

To derive the Equation of Time (EoT), two main effects that cause discrepancies between solar time (the temporal measure derived from the Sun's true celestial position) and average solar time (the standardized time kept by a chronometric device, operating uniformly without irregularities) must be accounted for.

1. The Earth's orbital eccentricity: Earth's heliocentric trajectory is elliptical, exhibiting a deviation from perfect circularity. Consequently, the planet's velocity increases as it approaches perihelion (the proximal point to the Sun) and decreases as it recedes to aphelion (the farthest point from the Sun).
2. The axial tilt (obliquity) of the Earth: Fluctuations in the Sun's apparent velocity along the ecliptic throughout the annual cycle are a consequence of Earth's axial tilt relative to its orbital plane. The Equation of Time emerges from the integration of these two dynamic influences.

**Step 1: The Orbital Eccentricity Component**

The Earth's elliptical orbit introduces a variation in the speed at which the Sun appears to move across the sky. This is described by the term involving the "mean anomaly," $M$:

$$M = 0.9856^0 \times (n-3) \quad (1.20)$$

where "$n$" signifies the ordinal day of the year, with $n = 1$ congruent to January 1.

The equation represents the mean anomaly of Earth's orbit around the Sun. The mean anomaly is an angular measure that indicates the position of the Earth in its elliptical orbit as if it were moving at a constant speed.

**Understanding the Mean Anomaly ($M$)**

Mean Anomaly ($M$): The average irregularity is the angle that the Earth would have moved if it were in a circular orbit moving at a constant speed. It is a linear measure of time since the Earth passed perihelion (the position in the orbit nearest to the Sun).

2. **Orbital Period and Mean Motion**

- A complete revolution of Earth around the Sun spans roughly 365.25 days. This corresponds to 360° of angular motion.
- Mean Motion ($n_m$): The average rate of angular motion, or mean motion, can be calculated as

$$n_m = \frac{360^{\circ}}{365.25} \approx 0.9856^0 \text{ per day} \tag{1.21}$$

This value (0.9856°) represents the average angular movement of the Earth per day in its orbit.

3. **Explanation of the Formula**

In this equation:

$$M = 0.9856^0 \times (n - 3) \tag{1.22}$$

"$M$" is the mean anomaly in degrees.

"$n$" denotes the day of the year, with $n = 1$ represents January 1.

The term $(n - 3)$ accounts for the fact that perihelion (when the Earth reaches its nearest point to the Sun) usually occurs around January 3. Thus, $(n - 3)$ represents the number of days since perihelion.

4. **Step-By-Step Derivation**

1. Daily Angular Motion: As calculated earlier, the Earth moves ~0.9856° per day on average.
2. Day Offset: The mean anomaly is typically measured from perihelion, which is around January 3. So, for any day "$n$" after January 1, the number of days since perihelion is represented by $(-3)$.
3. Total Mean Anomaly: By multiplying the daily angular motion by the number of days since perihelion, the mean anomaly is obtained: $M = 0.9856^0 \times (n - 3)$

This equation tells how far, in degrees, the Earth has moved along its orbit since perihelion, assuming a constant speed.

The contribution to the EoT due to orbital eccentricity is approximately

$$\text{Eccentricity component} = 1.914 \times \sin(M) \tag{1.23}$$

**Step 2: The Axial Tilt (Obliquity) Component**

Seasonal fluctuations in the Sun's apparent motion and radiative intensity along the celestial sphere are driven by the axial tilt of the Earth. This is captured by the equation:

$$\text{Obliquity component} = 7.5 \times \sin(2L) \tag{1.24}$$

where $L$ represents the Sun's mean longitude in heliocentric orbit, which is approximately

$$L = M + \text{perigee of Earths orbit} \tag{1.25}$$

However, in this specific approximation, it can be simplified to

$$L \approx 0.9856^0 \times (\mathrm{n} - 3) + \sin(M) \tag{1.26}$$

**Step 3: Combine the Two Components**

The total Equation of Time is the sum of the two effects:

$$\text{EoT} = 7.5 \sin(0.9856 \times (n - 3) + 1.914 \times \sin(0.9856 \times (n - 2))) \tag{1.27}$$

This equation approximately represents the EoT by combining the contributions of both orbital eccentricity and axial tilt.]

**Local Solar Time (LST)**

To transform from standard clock time to local solar time, you need to adjust for the local time zone and the correction factor known as the Equation of Time:

$$\text{LST} = \text{Standard Time} + \frac{\text{Longitude} - \text{Standard Meridian}}{15} + \frac{\text{EoT}}{60} \tag{1.28}$$

where

**Standard Time**: The local clock time.

**Longitude**: The geographic coordinate of longitude location (in degrees).

**Standard Meridian**: The central meridian is defined for the time zone (in degrees, e.g., for UTC-5, the standard meridian is $-75°$).

**EoT**: Equation of Time in minutes.

**Note**

To derive the formula for Local Solar Time (LST), the components will be broken down and their relationships understood.

Standard Time (ST): The time indicated by the clock according to the applicable time zone.

Longitude ($\lambda$): The location's longitude, expressed in degrees.

Standard Meridian (SM): The reference longitude for standard time zone determination, typically a multiple of 15° (e.g., 0°, 15°, and 30°).

Equation of Time (EoT): The adjustment coefficient that compensates for the irregularities in Earth's orbit and axial tilt, which influence solar time.

1. **Difference in Longitude**

In a 24-h period, the Earth undergoes a complete 360° rotation on its axis, meaning it rotates 15° every hour.

The time difference caused by a difference in longitude (from the standard meridian) is given by

$$\text{Time Difference} = \frac{\text{Longitude} - \text{Standard Meridian}}{15} \tag{1.29}$$

This difference is positive if the location is east of the standard meridian and adverse if west.

2. **Correction for Equation of Time (EoT)**

The Equation of Time delineates the temporal discrepancy between apparent and mean solar time. Since this correction is given in minutes, converting it to hours gives

$$\text{EoT Correction} = \frac{\text{EoT}}{60} \tag{1.30}$$

3. **Combining These Elements**

Local Solar Time (LST) is the sum of the Standard Time, the modification due to the correction for the Equation of Time, and the longitudinal difference:

$$\begin{aligned} \text{LST} = \text{Standard} + \frac{\text{Longitude} - \text{Standard Meridian}}{15} \\ + \frac{\text{EoT}}{60} \end{aligned} \tag{1.31}$$

The Equation of Time delineates the temporal divergence, the discrepancy between solar time and mean time, a consequence of Earth's oval-shaped orbit and axial obliquity, inducing variations in the length of the solar day. The Equation of Time is used to adjust clock time to solar time, ensuring accurate timekeeping.

This adjustment is important for applications that require precise timing, such as astronomical observations and solar energy units.

**Application in Solar Energy Systems**

In solar energy units, solar time and solar declination are used to calculate the preferred angles for solar panels and to predict solar energy availability. By accounting for these factors, engineers can design systems that maximize energy capture and improve overall efficiency. Additionally, understanding solar time helps in scheduling maintenance and optimizing energy usage determined by the Sun's position.

**Applications and Implications**

**In Solar Energy Technology**

Understanding the different types of solar radiation and their direction is essential for designing and optimizing solar energy systems. By accounting for direct, diffuse, and reflected radiation, engineers can improve the efficiency of solar panels as well as other solar technologies. Accurate measurement and prediction of solar radiation help in forecasting energy production, scheduling maintenance, and optimizing energy usage. Technologies such as photovoltaic arrays, solar thermal collectors, and concentrated solar power systems are critically dependent on a profound understanding of solar radiation and its intricate characteristics.

**In Climate Science**

Variations in solar radiation serve as a fundamental catalyst for Earth's climate system, intricately influencing thermal dynamics, meteorological phenomena, and long-term climatic trajectories. By studying solar radiation, scientists can develop climate models that predict the impacts of solar variability on global and local climates. The analysis of solar radiation's impact on Earth's energy equilibrium is indispensable for assessing climatic shifts and devising strategies to counteract their consequences.

**In Agriculture and Urban Planning**

Solar radiation knowledge is key for energy, climate, and agriculture, optimizing agricultural practices as well as urban planning. In agriculture, understanding solar radiation patterns helps in managing crops and planning planting schedules. In urban planning, knowledge of solar angles and radiation intensity can be used to design energy-efficient buildings and landscapes that maximize natural light and reduce energy consumption.

**Case Studies and Examples**

Real-world applications of solar radiation principles include solar energy projects, climate research, and architectural design. Case examples demonstrate effective solar energy implementations, innovative technologies, and research findings provide valuable insights into the practical applications of solar radiation concepts.

**Future Directions and Research**

The field of solar radiation research is continually evolving, with emerging technologies and advancements in measurement techniques. Future research will focus on improving solar energy efficiency, developing new technologies, and enhancing our understanding of solar radiation's impact on climate and environmental systems.

## 1.8 Introduction to Solar Energy in Industry

Solar energy has quickly become a major force in global energy due to its innovation and sustainability, especially within the industrial sector. As the world increasingly shifts from fossil fuels to sustainable energy sources, solar power has gained prominence due to its scalability, declining costs, and minimal environmental impact. The industrial sector, constituting ~24% of global energy consumption, represents a formidable segment of the worldwide energy demand matrix, and is progressively incorporating solar energy into its operations to reduce carbon footprints, lower energy costs, and comply with more stringent environmental regulations. This transformation to solar energy is driven not just by financial considerations but also by the global push toward sustainability. In a concerted effort to drive solar energy adoption, governments worldwide have instituted an array of policies and incentives, encompassing tax credits, subsidies, and renewable energy mandates. These efforts have led to significant growth in solar installations across various industries, making solar energy a critical component of industrial energy strategies.

**Global Solar Energy Adoption**

The uptake of solar energy exhibits significant regional variation, with Europe, North America, and Asia–Pacific leading the way. Europe has been a leader in adopting renewable energy, with countries like Germany and Spain taking the lead in solar capacity due to the European Union's ambitious climate goals. In North America, federal and state-level incentives, coupled with the falling costs of solar technology, have driven significant adoption, particularly in the United States. States like California and Texas have seen widespread industrial solar installations. Meanwhile, the Asia–Pacific region, led by China and India, has experienced the fastest growth in solar energy adoption. China's dominance in solar panel manufacturing has made solar energy more accessible and affordable, leading to its widespread use in various industries. This expansion has been driven by technological innovations, economies of scale, and favorable government policies [38]. Figure 1.18 shows the annual energy capacity for concentrated solar power and solar photovoltaic.

**Industrial Dependence on Solar Energy**

Solar energy adoption has been growing across a variety of industrial sectors, each with distinct energy needs and operational challenges. The manufacturing sector, particularly in energy-intensive industries like steel, cement, and chemicals, has been

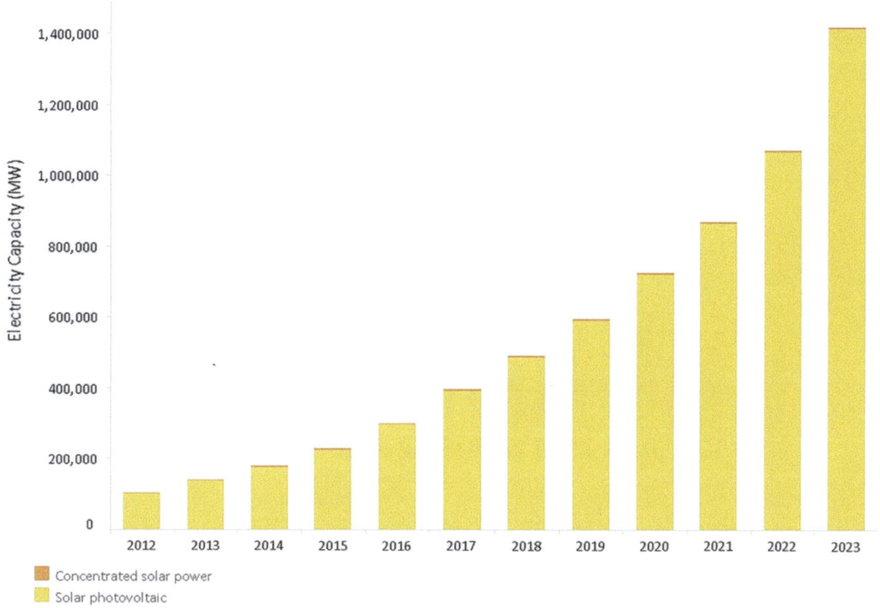


**Fig. 1.18** Annual energy capacity for solar photovoltaic as well as concentrated solar power. *Source* https://www.irena.org/Energy-Transition/Technology/Solar-energy

at the forefront of this shift. These industries have embraced solar energy as a way to reduce operational costs and carbon emissions, with many companies investing in large-scale solar installations that are often combined with energy storage solutions. The agricultural sector is also increasingly using solar energy for applications such as irrigation, greenhouse heating, and crop drying. Solar-powered water pumps and solar greenhouses are becoming more common, especially in sun-rich regions. The technology sector, including data centers and semiconductor manufacturing, is another major consumer of solar energy. Companies like Google, Apple, and Microsoft have invested heavily in large solar farms to power their operations. The transportation sector is exploring the use of solar energy for electric vehicle charging stations and solar-powered public transport systems, a trend that is expected to grow as electric vehicles become more widespread. Additionally, the construction industry is integrating solar energy into building designs, with solar panels being incorporated into rooftops, facades, and even windows, thus enhancing the sustainability of buildings [39–41]. Figure 1.19 represents the expected solar power market scale from 2022 to 2032.

**Economic Impact**

The economic implications of solar resources adoption in the industrial domain are profound. One of the most significant advantages is the decreasing cost of solar energy, which has made it compete with traditional energy sources such as coal,

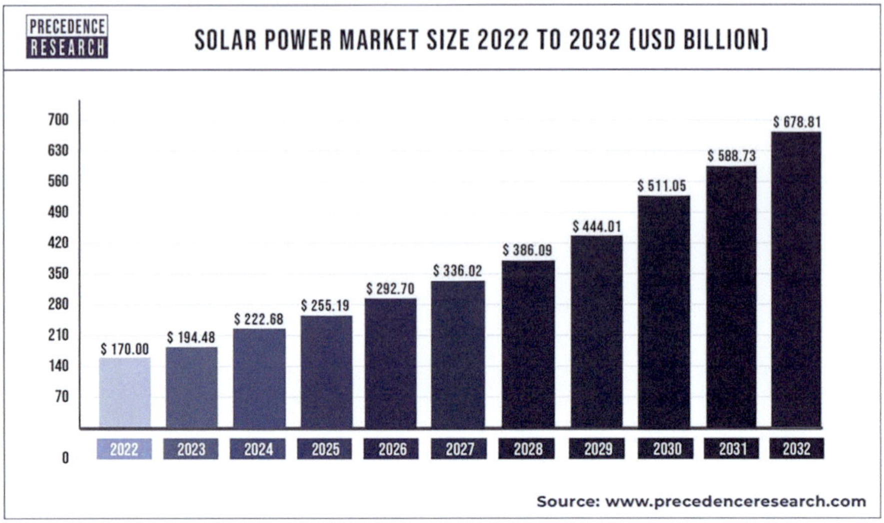


**Fig. 1.19** Expected solar power market scale 2022–2032

natural gas, and oil. Since 2010, the levelized cost of electricity (LCOE) from the Sun's energy has experienced a dramatic reduction, plummeting from around $250/MWh to a striking $40/MWh in 2023. By significantly reducing costs, solar is emerging as one of the most cost-effective energy alternatives available. Additionally, as solar energy adoption grows, the return on investment (ROI) for industrial sector projects is becoming increasingly compelling. The initial capital expenditure is often recouped within 5–15 years, depending on the scale of the installation and local energy prices [42]. This financial viability is further enhanced by various government incentives and the potential for generating revenue by selling surplus energy back to the grid. Furthermore, the solar industry has played a crucial role in job creation, with millions of jobs generated globally in areas including maintenance, manufacturing, installation, and research and development. In the United States alone, the solar industry employs over 250,000 people, with the industrial sector contributing significantly to these employment figures [43]. Presented in Fig. 1.20 is a pie chart that intricately quantifies the relative share of diverse energy sources in fulfilling the global energy demand.

**Technological Advances and Challenges**

Technological innovations in solar energy have significantly boosted the efficiency and flexibility of solar units, making them increasingly well-suited for industrial applications. Recent breakthroughs in photovoltaic cell technology have significantly improved conversion efficiencies, with some cells now converting over 25% of sunlight into electricity. Advancements like bifacial solar panels, designed to capture sunlight from both sides, represent significant innovations, and are further boosting energy production. In addition to photovoltaic technology, solar thermal systems

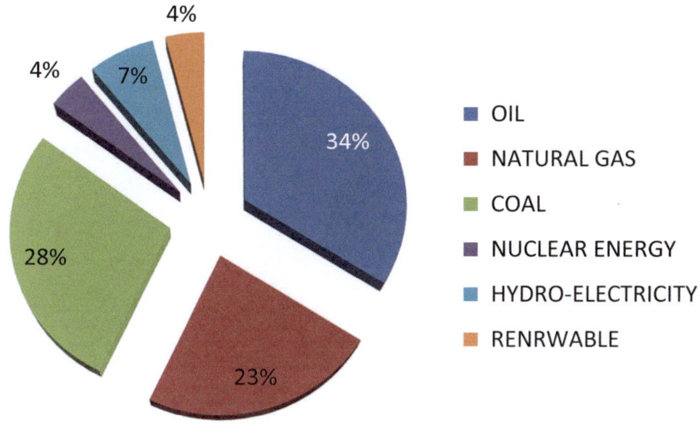


**Fig. 1.20** Proportional distribution of energy sources in global energy demand fulfillment

are gaining traction in industrial settings. By using mirrors or lenses to concentrate sunlight, these systems generate heat for energy production, which can be directly utilized in industrial processes or converted into electricity. Despite these advancements, challenges remain in scaling solar energy for widespread industrial use. To ensure a dependable power supply, the intermittency of solar power necessitates effective energy storage solutions. Moreover, the upfront capital expenses for large-scale solar installations can be a barrier, although these costs are steadily decreasing as technology improves and economies of scale are realized [44–46].

**Future Outlook**

Solar energy's future in the industrial sector appears promising, with projections indicating continued growth in adoption and capacity. As solar technology continues to advance and costs decline further, it is expected that more industries will make solar energy the primary power source. The increasing pressure from governments and consumers encouraging industries to lower their carbon footprints is expected to accelerate this trend. Solar energy holds vast potential to satisfy global energy requirements which is significant, especially as more regions adopt ambitious renewable energy targets. Combining solar energy with other renewable sources like wind and hydroelectric power forms a comprehensive and sustainable energy solution. Hydro enhances energy reliability and sustainability, along with advances in energy storage technologies, and could foster a more resilient and sustainable energy framework for the industrial sector. Additionally, policies addressing climate change, like carbon pricing and emissions trading schemes, will be essential in driving the global adoption of solar energy in industries.

The global industrial field's reliance on solar energy is growing rapidly, driven by the need for inexhaustible as well as cost-effective energy solutions. The implementation of solar energy across various industries—from manufacturing to technology and agriculture—demonstrates the viability and potential of this sustainable energy source. As solar technology continues to evolve and costs reduce, the industrial sector is poised to increase its dependence on solar energy, with its economic and environmental advantages; solar energy plays a vital role in worldwide initiatives to address climate change and shift toward a low-carbon economy, making it an increasingly attractive option for industries worldwide. Moving forward, continued support from governments and the private sector will be essential in realizing the complete solar energy in the industrial sector.

## References


1. Schrijver CJ, Siscoe GL (2009) The sun as a variable star: solar and stellar irradiance variability. Astrophys Space Sci 302:1–15
2. Zirker JB, Engvall OF (2019) Chapter 1—discoveries and concepts: the sun's role in astrophysics. Sun Guide Stellar Phys 15:1–26
3. Frohlich C, Huber MCE, Solanki SK, Steiger R (1998) Solar composition and its evolution: from Core to Corona. In: Proceedings of an ISSI workshop 26–30 January 1998, Bern
4. George C (1999) Solar variability and its implications for the human environment. J Atmosp Solar Terrest Phys 61:3–14
5. Adelberger EG, García A, Robertson RGH et al (2011) Solar fusion cross sections II: the pp chain and CNO cycles. Rev Mod Phys 4:2318
6. Owusu PA, Asumadu-Sarkodie S, Dubey S (2016) A review of renewable energy sources, sustainability issues and climate change mitigation. Cogent Eng 3(1):1167990
7. Zohuri B, McDaniel P (2021) Chapter 1—population growth driving energy demand. In: Introduction to energy essentials. Academic Press, pp 1–42
8. Chakraborty O, Das B, Gupta R (2022) Effect of spherical balls insert in collector tube of parabolic trough collector by considering $Al_2O_3$ based water. J Brazil Soc Mech Sci Eng 44:1–17
9. Rathore PKS, Rathore S, Singh RP, Agnihotri S (2018) Solar power utility sector in India: challenges and opportunities. Renew Sustain Energy Rev 81:2703–2713
10. Gharzi M, Kermani AM, Shamsabadi HT (2023) Experimental investigation of a parabolic trough collector-thermoelectric generator (PTC-TEG) hybrid solar system with a pressurized heat transfer fluid. Renew Energy 202:270–279
11. Negi S, Jaswal G, Dass K, Mazumder K, Elumalai S, Roy JK (2020) Torrefaction: a sustainable method for transforming of agri-wastes to high energy density solids (biocoal). Rev Environ Sci Bio/Technol 19(2):463–488
12. Ahmadi A, Das B, Ehyaei MA, Esmaeilion F, Assad MEH, Jamali DH, Koohshekan O, Kumar R, Rosen MA, Negi S, Bhogilla SS, Safari S (2021) Energy, exergy, and techno-economic performance analyses of solar dryers for agro products: a comprehensive review. Sol Energy 228:349–373
13. Ajadi T, Boyle R, Strahan D, Kimmel M, Collins B, Cheung A, Becker L (2019) Global trends in renewable energy investment 2019. In: Frankfurt school-UNEP centre. BNEF, Frankfurt. http://www.fs-unep-centre.org. Accessed 16 Feb 2020
14. Kastanaki E, Giannis A (2022) Energy decarbonisation in the European Union: assessment of photovoltaic waste recycling potential. Renew Energy 192:1–13
15. Eurostat EU (2022) Overachieves 2020 renewable energy target. https://ec.europa.eu/eurostat/web/products-eurostat-news/-/ddn-20220119-1. Accessed 14 Jan 2023
16. Rabbi MF, Popp J, Mate D, Kovacs S (2022) Energy security and energy transition to achieve carbon neutrality. Energies 15(21):8126
17. Cifuentes-Faura J (2022) European Union policies and their role in combating climate change over the years. Air Qual Atmos Health 15:1333–1340

18. Yao L (2022) Domestic dynamics and china's engagement in global renewable energy governance. Chin J Int Polit 15(1):69–86
19. Yang H, Huang X, Hu J, Thompson JR, Flower RJ (2022) Achievements, challenges and global implications of China's carbon neutral pledge. Front Environ Sci Eng 16(8):1–4
20. Lal SR, Herbert GM, Arjunan P, Suryan A (2022) Advancements in renewable energy transition in India: a review. Energy Sour A Recov Utiliz Environ Effects 2022(14):1–31
21. Yarlagadda B, Smith SJ, Mignone BK, Mallapragada D, Randles CA, Sampedro J (2022) Climate and air pollution implications of potential energy infrastructure and policy measures in India. Energy Clim Change 3:100067
22. Chakraborty O, Roy S, Debnath BK, Negi S, Rosen MA, Safari S, Assad MEH, Gupta R, Das B (2023) Energy, exergy, environment and techno-economic analysis of parabolic trough collector: a comprehensive review. Energy Environ 35:1118–1181
23. Lata T, Mishra AK, Dubey AK, Tripathi CB, Baredar P (2016) Renewable energy: an overview on its contribution in current energy scenario of India. Renew Sustain Energy Rev 60:226–233
24. Jha SK, Puppala H (2017) Prospects of renewable energy sources in India: prioritization of alternative sources in terms of energy index. Energy 127:116–127
25. Mao H, Chen X, Luo Y, Deng J, Tian Z, Yu J, Xiao Y, Fan J (2023) Advances and prospects on estimating solar photovoltaic installation capacity and potential based on satellite and aerial images. Renew Sustain Energy Rev 179:113276
26. Liu S, Biju VP, Qi Y, Chen W, Liu Z (2023) Recent progress in the development of high-efficiency inverted perovskite solar cells. NPG Asia Mater 15:27
27. Manokar AM, Winston DP, Vimala M (2016) Performance analysis of parabolic trough concentrating photovoltaic thermal system. Proced Technol 24:485–491
28. Kumar KH, Daabo AM, Karmakar MK, Hirani H (2022) Solar parabolic dish collector for concentrated solar thermal systems: a review and recommendations. Environ Sci Pollut Res 29:32335–32367
29. Sharma A (2011) A comprehensive study of solar power in India and world. Renew Sustain Energy Rev 15:1767–1776
30. Lucena JAY (2021) Chapter 5—recent advances and technology trends of wind turbines. Recent Adv Renew Energy Technol 1:177–210
31. Diaz H, Soares CG (2020) Review of the current status, technology and future trends of offshore wind farms. Ocean Eng 209:107381
32. Sari MA, Badruzzaman M, Cherchi C, Swindle M, Ajami N, Jacangelo JG (2018) Recent innovations and trends in in-conduit hydropower technologies and their applications in water distribution systems. J Environ Manag 228:416–428
33. Paish O (2002) Small hydro power: technology and current status. Renew Sustain Energy Rev 6:537–556
34. Gong L, Han D, Chen Z, Wang D, Jiao K, Zhang X, Yu B (2023) Research status and development trend of key technologies for enhanced geothermal systems. Natl Gas Ind B 10:140–164
35. Zarrouk SJ, Moon H (2014) Efficiency of geothermal power plants: a worldwide review. Geothermics 51:142–153
36. Usman IMT, Ho YC, Baloo L, Lam MK, Sujarwo W (2022) A comprehensive review on the advances of bioproducts from biomass towards meeting net zero carbon emissions (NZCE). Bioresour Technol 366:128167
37. Adeniyi OM, Azimov U, Burluka A (2018) Algae biofuel: current status and future applications. Renew Sustain Energy Rev 90:316–335
38. Fabra N (2021) The energy transition: an industrial economics perspective. Int J Ind Organ 79:102734
39. Bolinger M, Seel J, Robson D (2019) Utility-scale solar: empirical trends in project technology, cost, performance, and PPA pricing in the United States—2019 edition. Energy Mark Policy 21:455
40. Mekhilef S, Saidur R, Safari A (2011) A review on solar energy use in industries. Renew Sustain Energy Rev 15:1777–1790

41. Tsoutsos T, Frantzeskaki N, Gekas V (2005) Environmental impacts from the solar energy technologies. Energy Policy 33:289–296
42. Waldau AJ (2018) Snapshot of photovoltaics—February 2018. In: European commission, joint research centre (JRC), directorate C: energy, transport and climate, energy efficiency and renewables unit, via E. Fermi 2749, 21027 Ispra, VA, Italy
43. Maurya PK, Mondal S, Kumar V, Singh SP (2021) Roadmap to sustainable carbon-neutral energy and environment: can we cross the barrier of biomass productivity? Environ Sci Pollut Res 28:49327–49342
44. Moro JH, Martinez-Duart JM (2013) Analytical model for solar PV and CSP electricity costs: present LCOE values and their future evolution. Renew Sustain Energy Rev 20:119–132
45. Kabir E, Kumar P, Kumar S, Adelodun AA, Kim KH (2018) Solar energy: potential and future prospects. Renew Sustain Energy Rev 82:894–900
46. Singh A, Vats G, Khanduja D (2016) Exploring tapping potential of solar energy: prioritization of Indian states. Renew Sustain Energy Rev 58:397–406

# Chapter 2
# Different Types of Solar Concentrating Collectors

This chapter delves into the diverse world of solar concentrating collectors—essential technologies that convert sunlight into usable thermal energy through optical concentration. It explores major types such as Parabolic Dish, Solar Tower, Parabolic Trough, and Fresnel Lens systems, explaining their unique working principles, thermal performance, and design characteristics. By tracing their historical development and technological evolution, the chapter highlights how each system has advanced in efficiency and application. These collectors are examined in terms of their advantages, limitations, and suitability for various industrial processes, electricity generation, and space heating. With each type offering distinct performance traits, this chapter provides a comparative framework for understanding their role in modern solar energy systems and their contribution to a sustainable energy future.

## 2.1 Parabolic Dish Collector (PDC)

The Parabolic Dish Collector (PDC) is a solar concentrator that has evolved significantly. Concentrating sunlight using reflective surfaces dates back to ancient times, with scientific foundations laid in the eighteenth and nineteenth centuries. Early experiments, such as Horace-Benedict de Saussure's first solar oven in 1767 [1] and Augustin Mouchot's solar-powered steam engine in the 1860s, demonstrated the potential of concentrated solar energy. Modern PDC development began in the early twentieth century, with American engineer Frank Shuman constructing a solar power plant in Egypt in 1913 using parabolic troughs, demonstrating large-scale solar concentrators' feasibility [2, 3]. Mid-twentieth-century advancements in materials like polished aluminum and coatings improved efficiency, and the integration of Stirling engines led to dish-Stirling systems that converted concentrated solar heat into mechanical energy and electricity. The 1970s energy crisis spurred further development, with innovations in dish curvature, receiver efficiency, and tracking

 
O. Chakraborty and B. Das, *Parabolic Solar Trough Collector*, Mechanical Engineering Series, https://doi.org/10.1007/978-3-032-11229-3_2

systems, improving commercial viability. Today, PDCs are recognized for their high efficiency in converting solar energy into electricity, with applications from residential to large-scale solar power plants. Ongoing research aims to reduce costs and improve performance. Volumetric collectors in CSP systems, integral to PDCs, transform concentrated solar energy into thermal energy, with studies focused on optimizing their design and minimizing heat losses [4–13].

Figure 2.1 illustrates the design and components of the Economic Solar Paraboloidal Concentrator (ESPC) developed by the Solar Energy Research Laboratory at the University of Rajasthan, Jaipur.

The Parabolic Dish Collector (PDC) is a highly advanced solar energy system, specifically designed for high-temperature applications to generate heat that can be converted into electricity. Its operation is based on parabolic reflection, where a parabolic reflector captures sunlight, which is then concentrated onto a single focal point. The parabolic shape of the dish ensures that sunlight, arriving in parallel rays from the sun, is efficiently reflected and focused onto the receiver, where the concentrated energy is converted into thermal energy [14]. This ability to focus solar energy makes PDCs ideal for industrial and residential use, particularly in systems requiring high temperatures.

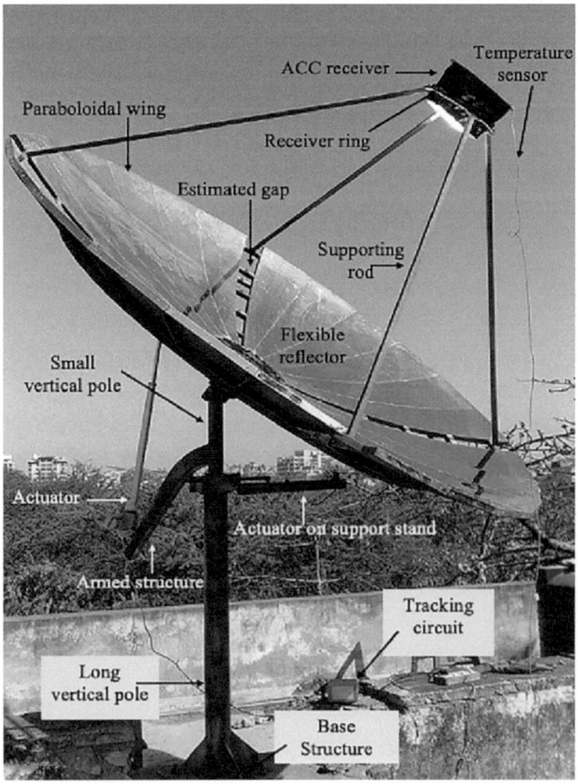


**Fig. 2.1** Design and components of the economic solar paraboloidal concentrator (ESPC) developed at the solar energy research laboratory, University of Rajasthan, Jaipur, India [3]

The design and construction of a PDC involve several key components. The parabolic reflector is made from highly reflective materials such as smooth aluminum, silver-coated glass, and advanced reflective films, selected for their high durability and reflectivity. Precision in the curvature of the dish is crucial, as even slight deviations can significantly reduce the efficiency of sunlight concentration [2]. The receiver absorbs the concentrated solar energy at the dish's focal point. The receiver is typically coated with materials that reduce heat loss and improve energy absorption. Its primary function is to transfer the captured heat to a working fluid, such as synthetic oils, molten salts, and air, which circulates through the receiver and absorbs the thermal energy [15]. The choice of working fluid depends on the required temperature range; synthetic oils are used for lower temperatures, while molten salts are preferred for higher temperatures and thermal storage due to their ability to store energy efficiently [14].

To maintain optimal sunlight focus, PDCs are mounted on support structures equipped with tracking systems that ensure the dish continuously aligns with the sun. These systems can be single-axis or dual-axis, with dual-axis tracking offering more precise alignment, maximizing the system's efficiency throughout the day [16]. Figures 2.2 and 2.3 illustrate the components of the PDC, highlighting the HTF pipes and energy flow within the system.

**Efficiency and Performance**

Several determinants govern the efficiency of a Parabolic Dish Collector, including the reflective quality of the surface, the precision in maintaining the parabolic shape,

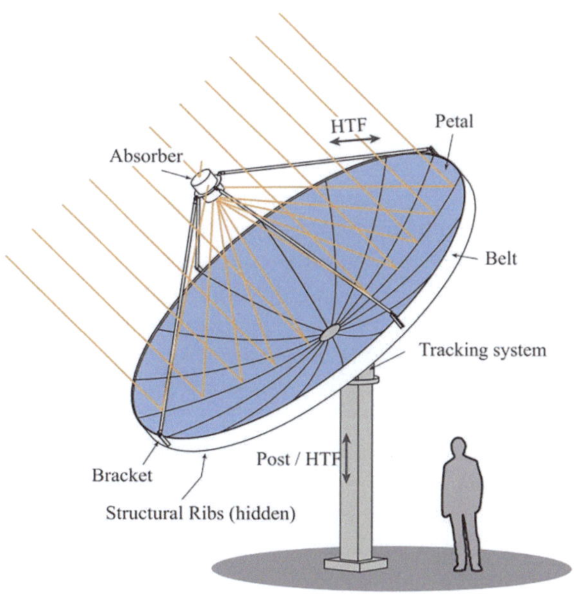


**Fig. 2.2** Overview of PDC components, featuring HTF pipes within the post- and petal-based mirror surface design [17]

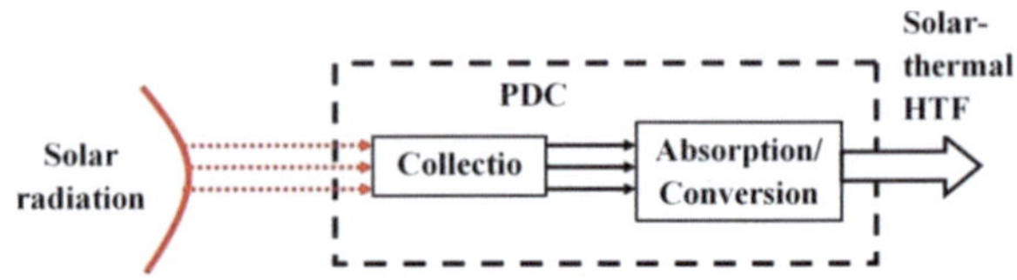


**Fig. 2.3** Block diagram and energy flow of the PDC, showing collection in the parabolic reflector and absorption in the receiver (HTF—heat transfer fluid)

and the operational efficacy of the tracking system. PDCs are known for their high efficiency in converting solar energy into thermal energy, with some systems achieving conversion efficiencies of over 30%. This high efficiency is due to the ability of the parabolic dish to concentrate sunlight to very high temperatures, often exceeding 1000 °C (1832 °F).

Relative to other solar concentration systems, such as parabolic troughs or solar power towers, Parabolic Dish Collectors (PDCs) offer several advantages, including their ability to achieve significantly higher thermal levels, making them ideal for applications that require extreme heat [14]. Additionally, PDCs are more compact due to their ability to concentrate light into a small focal area, which makes them well-suited for locations with limited space [1]. However, PDCs face challenges, including the complexity and higher cost associated with the precision required for manufacturing and aligning the dish, making them more expensive than simpler systems [14]. Moreover, the tracking systems and reflective surfaces of PDCs necessitate regular maintenance to ensure continued optimal performance [14].

### Applications

Parabolic Dish Collectors (PDCs) are highly versatile and can be used in various applications, making them an essential technology in solar energy systems [18, 19]. One of the primary usage of PDCs is electricity generation, where they are paired with Stirling engines or other thermodynamic cycles to convert thermal energy into electricity. This application is particularly beneficial in regions with high solar insolation, where PDCs are a key component of solar power plants, enabling efficient conversion of concentrated solar energy into usable power. In addition, PDCs can be integrated with thermal storage systems to store energy for use when solar radiation is unavailable. These systems often use materials like molten salts to store thermal energy, allowing for a more reliable energy supply, especially during cloudy days or nighttime. Another critical application of high-temperature PDCs is in industrial processes, such as chemical manufacturing, metal melting, and steam generation, where they provide the necessary high temperatures to support these energy-intensive tasks.

Recent advancements in PDC technology have focused on improving the efficiency, reducing costs, and expanding the range of applications [20–22]. Advanced materials are being developed for reflective coatings and dish structures, improving performance and durability. Innovations in tracking systems are also being made to enhance the accuracy and reliability of solar tracking, minimizing the need for manual adjustments and lowering maintenance costs. Additionally, there is increasing integration of PDCs with other renewable energy technologies, such as photovoltaic units, to create hybrid systems. These hybrid systems offer greater overall efficiency and flexibility, allowing for more consistent and adaptable energy generation, making PDCs a key player in the future of renewable energy solutions.

**Challenges and Future Directions**

Despite their advantages, Parabolic Dish Collectors (PDCs) face several challenges that hinder their widespread adoption [2, 14, 20]. One of the main obstacles is **cost**, as the high initial investment and ongoing maintenance expenses can make PDCs less economically viable for many applications. Reducing these costs through continued research and innovation is essential to making PDC technology more accessible. Another challenge is scalability; scaling up PDC systems for large-scale applications presents technical and economic difficulties, and finding practical solutions to improve scalability is crucial for broader implementation. Future research and development in PDC technology will likely focus on overcoming these challenges by enhancing system performance, optimizing cost-efficiency, and exploring new applications. Advancements in materials, design, and integration with other technologies will play a critical role in advancing the capabilities of PDCs and expanding their impact in the renewable energy sector.

**Mathematical Formulation**

The efficiency of a PDC is primarily determined by its concentration ratio, which is the ratio of the dish's aperture area to the receiver's surface area. Mathematically, the concentration ratio "C" can be expressed as

$$C = \frac{A_{\text{dish}}}{A_{\text{receiver}}} \tag{2.1}$$

where $A$ dish is the aperture area of the parabolic dish and $A_{\text{receiver}}$ is the area of the absorbing surface. An elevated concentration ratio signifies the concentration of a larger volume of solar irradiance onto a significantly smaller spatial domain, leading to higher temperatures at the receiver and, consequently, greater efficiency in converting solar energy into heat. However, achieving a high concentration ratio also requires precise manufacturing and alignment of the dish and receiver, as even slight deviations can significantly impact the system's performance.

The thermal efficiency of a PDC is another critical parameter and can be described by the equation:

$$\eta = \frac{\text{Useful thermal energy output}}{\text{Solar energy input}} \tag{2.2}$$

$$\eta = \frac{Q_u}{A_c \times G_{in}} \tag{2.3}$$

where

$\eta$ = Thermal efficiency of the collector.

$Q_u$ = Effective thermal energy captured by the collector (W).
$A_c$ = Area of the collector's aperture ($m^2$).
$G_{in}$ = Incoming solar radiation on aperture area ($W/m^2$).

$$Q_u = m \times c_p \times \Delta T \tag{2.4}$$

where
$\eta_o$ = Optical efficiency of the system (dimensionless).
$Q_{loss}$= Thermal losses from the collector to the environment (W).
The thermal efficiency is then

$$\eta = \frac{\eta_o \times A_c \times G_{in} - Q_{loss}}{A_c \times G_{in}} = \eta_0 - \frac{Q_{loss}}{A_c \times G_{in}} \tag{2.5}$$

This efficiency is affected by various factors, such as the optical efficiency of the system, which accounts for losses due to reflection, absorption, and scattering. Optical efficiency, denoted as $\eta_o$, measures how effectively the system concentrates and directs sunlight onto the receiver. It is affected by the quality of the reflective interface, the precision of the parabolic shape, and the alignment of the dish with the sun. The optical efficiency can be further expressed in terms of the system's ability to minimize optical losses, such as those from imperfect reflection or the dispersion of light due to surface roughness.

Note:

[The optical efficiency $\eta_o$ of a parabolic solar dish collector measures how effectively the collector focuses and transmits solar radiation onto the receiver. It can be computed through the following formula:

$$\eta_o = \rho \times \tau \times \alpha \times \gamma \times \eta_c \tag{2.6}$$

where
$\rho$: Reflectivity of the dish surface (dimensionless), representing the fraction of incident Sunlight reflected from the dish. This value depends on the material and condition of the dish surface.

$\tau$: Transmissivity of the cover glass or any transparent material (dimensionless), representing the fraction of solar radiation that passes through the cover to reach the receiver. This applies with a transparent cover; otherwise, it can be taken as one if no cover is used.

$\alpha$: Absorptivity of the receiver (dimensionless), representing the fraction of incident radiation on the receiver that is absorbed. This depends on the material and coating of the receiver.

$\gamma$: Intercept factor (dimensionless) represents the fraction of reflected solar radiation the receiver intercepts. It is usually $<1$ and depends on the geometric alignment and the accuracy of the dish in focusing sunlight onto the receiver.

$\eta_c$: Concentration ratio efficiency, accounting for the errors in the concentration process, such as surface imperfections or misalignment (dimensionless). It is related to the optical quality of the dish and how well it concentrates the sunlight.]

The next crucial step involves managing the heat losses from the receiver, which can occur through conduction, convection, and radiation. Heat losses reduce the amount of thermal energy utilized for power generation or other applications. Therefore, the energy balance within the system is crucial for maintaining high efficiency. The sound thermal energy output, denoted as $Q_{\text{useful}}$, can be calculated by subtracting the total heat losses from the absorbed energy:

$$Q_{\text{useful}} = Q_{\text{absorbed}} - Q_{\text{losses}} \tag{2.7}$$

where $Q_{\text{losses}}$ includes all forms of heat dissipation, including conductive losses through the receiver material, convective losses to the surrounding air, and radiative losses back into the environment.

In practical applications, the thermal energy generated by the Parabolic Dish Collector (PDC) is often used to drive a Stirling engine, which converts thermal energy into mechanical work by cyclically compressing and expanding air or gas at different temperatures. The engine's efficiency depends on the temperature difference between the receiver (heat source) and the heat sink (ambient environment). Higher receiver temperatures enhance engine efficiency. Factors such as the receiver's operating temperature, heat loss minimization, and precise alignment of the dish with the sun are crucial to maximizing the PDC system's efficiency. Advanced dual-axis tracking systems ensure continuous alignment with the sun to maximize solar energy capture. The PDC's performance is influenced by its concentration ratio, optical and thermal efficiency. It can be optimized through careful design and operational management, making PDCs an essential technology for solar energy applications.

## 2.2 Parabolic Solar Trough Collector

The development of parabolic solar trough collectors represents a significant leap in solar energy technology, built on over a century of innovation. The concept dates back to the mid-nineteenth century when French inventor Augustin Mouchot first demonstrated the potential of concentrating solar energy using a parabolic reflector to generate steam. Although this early work was groundbreaking, it was overshadowed by the dominance of fossil fuels. Further advancements came in the early twentieth century when American engineer Frank Shuman built one of the first practical solar power plants in Maadi, Egypt, using parabolic troughs to concentrate sunlight and generate steam for powering a water pump. Despite setbacks, including World War I, interest in solar energy revived during the 1970s energy crisis, prompting improvements in concentrating solar power (CSP) technologies, such as parabolic troughs, to enhance efficiency for large-scale use. In the 1980s, Luz International's Solar Energy Generating Systems (SEGS) in the Mojave Desert utilized

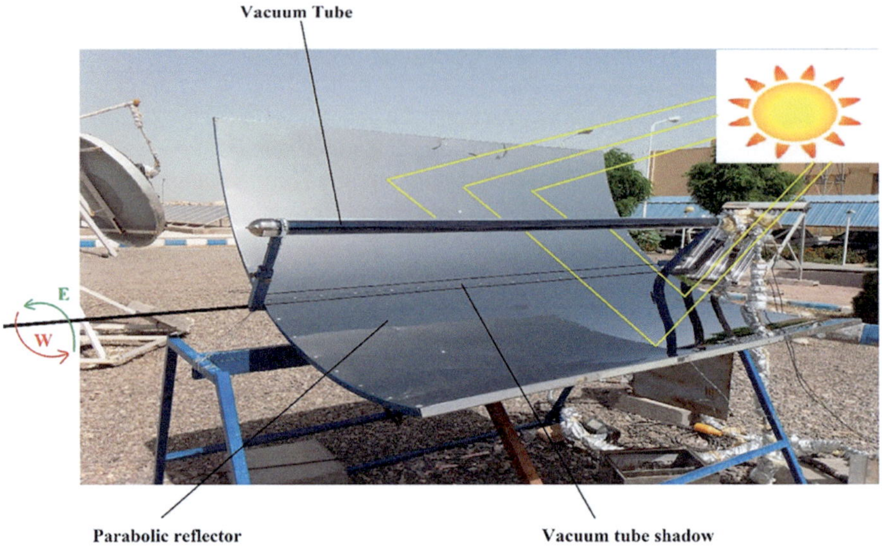


**Fig. 2.4** Experimental setup of a parabolic trough solar collector (PTSC) [23]

parabolic troughs to produce electricity on a commercial scale, demonstrating their viability. Today, parabolic trough collectors are a well-established CSP technology that continuously improves through materials, heat transfer fluids, and thermal energy storage advances. These innovations strengthen their role in the global transition to sustainable energy, as seen in the evolution depicted in Fig. 2.4.

Recent advancements in parabolic trough solar collectors (PTSCs) have focused on improving heat transfer performance through innovations in nanofluids and specialized inserts. Javidan et al. [24] investigated hybrid nanofluids, specifically $TiO_2$–Cu, Ag–MgO, and $Al_2O_3$–CuO particles in Syltherm-oil-800. They found that adding $TiO_2$–Cu particles at a 0.03 mass fraction improved the heat transfer coefficient by 21%. Vahidinia et al. [25] tested a hybrid nanofluid with MWCNT and $Fe_3O_4$ particles in Therminol VP-1 and combined it with an internally finned receiver tube, finding increased energetic efficiency with higher inlet temperatures. Chakraborty [26] introduced spinning flower-shaped inserts in a PTSC tube using a ternary nanofluid mixture of $Al_2O_3$, GO, and CuO in water, achieving a 30.87% improvement in thermal efficiency and a 98.34% increase in the heat transfer coefficient at 15 rad/s. Rawat et al. [27] compared ternary hybrid nanofluids ($Al_2O_3$–Cu–CNT) with mono and hybrid nanofluids, showing superior performance of the ternary blend.

Chakraborty et al. [28] also explored hybrid nanofluids with oval-shaped inserts in spinning receiver tubes, leading to a 29.21% increase in thermal efficiency and a 44.21% improvement in the heat transfer coefficient. Adun et al. [29] found that CuO–MgO–$TiO_2$ nanofluids provided the highest energetic efficiency at 25.1% in a Kalina cycle integrated with a PTSC. Chakraborty et al. [30] reported improvements of 16% in the thermal performance index and 13.3% in thermal efficiency with elliptical inserts using CuO and $Al_2O_3$ nanoparticles. Menbari et al. [31, 32] observed that binary nanoparticles in ethylene glycol–water mixtures achieved superior thermal efficiency compared to other nanofluid combinations. Minea and El-Maghlany [33] highlighted that Ag–MgO nanoparticles in distilled water offered the highest thermal efficiency for PTSCs, reaching 60%. Bellos and Tzivanidis [34] noted a 1.8% improvement in thermal efficiency with hybrid nanofluids compared to 0.7% with mono nanofluids. Chakraborty et al. [35] achieved a 50.90% increase in the Nusselt number and a 38.03% improvement in thermal efficiency using spinning star inserts and a binary hybrid nanofluid of copper oxide and aluminum oxide in water. Last, Dou et al. [36] found significant enhancements in convective heat transfer with Cu nanoparticles in synthetic oil. Zhang et al. [37] reported a 22.34% improvement with a quaternary mixture of molten salt and $SiO_2$ nanoparticles. Mwesigye et al. [38] observed a 13.9% increase in thermal performance with Ag nanoparticles in Therminol®VP-1. These studies underscore the crucial role of nanofluids and inserts in enhancing PTSC efficiency.

A parabolic solar trough collector is a concentrating solar power (CSP) system that uses parabolic-shaped mirrors to focus sunlight onto a receiver tube at the parabola's focal line. The principle of operation involves the following steps.

### Principle of Operation of Parabolic Solar Trough Collector

The parabolic trough is a curved mirror shaped like a parabola, designed to track the sun's movement across the sky using either one-axis or two-axis tracking systems, ensuring that sunlight is continuously focused onto the receiver tube throughout the day. As sunlight strikes the parabolic mirror, it is reflected and converges at the focal point of the receiver tube. The parabolic shape guarantees that all incoming parallel rays of sunlight are concentrated onto a narrow line, maximizing efficiency. The receiver tube, coated with a selective material with high absorptivity for solar radiation and low emissivity to minimize heat loss, absorbs the concentrated sunlight, heating the fluid inside, typically thermal oil or molten salt. This heated fluid is then circulated through a heat exchanger or directed to a power generation system, where the thermal energy is used to produce steam. The steam drives a turbine connected to an electrical generator, generating electricity. In some systems, excess heat is stored in thermal storage systems, allowing for energy use when sunlight is unavailable. After transferring heat to the steam generator or heat exchanger, the fluid is cooled and recirculated back through the receiver tube to be reheated by the concentrated sunlight, maintaining the system's continuous operation.

Note:

**Concentration Ratio:** The parabolic trough achieves a high concentration ratio, which increases the efficiency of the heat absorption process.

**Tracking System:** A tracking mechanism ensures the trough follows the sun, optimizing the amount of sunlight collected and focused on the receiver.

This system is widely used in solar thermal power plants and can achieve high temperatures, making it suitable for generating electricity at a commercial scale.

Figure 2.5 illustrates the different modes of heat interaction, beginning from the working atmosphere (denoted as 1 in the figure), passing to the working fluid (denoted as 3 in the figure), and returning to the atmosphere (also denoted as 1 in the figure). During these modes of heat interaction, while the working fluid is flowing through the tube, it absorbs reflected heat from the trough.

### Design and Construction of a Parabolic Solar Trough Collector

Designing and constructing a parabolic solar trough collector involves several key components and considerations to optimize the system's efficiency. Below is a step-by-step overview of the design and construction process.

### Design Considerations

The solar trough needs to be placed in an area with maximum sunlight exposure, generally oriented north–south for daily east–west sun tracking. Its design should aim for a high concentration ratio, directing more sunlight onto the receiver tube

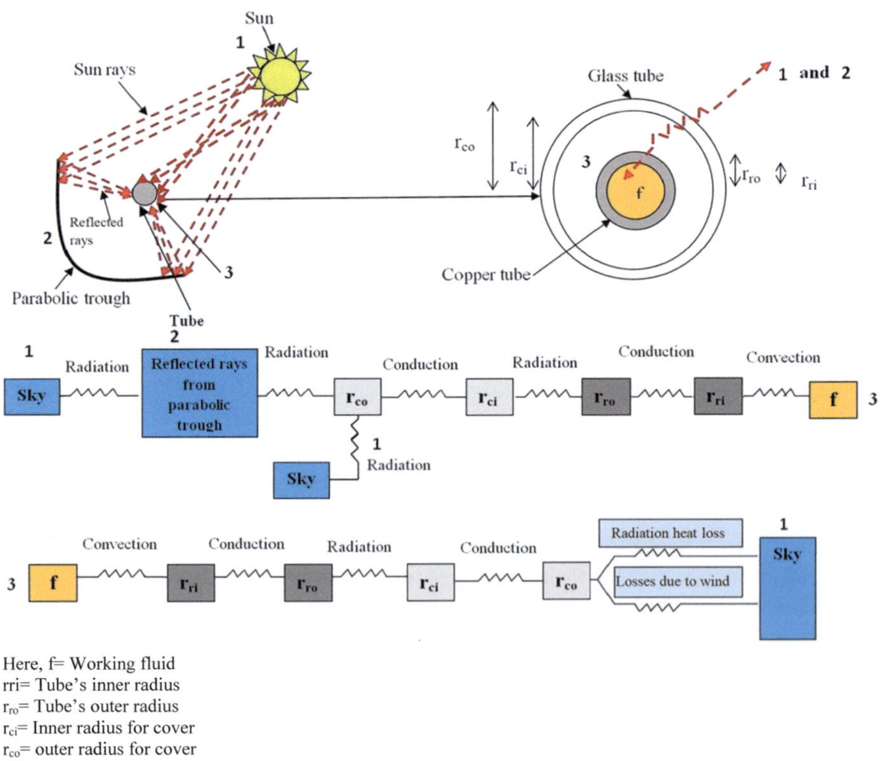


**Fig. 2.5** Different modes of heat interaction for parabolic solar trough collector (PTC)

to boost thermal efficiency. The tracking system can be single- or dual-axis, with single-axis being the more common choice for parabolic troughs. The main part of the system is the parabolic reflector, made from highly reflective materials like aluminum or silver-coated glass, designed to focus sunlight onto the receiver tube positioned at the focal line. The receiver tube, usually a metal tube with a selective surface, absorbs solar energy efficiently while minimizing heat loss. The support structure, made from strong metals like steel or aluminum, keeps the reflector and receiver tube aligned precisely and is built to endure harsh weather and thermal stress. The tracking mechanism, typically motorized, adjusts the trough's angle throughout the day using a sensor-based system. A heat transfer fluid (HTF), such as thermal oil or molten salt, flows through the receiver tube, absorbing concentrated solar energy, and should have a high boiling point and excellent thermal conductivity for effective energy transfer.

**Design Process**

The parabola geometry calculation begins by defining the parabola's dimensions based on the desired focal length, where the equation $y = \frac{x^2}{4f}$ is used, with "$f$" representing the focal length, which determines the distance from the parabola to the receiver tube. The parabolic reflector design follows, ensuring the shape matches the calculated geometry, often segmented for easier construction and assembly. The receiver tube is then designed with materials that optimize solar absorption and minimize heat loss. Its diameter is selected to match the width of the focal line for efficient energy collection. A tracking system is designed for precision, capable of maintaining the focus of the sun's rays on the receiver tube with durability, weather resistance, and energy efficiency in mind. The thermal system design involves choosing an appropriate heat transfer fluid (HTF), laying out the fluid circuit, and integrating it with a heat exchanger or power generation system. During construction, the reflector is fabricated from polished aluminum or reflective glass, ensuring the parabolic shape adheres to design specifications. The receiver tube is positioned at the focal line of the parabola and secured with a support structure that minimizes thermal expansion. The support structure must be sturdy, level, and able to hold the mirror and receiver tube in precise alignment. The tracking system, including motors, sensors, and control units, is installed and tested for accuracy in following the sun. The HTF circuit is connected, including the pump, heat exchanger, and storage system if necessary, with all connections sealed and insulated. Finally, the system undergoes testing and calibration to ensure optimal performance, with checks for tracking accuracy, mirror and receiver alignment, and HTF flow rate.

**Maintenance and Optimization**

Regular cleaning of the parabolic mirrors is crucial to maintaining their reflectivity, as dust, dirt, and other debris can reduce their ability to focus sunlight efficiently, directly impacting the system's thermal performance. Cleaning should be done with non-abrasive materials and techniques to avoid damaging the reflective surface, and the frequency will depend on the environmental conditions—more frequent cleaning may be necessary in dusty or polluted areas. The tracking system requires periodic maintenance, including inspecting the motors, gears, and sensors to ensure they function properly. Lubrication of moving parts is essential to preventing friction and wear, ensuring the system operates smoothly and accurately tracks the sun's movement from east to west. Over time, components such as bearings and actuators may wear down, so these should be checked and replaced if necessary. The heat transfer fluid (HTF) should be inspected regularly for signs of degradation, such as changes in viscosity or contamination, which can reduce its thermal efficiency. Additionally, the entire HTF circuit must be checked for leaks, as any loss of fluid can lead to reduced performance or even system failure. It's also essential to monitoring the pressure and temperature within the system to ensure the HTF is circulating properly and not subject to overheating or excessive wear, which could compromise the overall efficiency of the solar trough system.

**Efficiency and Performance**

This design and construction process ensures the parabolic solar trough collector efficiently captures and converts solar energy into thermal energy for various applications, such as electricity generation or industrial heating.

The efficiency and performance of a parabolic solar trough collector are determined by various factors, including optical, thermal, and system efficiencies, as well as external conditions like location and weather. These factors work together to define how effectively the system converts solar energy into usable thermal energy.

**Optical Efficiency**

Optical efficiency is crucial in determining how much sunlight is captured and focused onto the receiver tube. The reflectivity of the parabolic mirror, typically made from polished aluminum or silver-coated glass, plays a significant role. High-quality mirrors with reflectivity over 90% ensure that most of the sunlight is concentrated onto the receiver. The accuracy of the parabolic shape is equally important; even minor deviations can cause significant losses in concentrated sunlight. Additionally, the receiver tube, which is coated with a selective material, should have high absorptivity (around 95% [11]) to maximize the conversion of sunlight into heat while minimizing radiative losses. The tracking system, which keeps the trough aligned with the sun, must be precise to maintain optimal focus throughout the day.

**Thermal Efficiency**

Thermal efficiency depends on the heat transfer fluid (HTF) properties and the design of the receiver tube. HTFs like thermal oils or molten salts are chosen for their ability to operate at high temperatures with minimal heat loss. The receiver tube's material, often copper or stainless steel, should have high thermal conductivity to efficiently transfer absorbed heat to the HTF. Heat losses from the receiver occur through conduction, convection, and radiation; minimizing these losses is essential for maintaining high thermal efficiency. Vacuum insulation around the receiver tube and the use of selective coatings help in reducing these losses. The temperature gradient between the receiver tube and the environment also influences efficiency, with higher temperatures typically leading to higher thermal efficiency but also increased heat losses.

**System Efficiency**

The overall system efficiency is a combination of optical and thermal efficiencies, along with the efficiency of the power generation process. The heat exchanger, which transfers thermal energy from the HTF to the working fluid (such as water for steam generation), must be efficient to minimize energy losses. The system's integration with power generation, including the turbine and generator, also affects overall efficiency. Thermal storage systems, if integrated, can optimize the use of collected energy, especially when sunlight is not available, thereby improving the system's capacity factor—the ratio of actual energy produced to the maximum possible energy output.

**Performance Metrics and Improving Efficiency**

Performance metrics for parabolic solar trough collectors include instantaneous efficiency, annual efficiency, thermal output, and capacity factor. Instantaneous efficiency measures system performance at a specific time, while annual efficiency accounts for variations over the year due to changes in sunlight and downtime. Thermal output, typically measured in kilowatts or megawatts, represents the total energy produced, and a high capacity factor indicates the system's reliability and effectiveness. Efficiency improvements can be achieved through advanced tracking systems that optimize alignment with the sun, higher quality reflector materials to enhance optical efficiency, and improved coatings on the receiver tube to boost absorptivity and reduce emissivity. Additionally, integrating efficient thermal storage systems ensures better energy use when sunlight is unavailable. In summary, optimizing the performance and efficiency of a parabolic solar trough collector requires precise design, high-quality materials, and careful component integration to maximize the conversion of solar energy into thermal energy.

**Applications of Parabolic Solar Trough Collectors**

Concentrated Solar Power (CSP), including parabolic solar trough collector plants, harnesses solar energy to generate electricity by heating a fluid to produce steam that drives turbines, while hybrid power plants integrate CSP with fossil fuel plants to reduce fuel consumption and emissions. CSP technology also plays a vital role in industrial processes, providing thermal energy for desalination, where seawater is distilled into fresh water in arid regions, and enhancing oil recovery (EOR) by supplying steam to improve oil viscosity during extraction. It is also used for industrial heating, meeting the needs of sectors like chemical manufacturing and food processing that require temperatures between 150 and 400 °C. In residential and commercial applications, solar thermal systems efficiently heat water and provide space heating, reducing dependence on conventional energy sources. Solar cooling, particularly through absorption cooling, powers chillers for cooling applications, especially in areas with high solar radiation. Thermal storage systems store excess heat for use during low sunlight periods, ensuring a continuous supply of energy.

In agriculture, parabolic solar troughs help heat greenhouses, extending growing seasons and boosting crop yields in colder climates. Overall, parabolic solar trough collectors are versatile and efficient, supporting everything from power generation to industrial applications, residential heating, and agricultural needs, promoting the adoption of renewable energy and reducing reliance on fossil fuels.

**Challenges of Parabolic Solar Trough Collectors**

Parabolic trough systems face high initial costs, with substantial capital investment required for construction and installation, making them financially prohibitive for smaller projects or developing regions. Their land requirements are equally challenging, as they demand expansive areas, posing significant obstacles in densely populated areas or regions with elevated land costs. Despite advanced insulation, these systems suffer from efficiency losses due to unavoidable heat losses through conduction, convection, and radiation, particularly at higher operating temperatures. Additionally, optical imperfections like misalignment, dirt, and degradation of mirrors degrade optical efficiency over time, diminishing their overall performance. Water usage is another limitation, as many CSP plants, including those utilizing parabolic troughs, depend on substantial water for cooling, creating a constraint in arid regions. The intermittency of solar power, tied directly to sunlight availability, further compounds the issue, with reduced efficiency on cloudy days or at night. However, future advancements could address these issues through cost reductions driven by material innovations, cheaper and more durable components for mirrors, receivers, and tracking systems, and economies of scale from expanded deployment and manufacturing. Efficiency improvements are also on the horizon, with research into advanced selective coatings for receiver tubes that reduce heat loss and enhance thermal efficiency, alongside improved tracking systems that maximize sunlight capture by increasing precision and reliability.

**Integration with Thermal Storage**

Enhancing energy storage technologies is critical for addressing the intermittency of parabolic trough systems, as it allows thermal energy to be stored for extended periods and used when sunlight is unavailable. This would provide a continuous and reliable energy supply, ensuring that power generation does not stop during the night or on cloudy days. Integrating hybrid systems, combining parabolic troughs with photovoltaic (PV) panels or other renewable energy sources, offers another significant improvement by ensuring more consistent and stable power output. By combining the strengths of different renewable technologies, these hybrid systems can compensate for the fluctuations in energy generation, making the overall power supply more reliable. Moreover, developing water-efficient cooling technologies, such as dry cooling systems, is essential for making parabolic troughs more viable in regions facing water scarcity. Dry cooling solutions minimize water usage by utilizing air instead of water to dissipate heat, thereby addressing the challenges of high water consumption associated with traditional cooling methods and making parabolic trough systems more sustainable in water-scarce environments.

## Mathematical Formulation

The mathematical formulation of a parabolic solar trough collector involves several equations that describe the geometry, optical efficiency, thermal performance, and overall energy balance of the system. Here's an overview of the key mathematical aspects.

### Geometry of the Parabola

The shape of the parabolic reflector is defined by the equation of a parabola:

$$y^2 = \frac{x^2}{4f} \tag{2.8}$$

where

$x$ is the horizontal distance from the parabola's vertex,

$y$ is the vertical distance from the parabola's axis, and

$f$ is the focal length (the distance from the vertex to the focal point where the receiver tube is located).

### Optical Efficiency ($\eta_o$)

Optical efficiency represents the fraction of incident solar energy that is concentrated onto the receiver:

$$\eta_o = \rho \times \tau \times \alpha \times \gamma \tag{2.9}$$

where

$\rho$—is the reflectivity of the mirror,

$\tau$—is the transmissivity of any cover material over the receiver tube,

$\alpha$—is the absorptivity of the receiver tube surface, and

$\gamma$—is the intercept factor, representing the fraction of reflected rays hitting the receiver.

Note:

Energy Reflected by the Mirror:

The mirrors reflect a fraction ρ of the incident solar energy. Thus, the energy reflected toward the receiver is

$$I_r = \rho \times I \tag{2.10}$$

$I$ is the total incident solar energy falling on the collector.

### Energy Passing Through the Cover Material

The reflected energy $I_r$, then passes through the cover material. The transmissivity $\tau$ of the cover material determines what fraction of this energy actually reaches the receiver tube. Therefore, the energy reaching the receiver is

$$I_{r,\text{after cover}} = \tau \times (\rho \times I) \tag{2.11}$$

Energy Absorbed by the Receiver Tube:

The receiver tube absorbs a fraction $\alpha$ of the energy that reaches it. Hence, the energy absorbed by the receiver tube is

$$I_{\text{absorbed}} = \alpha \times I_{r,\text{after cover}} = \alpha \times (\tau \times \rho \times I) \tag{2.12}$$

### Adjust for the Intercept Factor

Not all of the reflected energy will hit the receiver tube due to geometrical and alignment factors. The intercept factor $\gamma$ accounts for this and represents the fraction of the reflected energy that actually reaches the receiver. The effective energy absorbed by the receiver, considering the intercept factor, is

$$I_{\text{absorbed}} = \gamma \times (\alpha \times \tau \times \rho \times I) \tag{2.13}$$

### Define Optical Efficiency

The optical efficiency $\eta_o$ is defined as the fraction of the incident solar energy that is effectively absorbed by the receiver. Thus:

$$\eta_o = \frac{\text{Energy absorbed by the receiver}}{\text{Total incident solar energy}} = \frac{I_{\text{absorbed}}}{I} \tag{2.14}$$

Substitute $I_{\text{absorbed}}$

$$\eta_o = \frac{\gamma \times (\alpha \times \tau \times \rho \times I)}{I} \tag{2.15}$$

The $I$ terms cancel out.

Thus, the formula for optical efficiency $\eta_o$ is

$$\eta_o = \rho \times \tau \times \alpha \times \gamma \tag{2.16}$$

### Thermal Efficiency ($\eta_t$)

Thermal efficiency is the ratio of useful thermal energy gained to the solar energy incident on the collector:

$$\eta_t = \frac{Q_u}{A_c \times G} \tag{2.17}$$

where

$Q_u$ is the useful thermal power (energy) output.

$A_c$ is the aperture area of the collector (the area through which sunlight enters).

$G$ is the solar irradiance (W/m$^2$).

4. Useful Thermal Power ($Q_u$)

The useful thermal power generated by the system is given by

$$Q_u = \eta_o \times A_c \times G - Q_{\text{loss}} \tag{2.18}$$

where

$Q_{\text{loss}}$ represents the heat losses due to conduction, convection, and radiation.

**Heat Losses ($Q_{\text{loss}}$)**

Heat losses from the receiver can be expressed as

$$Q_{\text{loss}} = h_{\text{loss}} \times A_r \times (T_r - T_a) \tag{2.19}$$

where

$h_{\text{loss}}$ is the overall heat transfer coefficient (W/m$^2$K),
$A_r$ is the surface area of the receiver tube,
$T_r$ is the temperature of the receiver tube, and
$T_a$ is the ambient temperature.

**Heat Transfer Fluid (HTF) Energy Gain**

The energy gained by the heat transfer fluid as it flows through the receiver tube is

$$Q_u \doteq \dot{m} \times c_p \times (T_{\text{out}} - T_{\text{in}}) \tag{2.20}$$

where

$\dot{m}$—is the mass flow rate of the HTF (kg/s),
$c_p$ is the specific heat capacity of the HTF (J/kg K), and
$T_{\text{out}}$ and $T_{\text{in}}$ are the outlet and inlet temperatures of the HTF, respectively.

These equations form the mathematical basis for analyzing the performance of a parabolic solar trough collector, allowing engineers to design, optimize, and predict the efficiency and output of the system under various conditions.

## 2.3 Solar Tower

The development of solar tower technology represents a remarkable evolution in solar energy harnessing, combining centuries of innovation with cutting-edge advancements. The concept of concentrating solar energy can be traced back to the early twentieth century when pioneering experiments, such as those by French inventor Augustin Mouchot in 1912, demonstrated the potential of using mirrors to concentrate sunlight for generating steam. Despite these early efforts, the technology saw limited application until the energy crises of the 1970s spurred renewed interest in renewable energy sources. The 1980s marked a significant turning point with the development of the Solar One project in California by Luz International. Solar One, a solar power tower using a central receiver to convert concentrated solar energy into heat, laid the groundwork for future innovations. Its successor, Solar Two, operational from 1995 to 1999, further advanced the technology by operating at higher temperatures and testing new materials. The 2000s saw commercial-scale implementations, with notable projects like the PS10 Solar Power Tower in Spain, which showcased the feasibility of large-scale solar tower plants. The Ivanpah Solar Electric Generating System, completed in 2014, became one of the largest solar tower installations globally, demonstrating the technology's capacity for significant power generation [39]. Modern advancements have focused on improving the efficiency and cost-effectiveness of solar towers. Key innovations include the development of advanced heliostats, which are mirrors that track the sun and reflect its light onto a central receiver with greater precision. The central receiver technology has evolved to use high-temperature materials and efficient heat exchangers, enhancing the overall performance of solar towers [40]. Additionally, thermal energy storage systems, such as molten salt storage, have been integrated into solar towers to enable continuous electricity production even when sunlight is not available, thereby improving the reliability of the technology. The integration of hybrid systems, combining solar towers with other renewable energy sources or backup power systems, has further enhanced the feasibility and economic viability of solar towers. Looking ahead, the future of solar tower technology is promising, with ongoing research aimed at enhancing efficiency, reducing costs, and expanding deployment in diverse regions. Innovations in materials, control systems, and hybrid configurations continue to push the boundaries of solar tower capabilities, making them a crucial component in the global transition to sustainable energy. The development of solar towers reflects a significant achievement in renewable energy technology, evolving from experimental concepts to a scalable and reliable solution for clean energy production, poised to play a key role in meeting the world's growing energy demands while reducing dependence on fossil fuels and mitigating climate change.

The solar power tower is a type of solar furnace that uses a central tower to concentrate sunlight. It employs an array of movable flat mirrors, called heliostats, which focus sunlight onto a receiver at the top of the tower. Figure 2.6 depicts the structure of a typical solar power tower collector. Concentrated Solar Power (CSP) technologies, including solar power towers, are gaining attention for their capacity to generate substantial electrical power at a relatively low cost [41]. CSP systems are projected to meet around 11% of the world's electricity needs by 2050 [42]. These systems can achieve temperatures over 1000 °C [43] and utilize thermal energy to drive thermodynamic cycles. Excess energy is stored in a Thermal Energy Storage (TES) system [44], allowing for electricity generation even after sunset and accommodating variations in solar irradiance, thereby improving the Capacity Utilization Factor (CUF) and energy output [45]. The heliostat field's size, defined by the Solar Multiple (SM), is critical for the plant's economic efficiency. Optimal design balances performance and cost; too small a field reduces thermal output, while too large a field increases inefficiency during peak solar times [46]. Recent

**Fig. 2.6** Solar tower collector [40]

studies have employed optimization techniques to enhance solar flux distribution and efficiency [47, 48].

The working principle of a solar tower, also known as a solar power tower, is centered around the concept of concentrating solar energy to generate electricity. Here's a detailed explanation of how it works.

**Solar Radiation Collection**

Heliostats are large, adjustable mirrors arranged around a solar tower, designed to track the sun's movement and reflect its light toward a central receiver located at the top of the tower. Each heliostat is mounted on a tracking system that follows the sun's path across the sky, ensuring that the sunlight remains concentrated on the receiver throughout the day. The central receiver, situated at the focal point of the heliostat array, captures the concentrated solar radiation. This energy heats up a heat transfer fluid, such as synthetic oils, molten salts, and other high-temperature fluids, inside the receiver. Molten salts are commonly used due to their ability to operate at high temperatures and store thermal energy effectively. Some solar tower systems are equipped with thermal storage tanks where the heat transfer fluid is stored, allowing the system to continue generating electricity even when the sun is not shining.

The heated fluid flows through a heat exchanger, transferring its thermal energy to a secondary fluid like water, which is heated to produce steam. This steam drives a turbine connected to a generator, converting mechanical energy into electricity that can be fed into the power grid. Sophisticated control systems manage the heliostats' orientation to ensure maximum solar energy collection throughout the day. Additionally, sensors and data acquisition systems continuously monitor parameters like solar intensity, temperature, and fluid flow rates, optimizing efficiency and maintaining

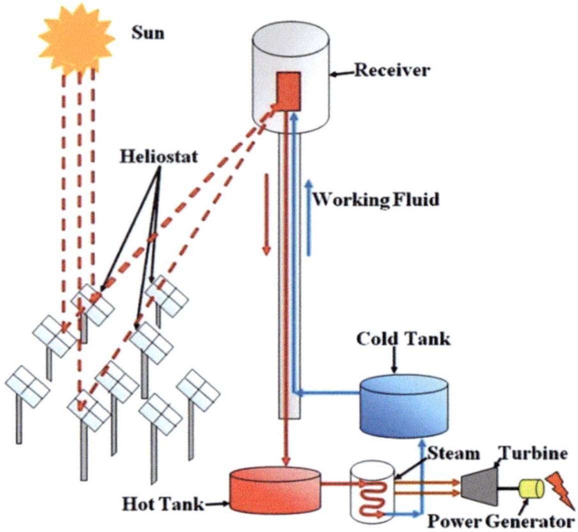


**Fig. 2.7** Overview of solar tower components

system performance. This integrated system effectively harnesses solar energy to generate clean, renewable electricity.

In essence, a solar tower system works by using a field of heliostats to concentrate sunlight onto a central receiver atop a tower. The concentrated solar energy heats a fluid, which is then used to produce steam. This steam drives a turbine connected to a generator, thereby converting solar energy into electricity. The integration of thermal energy storage allows the system to provide power even when sunlight is not available, enhancing its reliability and efficiency. Figure 2.7 illustrates the working of a solar tower.

### Effectiveness and Performance

Solar towers are highly efficient in energy conversion, primarily due to their ability to achieve higher operating temperatures than other solar technologies, which significantly improves thermal-to-electricity conversion efficiency. The use of advanced receivers and thermal storage systems further enhances their performance by efficiently capturing and storing solar energy. Thermal storage systems, such as molten salt tanks, play a crucial role in providing a reliable energy supply by allowing solar towers to generate power even when sunlight is not available, making them more versatile and dependable. This capability is particularly valuable for meeting energy demands during periods of low solar availability. Solar towers can be scaled up to accommodate large-scale power generation, making them an excellent option for utility-scale applications. However, the performance of solar towers is highly dependent on solar irradiance, meaning they are most effective in areas with high direct sunlight and clear skies, such as deserts. To maximize energy capture, solar towers rely on sophisticated tracking systems that ensure mirrors remain aligned with the sun. Any misalignment or failure in these systems can reduce efficiency. Regular maintenance, including mirror cleaning and tracking system checks, is essential to ensuring optimal performance and minimizing downtime. The land area required for solar tower installations can be significant, which may impact local ecosystems, requiring careful planning to balance land use and environmental considerations. When managed properly with advanced technology, solar towers perform exceptionally well and are a valuable solution for renewable energy generation.

**Comparative Performance**

Solar towers are a highly efficient renewable energy technology that converts sunlight into electricity using large arrays of mirrors (heliostats) and a central receiver. The system's performance depends on factors like solar irradiance, tracking precision, and maintenance. Below is a detailed breakdown of their energy conversion efficiency, scalability, performance, and environmental impact, compared to other solar technologies. Table 2.1 represents a comparison of solar tower with other solar technologies.

**Applications**

Solar towers are highly efficient for large-scale energy production, capable of generating substantial amounts of electricity that can be fed directly into the national grid. This makes them ideal for supporting large populations and meeting significant energy demands. Their ability to produce high temperatures also makes them suitable for industrial heat applications, such as chemical production and metal processing, including processes like refining and smelting. In addition to these uses, solar towers incorporate advanced thermal energy storage systems, such as molten salt, to store excess heat collected during the day. This stored energy can be used to generate electricity during nighttime or cloudy periods, ensuring a continuous power supply even when solar energy is not available. Solar towers can also be integrated into hybrid systems that combine solar energy with other sources, such as natural gas. This integration enhances reliability and efficiency, especially during periods of low solar availability, making it a versatile solution for diverse energy needs. Furthermore, solar towers are especially beneficial in remote and off-grid areas, providing self-sufficient energy in locations where traditional infrastructure is lacking. In addition to power generation, the heat from solar towers can be used for solar-powered desalination, addressing water scarcity in arid regions. By harnessing the sun's energy to produce electricity and fresh water, solar towers offer a sustainable and reliable solution to some of the world's most pressing challenges, such as energy shortages and water scarcity.

**Challenges**

Solar towers, while highly efficient, come with high initial costs due to significant capital investment required for construction and the use of advanced materials and technology, such as mirrors and tracking systems. Their complex construction and operation demand precision, as mirrors must be accurately aligned to focus sunlight onto the receiver, necessitating meticulous construction and ongoing maintenance to ensure optimal performance. Additionally, solar towers are highly dependent on

**Table 2.1** Solar tower comparison with other solar technologies

| Aspect | Solar towers | Comparison with other solar technologies |
|---|---|---|
| Energy conversion efficiency | Achieves higher temperatures, improving thermal-to-electricity conversion efficiency. Advanced receivers and thermal storage enhance energy generation | Concentrated solar power (CSP) systems in towers outperform flat-panel photovoltaic (PV) systems in terms of efficiency at higher temperatures. PV systems have lower operational temperatures |
| Thermal storage | Molten salt and other thermal storage systems allow energy generation even when sunlight is unavailable, ensuring reliability | Other solar technologies, such as PV and CSP with parabolic troughs, may require backup systems or have limited storage capacity |
| Scalability | Can be scaled up to meet large-scale energy demands, making them ideal for utility-scale power generation | Solar PV can also be scaled, but utility-scale solar towers are more efficient in areas with high direct sunlight, such as deserts |
| Solar irradiance dependency | Best in regions with high direct sunlight and clear skies (e.g., deserts) | Solar PV works in a wider range of locations, though it produces more energy in sunnier regions, similar to solar towers |
| Tracking systems | Uses sophisticated tracking systems to align heliostats with the sun for maximum efficiency | Fixed-position PV panels lack tracking systems and generally produce less energy than tracked solar towers |
| Maintenance and operations | Regular maintenance is required for cleaning mirrors and ensuring proper system function | Both solar towers and PV systems require maintenance, but solar towers' complex tracking systems may demand more frequent upkeep |
| Land and environmental impact | Requires a significant land area, which may impact local ecosystems. Efficient land use planning is crucial | PV systems, particularly on rooftops, have a smaller land footprint, but utility-scale PV installations also require large areas |

weather conditions, with performance being affected by sunlight variability, especially in regions with frequent cloud cover or rain. Maintenance is also a significant consideration, as keeping mirrors clean and ensuring the proper functioning of tracking systems are crucial for maintaining high efficiency. The large land area required for solar towers can be a constraint, particularly in densely populated or ecologically sensitive areas, and the construction and operation of these systems may impact local ecosystems and wildlife. Operational costs, including maintenance and the high initial investment, can be comparatively higher than other renewable energy technologies, potentially limiting their adoption despite their efficiency. These factors together present challenges in the widespread implementation of solar tower technology, even as it offers promising solutions for clean energy production.

**Mathematical Formulation**

The mathematical formulation of solar towers involves several key aspects related to energy collection, conversion, and performance. Here's a simplified overview of the key formulas and calculations.

**Solar Irradiance Collection**

Solar Irradiance ($G$) is the power per unit area received from the Sun. It is typically measured in watts per square meter ($W/m^2$):

$$G = \frac{E}{A} \tag{2.21}$$

where

$E$ = Energy received (Joules, J).
$A$ = Area of the receiver ($m^2$).

**Concentration Ratio**

The Concentration Ratio (CR) is the ratio of the area of the field of mirrors (or lenses) to the area of the receiver:

$$\text{CR} = \frac{A_{\text{mirror}}}{A_{\text{receiver}}} \tag{2.22}$$

where

$A_{\text{mirror}}$ = Area of the mirrors ($m^2$).
$A_{\text{receiver}}$ = Area of the receiver ($m^2$).

**Heat Transfer to Working Fluid**

Heat Input ($Q$) to the working fluid is given by

$$Q = G \times A_{\text{mirror}} \times \text{CR} \tag{2.23}$$

where

$G$ = Solar irradiance ($W/m^2$).
$A_{\text{mirror}}$ = Area of the mirrors ($m^2$).
CR = Concentration Ratio.

**Thermal Efficiency**

Thermal Efficiency ($\eta_{\text{thermal}}$) of the system measures how effectively the collected solar energy is converted into useful heat:

$$\eta_{\text{thermal}} = \frac{Q_{\text{useful}}}{Q_{\text{incident}}} \tag{2.24}$$

where
$Q_{useful}$ = Useful heat energy extracted (W).
$Q_{incident}$ = Incident solar energy (W).

**Conversion to Electrical Energy**

Electric Power Output ($P_{elec}$) can be calculated from the useful thermal energy and the efficiency of the thermal-to-electric conversion system (such as a steam turbine):

$$P_{elec} = Q_{useful} \times \eta_{conversion} \tag{2.25}$$

where
$\eta_{conversion}$ = Efficiency of converting thermal energy to electrical energy.

**Overall System Efficiency**

Overall Efficiency ($\eta_{overall}$) of the solar tower system considers both thermal and conversion efficiencies:

$$\eta_{overall} = \eta_{thermal} \times \eta_{conversion} \tag{2.26}$$

**Capacity Factor**

Capacity Factor (CF) is a measure of the actual output of a power plant compared to its maximum potential output over a period of time:

$$\text{CF} = \frac{P_{actual}}{P_{max}} \tag{2.27}$$

where
$P_{actual}$ = Actual power output (W).
$P_{max}$ = Maximum possible power output (W).

## 2.4 Fresnel Lens Collector

The Fresnel lens, named after the French physicist Augustin-Jean Fresnel, revolutionized optical technology when it was first developed in the early nineteenth century. Fresnel's innovation aimed to create a compact, lightweight lens with the same focusing power as a conventional lens. His design involved segmenting the lens into a series of concentric rings, each of which refracts light in a specific way, thus significantly reducing the lens's thickness and weight.

In the mid-nineteenth century, Fresnel lenses found practical applications in lighthouse technology, where their efficiency in directing light over long distances was invaluable. Their use in lighthouses became widespread, improving navigation and safety at sea.

The concept of the Fresnel lens was later adapted for solar energy collection. By the twentieth century, engineers began utilizing Fresnel lenses to concentrate sunlight onto small, highly efficient photovoltaic cells or thermal collectors. This adaptation allowed for more compact and cost-effective solar energy systems, as the Fresnel lens could focus sunlight with high precision while remaining relatively thin and light [49].

Today, Fresnel lenses are employed in various applications, including solar energy, automotive headlights, and even virtual reality, demonstrating their enduring impact and versatility. Fresnel lenses are used in solar energy systems primarily in two types of collectors: point-focusing and line-focusing. Point-focusing collectors concentrate sunlight to a single focal point, making them suitable for high-temperature applications, but they require precise two-axis tracking. Conversely, line-focusing collectors direct sunlight along a line, ideal for medium-temperature applications and requiring only one-axis tracking.

Research into Fresnel lens collectors has revealed a range of efficiencies. For instance, Wang et al. [50] achieved a high concentration ratio of 500 with an optical efficiency exceeding 71.6%. Lin et al. [51] explored different receiver designs, finding that triangular cavities had an efficiency of 69.8%, while rectangular ones achieved 61.8%. Comparative studies, such as those by Imtiaz Hussain and Lee [52], showed that point-focusing collectors generally offered a 7% higher average efficiency compared to line-focusing systems. Recent investigations have continued to advance the technology, with Ibrahim and Sakhrieh [53] reporting a peak efficiency of 77.3% and Zhai et al. [54] achieving an overall efficiency of 56.9%. These findings highlight the ongoing improvements and effectiveness of Fresnel lenses in enhancing solar energy collection. Figure 2.8 showcases the operational Fresnel lens solar collector field located in Lendemarke, Denmark.

**Fig. 2.8** Operation of the fresnel lens solar collector field at Lendemarke, Denmark [55]

**Working Principle of the Fresnel Lens Collector**

The working principle of a Fresnel lens collector revolves around its segmented optical structure, which allows it to efficiently focus light while maintaining a lightweight, cost-effective design. A Fresnel lens consists of concentric rings or parallel segments, each acting as a small prism that refracts light at a specific angle. This segmented design, which mimics the functionality of a traditional lens, dramatically reduces the lens's thickness and weight. The result is a more compact and less expensive alternative to conventional lenses, which is especially advantageous for large-scale solar energy systems. When sunlight strikes the Fresnel lens, the light passes through its segmented rings or segments, each of which bends (or refracts) the light toward a common focal point or line. The precise arrangement of these segments ensures that sunlight is concentrated effectively. In point-focusing collectors, the Fresnel lens directs the sunlight to a single focal point, where it can be used to heat a receiver or power a photovoltaic cell. In contrast, line-focusing collectors focus sunlight along a linear focal line, making them ideal for applications such as heating long receivers or generating electricity through a series of photovoltaic cells. In thermal applications, the concentrated sunlight heats a receiver placed at the focal point or along the focal line, which can then be used to generate steam, heat fluids, or directly provide thermal energy for industrial processes. In photovoltaic applications, the concentrated sunlight increases the intensity of light that falls on the photovoltaic cells, enhancing their efficiency and improving their ability to convert sunlight into electricity. To optimize performance, particularly in point-focusing collectors, tracking mechanisms are often incorporated. These systems adjust the position of the Fresnel lens or receiver throughout the day to follow the sun's movement, ensuring that the lens continuously captures and concentrates sunlight. The use of tracking systems maximizes the energy output by maintaining an optimal angle of incidence between the lens and the sunlight. This combination of segmented design, light concentration, and tracking mechanisms allows Fresnel lens collectors to provide an efficient and cost-effective solution for solar thermal and photovoltaic applications, offering a lightweight alternative to traditional solar concentrators (Fig. 2.9).

**Effectiveness and Performance of the Fresnel Lens Solar Collector**

Fresnel lens solar collectors are known for their efficiency in harnessing solar energy, utilizing a flat Fresnel lens to focus sunlight onto a small receiver, which enhances space efficiency. These collectors are cost-effective, as they typically cost less to manufacture and install compared to traditional parabolic collectors, thanks to reduced material requirements. The use of lightweight materials for both the lens and supporting structure makes them easier to handle and transport. Fresnel lens solar collectors also offer impressive thermal performance, achieving efficiencies of around 60–70%, depending on the system design and environmental conditions. Their compact design allows for installation in limited spaces or on existing infrastructure, making them ideal for areas where space is at a premium. To maintain optimal performance, regular cleaning and proper alignment are essential, ensuring

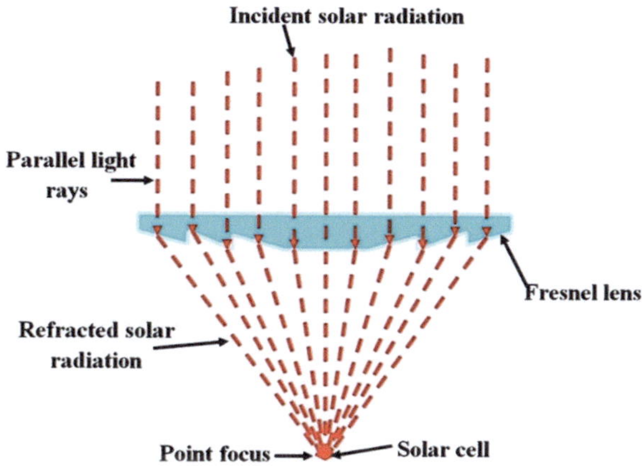


**Fig. 2.9** Overview of fresnel lens solar collector components

the mirrors remain effective in focusing sunlight. Due to their versatility, these collectors are suitable for a range of applications, from residential solar thermal systems to large-scale industrial processes. They can generate significant amounts of heat for various uses, such as water heating or industrial heat applications. Overall, Fresnel lens solar collectors provide an efficient, cost-effective, and space-saving solution for both residential and industrial solar thermal energy needs.

## Challenges

Fresnel lens solar collectors face several challenges, primarily related to tracking accuracy, optical efficiency, thermal management, maintenance, cost, and scalability. One of the key challenges is maintaining precise tracking of the sun, as point-focusing systems require two-axis tracking mechanisms to keep the focal point aligned with the receiver throughout the day. Even minor misalignment can drastically reduce the system's efficiency, making accurate tracking systems essential to maximizing energy collection and minimizing losses. Achieving high optical efficiency is also crucial, as the lenses must be precisely designed and manufactured to prevent light loss from imperfections or misalignment. Environmental contaminants like dust and dirt can further hinder performance, necessitating regular cleaning and maintenance, which adds to operational costs. Thermal management is another concern, as the concentrated sunlight generates high temperatures at the focal point, potentially causing thermal stresses and material degradation in the receiver. Effective thermal management, including advanced materials and cooling systems, is required to ensure the system's longevity, but can increase complexity and costs. Regular maintenance is essential to keeping the lenses clean and ensuring the mechanical components of the tracking system function properly. Additionally, the tracking systems themselves, while crucial for accurate alignment, can be costly and represent a significant portion of the overall investment. Finally, while Fresnel lens collectors are ideal for large-scale solar power plants, their scalability to smaller or residential systems is challenging. The need for precise tracking and large surface areas for sunlight concentration makes it difficult to adapt these systems for smaller applications, requiring innovative solutions to overcome scalability issues.

**Applications**

Fresnel lenses are a crucial component in a wide range of solar energy applications due to their ability to efficiently concentrate sunlight. In solar thermal power plants, Fresnel lenses focus sunlight onto a receiver, which absorbs the concentrated heat and transfers it to a fluid, typically water or oil. This fluid is then used to generate steam, which drives turbines connected to generators to produce electricity. This method is particularly advantageous for large-scale energy production and can be easily integrated into existing power grids. In concentrated photovoltaic (CPV) systems, Fresnel lenses focus sunlight onto high-efficiency photovoltaic (PV) cells, enhancing their ability to generate electricity. By concentrating the sunlight, CPV systems generate more power from a smaller area of PV material, reducing the cost of the materials and increasing the overall energy output. Fresnel lenses are also widely used in solar water heating systems, where they concentrate sunlight onto a thermal receiver to heat water. This method provides a sustainable and cost-effective solution for both residential and industrial water heating needs. Furthermore, Fresnel lens technology is employed in solar desalination plants, where concentrated solar heat is used to evaporate seawater, which then condenses to produce fresh water, an essential process in arid regions with limited freshwater resources. Fresnel lenses can also be utilized in solar cookers, where they focus sunlight onto cooking vessels, offering an eco-friendly and efficient alternative to conventional cooking methods. In research and development, Fresnel lenses are employed to test and refine solar technologies, aiding in the creation of new materials and innovations. Despite challenges in tracking accuracy, optical efficiency, thermal management, and maintenance, Fresnel lenses continue to play a significant role in advancing solar energy solutions, with applications ranging from power generation to water desalination and cooking.

**Mathematical Formulation of Fresnel Lens Solar Collectors**

The mathematical formulation of Fresnel lens solar collectors involves several key aspects, including the lens design, light concentration, and efficiency calculations. Here's an overview of the key mathematical concepts and equations used in the analysis and design of Fresnel lens collectors.

**Lens Design and Focusing**

**Lens Geometry**

The Fresnel lens is designed with concentric rings or segments, where each segment acts as a small prism to refract light. The basic geometric equation for a Fresnel lens can be derived from the lens's focal length ($f$) and the distance ($d$) of the lens from the focal point.

### Lens Focal Length

The focal length of a Fresnel lens can be estimated using the lens maker's equation, though the actual design is more complex due to the segmented structure. The general lens equation is

$$\frac{1}{f} = \frac{1}{d_o} + \frac{1}{d_i} \tag{2.28}$$

where

- $f$—is the focal length,
- $d_o$—is the object distance (distance from the lens to the sunlight), and
- $d_i$—is the image distance (distance from the lens to the focal point).

### Fresnel Lens Segments

The segments of a Fresnel lens can be approximated using simple prisms, where the angle of refraction $\theta$ for each segment is given by Snell's Law:

$$n_1 \sin(\theta_1) = n_2 \sin(\theta_2) \tag{2.29}$$

where

- $n_1$—is the refractive index of the medium from which light is coming (air),
- $n_2$—is the refractive index of the lens material,
- $\theta_1$—is the angle of incidence, and
- $\theta_2$—is the angle of refraction.

### Light Concentration

### Concentration Ratio

The concentration ratio (CR) of a Fresnel lens collector is defined as the ratio of the area of the lens to the area of the focal point (or receiver). It can be calculated as

$$\mathrm{CR} = \frac{A_{\text{lens}}}{A_{\text{receiver}}} \tag{2.30}$$

where

- $A_{\text{lens}}$—is the area of the Fresnel lens.
- $A_{\text{receiver}}$—is the area of the focal spot.

### Intensity of Concentrated Light

The intensity ($I_{\text{focus}}$) of the concentrated light at the focal point is

$$I_{\text{focus}} = \text{CR} \times I_{\text{solar}} \tag{2.31}$$

where

- $I_{\text{solar}}$—is the incident solar irradiance (W/m$^2$).
- CR is the concentration ratio.

## Efficiency Calculations

### Optical Efficiency

The optical efficiency ($\eta_{\text{opt}}$) of a Fresnel lens collector is the ratio of the actual amount of concentrated solar energy captured to the theoretical maximum. It is given by

$$\eta_{\text{opt}} = \frac{E_{\text{collector}}}{E_{\text{incident}}} \tag{2.32}$$

where

- $E_{\text{collected}}$—is the energy collected by the receiver.
- $E_{\text{incident}}$ is the incident solar energy.

### Thermal Efficiency

The thermal efficiency ($\eta_{\text{thermal}}$) is the ratio of useful thermal energy output to the solar energy input:

$$\eta_{\text{thermal}} = \frac{Q_{\text{useful}}}{I_{\text{solar}} \times A_{\text{lens}}} \tag{2.33}$$

where

- $Q_{\text{useful}}$—is the useful thermal energy output (e.g., heat transferred to the fluid),
- $I_{\text{solar}}$—is the solar irradiance, and
- $A_{\text{lens}}$—is the area of the Fresnel lens.

## Tracking and Alignment

### Tracking Accuracy

To maintain focus throughout the day, the tracking system must adjust the angle of the Fresnel lens. The angular adjustment required ($\Delta\theta$) can be calculated based on the latitude of the location and the time of year. The equation for the solar angle adjustment is

$$\Delta\theta = \text{SolarDeclination} + \text{HourAngle} \tag{2.34}$$

where

- **Solar Declination** is the angle between the rays of the sun and the plane of the Earth's equator.
- **Hour Angle** represents the time difference from solar noon, measured in degrees.

In summary, the mathematical formulation of Fresnel lens solar collectors involves geometric design considerations, light concentration calculations, efficiency metrics, and tracking adjustments. These equations help in optimizing the design and performance of the collectors to enhance solar energy capture and utilization.

## References


1. Duffie JA, Beckman WA (2013) Solar engineering of thermal processes, 4th edn. Wiley, Amsterdam
2. Lovegrove K, Stein W (eds) (2012) Concentrating solar power technology: principles, developments and applications. Woodhead Publishing, Sawston
3. Kumar S, Mahavar S, Mawire A, Ruivo CR (2023) Design development of an economic solar paraboloidal concentrator for industrial process heat generation. Energy Conver Manag 292:117380
4. Lovegrove K, Pye J (2012) 2—fundamental principles of concentrating solar power (CSP) systems. Concentrating solar power technology. Woodhead Publishing series in energy. Woodhead Publishing, Sawston, pp 16–67
5. de la Beaujardiere JFPP, Reuter HC (2018) A review of performance modelling studies associated with open volumetric receiver csp plant technology. Renew Sustain Energy Rev 82:3848–3862
6. Lee HJ, Kim JK, Lee SN, Kang YH (2012) Consistent heat transfer analysis for performance evaluation of multichannel solar absorbers. Sol Energy 86(5):1576–1585
7. Zaversk F, Aldaz L, Sanchez M, Ávila-Marín AL, Roldan MI, Fernandez-Reche J, Fussel A, Beckert W, Adler J (2018) Numerical and experimental evaluation and optimization of ceramic foam as solar absorber—single-layer versus multi-layer configurations. Appl Energy 210:351–375
8. Avila-Marin A, Fernandez-Reche J, Martinez-Tarifa A (2019) Modelling strategies for porous structures as solar receivers in central receiver systems: a review. Renew Sustain Energy Rev 111:15–33
9. Chen X, Xia XL, Liu H, Li Y, Liu B (2016) Heat transfer analysis of a volumetric solar receiver by coupling the solar radiation transport and internal heat transfer. Energy Conver Manag 114:20–27
10. Avila-Marin AL, Caliot C, de Lara MA, Fernandez-Reche J, Montes MJ, Martinez-Tarifa A (2019) Homogeneous equivalent model coupled with P1-approximation for dense wire meshes volumetric air receivers. Renew Energy 135:908–919
11. Barreto G, Canhoto P, Collares-Pereira M (2019) Three-dimensional CFD modelling and thermal performance analysis of porous volumetric receivers coupled to solar concentration systems. Appl Energy 252:113433
12. Roldan M, Fernandez-Reche J, Ballestrín J (2016) Computational fluid dynamics evaluation of the operating conditions for a volumetric receiver installed in a solar tower. Energy 94:844–856
13. Gu R, Ding J, Wang Y, Yuan Q, Wang W, Lu J (2019) Heat transfer and storage performance of steam methane reforming in tubular reactor with focused solar simulator. Appl Energy 233–234:789–801
14. Kalogirou SA (2004) Solar thermal collectors and applications. Progr Energy Combust Sci 30(3):231–295
15. Stine WB, Harrigan RW (1985) Solar energy systems design. Wiley, Amsterdam

16. Guven HM, Bannerot RB (1986) Determination of the optical performance of parabolic dish solar concentrators. Sol Energy 37(1):3–11
17. Lopez O, Banos A, Arenas A (2020) On the thermal performance of flat and cavity receivers for a parabolic dish concentrator and low/medium temperatures. Sol Energy 199:911–923
18. Turchi C, Mehos M, Ho C, Kolb G (2010) Current and future costs for parabolic trough and power tower systems in the US market. Solar PACES 2010, Perpignan
19. Twidell J, Weir T (2015) Renewable energy resources, 3rd edn. Routledge, New York
20. Tian Y, Zhao CY (2013) A review of solar collectors and thermal energy storage in solar thermal applications. Appl Energy 104:538–553
21. Zhai R, Wang J (2014) Analysis and design of a solar-thermal parabolic dish system. J Energy Eng 140(4):04014020
22. Ho CK, Iverson BD (2014) Review of high-temperature central receiver designs for concentrating solar power. Renew Sustain Energy Rev 29:835–846
23. Hosseini SMS, Dehaj MS (2021) An experimental study on energetic performance evaluation of a parabolic trough solar collector operating with $Al_2O_3$/water and GO/water nanofluids. Energy 234:121317
24. Javidan M, Gorji-Bandpy M, Al-Araji A (2023) Investigation and simulation of parabolic trough collector with the presence of hybrid nanofluid in the finned receiver tube. Theor Appl Mech Lett 45:100465. https://doi.org/10.1016/j.taml.2023.100465
25. Vahidinia F, Khorasanizadeh H, Aghaei A (2023) Energy, exergy, economic and environmental evaluations of a finned absorber tube parabolic trough collector utilizing hybrid and mono nanofluids and comparison. Renew Energy 205:185–199. https://doi.org/10.1016/j.renene.2023.01.085
26. Chakraborty O (2023) Influence of spinning flower structure inserts in the thermal performance of LS-2 model of parabolic trough collector with ternary hybrid nanofluid. Renew Energy 210:215–228. https://doi.org/10.1016/j.renene.2023.03.112
27. Rawat SK, Yaseen M, Pant M, Ujarari CS, Joshi DK, Chaube S, Negi AS, Kumar M (2023) Designing soft computing algorithms to study heat transfer simulation of ternary hybrid nanofluid flow between parallel plates in a parabolic trough solar collector: case of artificial neural network and particle swarm optimization. Int Commun Heat Mass Transf 148:107011. https://doi.org/10.1016/j.icheatmasstransfer.2023.107011
28. Chakraborty O, Das B, Gupta R (2022) Performance of parabolic trough collector having rotating receiver: effect of elliptical insert and hybrid nanofluid. Int J Energy Res 46:15254–15275. https://doi.org/10.1002/er.8225
29. Adun H, Adedeji M, Adebayo V, Shefik A, Bamisile O, Kavaz D, Dagbasi M (2021) Multi-objective optimization and energy/exergy analysis of a ternary nanofluid based parabolic trough solar collector integrated with kalina cycle. Solar Energy Mater Solar Cells 231:111322
30. Chakraborty O, Das B, Gupta R (2023) Thermal performance evaluation of parabolic trough collector having different inserts and working with hybrid nanofluid. Energy Environ 305:56407. https://doi.org/10.1177/0958305X231156407
31. Menbari A, Alemrajabi AA, Ghayeb Y (2016) Experimental investigation of stability and extinction coefficient of $Al_2O_3$–CuO binary nanoparticles dispersed in ethylene glycol–water mixture for low-temperature direct absorption solar collectors. Energy Convers Manag 108:501–510
32. Menbari A, Alemrajabi AA, Rezaei A (2017) Experimental investigation of thermal performance for direct absorption solar parabolic trough collector (DASPTC) based on binary nanofluids. Exp Therm Fluid Sci 80:218–227
33. Minea AA, El-Maghlany WM (2018) Influence of hybrid nanofluids on the performance of parabolic trough collectors in solar thermal systems: recent findings and numerical comparison. Renew Energy 120:350–364
34. Bellos E, Tzivanidis C (2018) Thermal analysis of parabolic trough collector operating with mono and hybrid nanofluids. Sustain Energy Technol Assess 26:105–115
35. Chakraborty O, Das B, Gupta R (2023) Performance of parabolic trough solar collector with spinning star inserts in receiver tube with hybrid nanofluid. Proceed Instit Mech Eng E J Process Mech Eng 11:58507. https://doi.org/10.1177/09544089231158507

36. Dou L, Ding B, Zhang Q, Kou G, Mu M (2023) Numerical investigation on the thermal performance of parabolic trough solar collector with synthetic oil/Cu nanofluids. Appl Therm Eng 227:120376. https://doi.org/10.1016/j.applthermaleng.2023.120376
37. Zhang C, Han S, Wu Y, Zhang C, Guo H (2022) Investigation on convection heat transfer performance of quaternary mixed molten salt based nanofluids in smooth tube. Int J Therm Sci 177:107534. https://doi.org/10.1016/j.ijthermalsci.2022.107534
38. Mwesigye A, Meyer JP (2017) Optimal thermal and thermodynamic performance of a solar parabolic trough receiver with different nanofluids and at different concentration ratios. Appl Energy 193:393–413
39. Sharma A (2011) A comprehensive study of solar power in India and World. Renew Sustain Energy Rev 15:1767–1776
40. Das P, Chandramohan VP (2023) A review on recent advances in hybrid solar updraft tower plants: challenges and future aspects. Sustain Energy Technol Assess 55:102978
41. Fuqiang W, Ziming C, Jianyu T, Yuan Y, Yong S, Linhua L (2017) Progress in concentrated solar power technology with parabolic trough collector system: a comprehensive review. Renew Sustain Energy Rev 79:1314–1328
42. Philibert C (2014) International energy agency (IEA), technology roadmap: solar thermal electricity
43. Cagnoli M, de la Calle A, Pye J, Savoldi L, Zanino R (2019) A CFD-supported dynamic system-level model of a sodium-cooled billboard-type receiver for central tower CSP applications. Sol Energy 177:576–594
44. Kim S, Lee I, Lee BJ (2017) Development of performance analysis model for central receiver system and its application to pattern-free heliostat layout optimization. Sol Energy 153:499–507
45. Avila-Marin AL, Fernandez-Reche J, Tellez FM (2013) Evaluation of the potential of central receiver solar power plants: configuration, optimization and trends. Appl Energy 112:274–288
46. Hussaini ZA, King P, Sansom C (2020) Numerical simulation and design of multi-tower concentrated solar power fields. Sustainability 12:1–22
47. Wang K, He YL, Xue XD, Du BC (2017) Multi-objective optimization of the aiming strategy for the solar power tower with a cavity receiver by using the non-dominated sorting genetic algorithm. Appl Energy 205:399–416
48. Farges O, Bezian JJ, El HM (2018) Global optimization of solar power tower systems using a Monte Carlo algorithm: application to a redesign of the PS10 solar thermal power plant. Renew Energy 119:345–353
49. Zhao Y, Hong H, Jin H (2017) Optimization of the solar field size for the solar–coal hybrid system. Appl Energy 185:1162–1172
50. Wang H, Huang J, Song M, Hu Y, Wang Y, Lu Z (2018) Simulation and experimental study on the optical performance of a fixed-focus Fresnel lens solar concentrator using polar-axis tracking. Energies 11(4):887
51. Lin M, Sumathy K, Dai YJ, Zhao XK (2014) Performance investigation on a linear Fresnel lens solar collector using cavity receiver. Sol Energy 107:50–62
52. Imtiaz Hussain M, Lee GH (2016) Thermal performance comparison of line- and pointfocus solar concentrating systems: experimental and numerical analyses. Sol Energy 133:44–54
53. Ibrahim A, Sakhrieh A (2016) Comparative study of Fresnel lenses and mirrors in concentrated solar applications. In: The fifth global conference on renewables and energy efficiency in desert regions, pp 1–7
54. Zhai H, Dai YJ, Wu JY, Wang RZ, Zhang LY (2010) Experimental investigation and analysis on a concentrating solar collector using linear Fresnel lens. Energy Convers Manag 51(1):48–55
55. Jensen AR, Sifnaios I, Caringal GP, Furbo S, Dragsted J (2022) Thermal performance assessment of the world's first solar thermal Fresnel lens collector field. Sol Energy 237:447–455

# Chapter 3
# Parabolic Solar Trough Collector Technology: Thermal and Optical Performance, Design, and Diverse Applications

This chapter focuses on the Parabolic Solar Trough Collector (PTC), a highly efficient technology for capturing solar thermal energy by concentrating sunlight onto a receiver tube. It begins by tracing the historical development of PTCs and their growing role in sustainable energy systems. A comprehensive literature review is presented, analyzing thermal and optical performance as well as highlighting the strengths and limitations of traditional designs. The technical aspects of PTC construction are explored in detail, with attention to key parameters that influence energy efficiency. The chapter also discusses a wide range of practical applications, including power generation, industrial process heat, refrigeration, and air-conditioning. With increasing global emphasis on clean energy, this chapter underscores how PTCs contribute meaningfully to reducing fossil fuel dependence and mitigating climate change.

## 3.1 Design and Operational Principles

Parabolic Solar Trough Collectors (PTCs) rely on a parabolic reflector to concentrate solar flux onto a tube. The reflector's shape ensures that parallel solar rays are focused on a single focal line, directing concentrated solar flux onto the receiver tube for optimal energy capture. Key components of a PTC system include (Fig. 3.1).

- **Parabolic Reflector**: Made from reflective materials, the parabolic reflector concentrates sunlight onto the receiver tube. Its curvature maximizes the solar energy focused in a linear area. Modern PTCs use tracking mechanisms (single-axis or dual-axis) to adjust the reflector's position throughout the day, ensuring the collector remains aligned with the sun for optimal energy absorption.



O. Chakraborty and B. Das, *Parabolic Solar Trough Collector*, Mechanical Engineering Series, https://doi.org/10.1007/978-3-032-11229-3_3

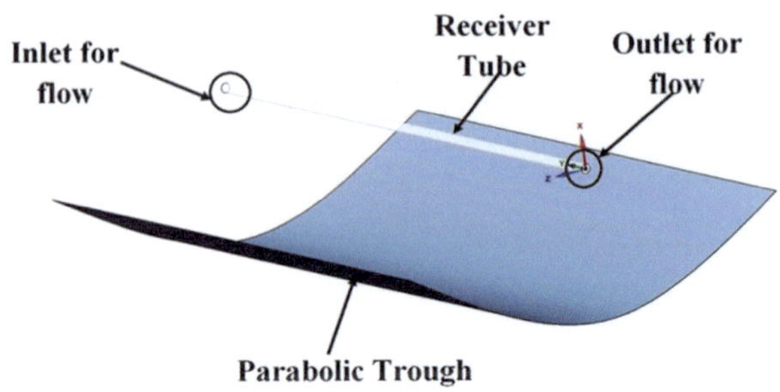


**Fig. 3.1** Parabolic solar trough collector

- **Receiver Tube**: Located on the focal line, the receiver tube assimilates concentrated solar radiation for thermal harvesting. It is constructed from high thermal absorptivity and low thermal emissivity materials, such as blackened metal or specialized coatings. Within the receiver tube, a heat transfer fluid absorbs concentrated thermal flux, enabling efficient energy transference.
- **Heat Transfer Fluid**: Fluids like synthetic oils or molten salts are used because they operate efficiently at extreme temperatures. The heated fluid is circulated through pipes to a thermal exchanger or storage system. In the thermal exchanger, the fluid transfers heat to water, which is converted into steam for electricity generation or industrial heating applications. The design of these components ensures the efficient capture and utilization of solar energy.

## 3.2 Applications

Parabolic Solar Trough Collectors offer several advantages that make them Adaptable for diverse operational applications. One of the primary applications of PTC technology is suited for large-scale and concentrated Solar Power (CSP) plants. In these plants, multiple PTC systems are deployed in an array to generate significant amounts of thermal energy, which is then utilized to produce electricity. The ability to store heat energy in high-temperature fluids allows these plants to provide power even when sunlight is not available, offering a level of reliability and dispatchability critical for meeting electricity demands.

In addition to power generation, PTCs are used in various industrial processes requiring substantial heat. For instance, they are employed in chemical manufacturing, food processing, and desalination plants. In these settings, the elevated-temperature heat created by PTCs can be utilized for heating, drying, and chemical reactions, contributing to more efficient and sustainable industrial operations.

## 3.3 Analysis of Traditional Parabolic Trough Collector (PTC) Designs

Parabolic trough collectors (PTCs) are an established and effective technology for harnessing solar thermal energy. Their design integrates several key features contributing to their efficiency and effectiveness in converting solar energy into usable thermal energy. This analysis will cover the fundamental design features of PTCs, focusing on their parabolic shape, tracking systems, and receiver tubes. Each component is crucial in maximizing the performance of PTC systems.

(i) **Design Features**

### Concentration of Solar Radiation

The parabolic shape of Parabolic Trough Collectors (PTCs) plays a pivotal role in their ability to concentrate solar radiation onto a small area, significantly enhancing the system's efficiency. This shape leverages a unique geometric property: parallel rays of sunlight striking a parabolic surface converge at a single focal point. In the case of PTCs, the focal point is extended into a linear region, known as the focal line, where the receiver tube is strategically positioned. The curvature of the parabolic reflector ensures that sunlight is reflected and concentrated onto this linear region, resulting in a high intensity of solar radiation directed at the receiver tube. This concentrated energy increases the amount of thermal energy captured, making the PTC system highly efficient at converting solar radiation into usable heat. The focused sunlight maximizes energy collection, essential for achieving high-temperature outputs necessary for various applications.

### Efficient Energy Collection

The parabolic shape also enhances energy collection efficiency by concentrating solar radiation over a larger area onto a much smaller, thermally conductive surface. By converging sunlight from a wide area, the parabolic reflector directs this energy toward the receiver tube, which absorbs the intensified radiation. This concentration process is crucial in optimizing the thermal energy transfer from the receiver tube to the heat transfer fluid (HTF) circulating within the system. The efficiency of this transfer is directly related to the geometry of the parabolic reflector. As the reflector converges sunlight, the amount of thermal energy collected increases, leading to higher operational temperatures in the receiver tube. This higher temperature allows for more effective heat conversion, making the PTC an ideal choice for applications requiring significant thermal energy output. By focusing sunlight, the parabolic reflector plays a critical role in improving the overall energy collection and system efficiency of PTCs.

### Thermal Performance

The parabolic shape not only concentrates solar radiation but also plays a significant role in improving the thermal performance of the PTC system. The concentrated solar energy heats the receiver tube to elevated temperatures, essential for applications requiring substantial thermal output, such as electricity generation and industrial heating. The receiver tube absorbs the high-intensity solar radiation, raising its temperature and increasing the heat transferred to the heat transfer fluid (HTF) circulating inside. This high temperature is crucial for optimizing the heat transfer process, as the HTF, now heated to a high temperature, can generate steam or provide industrial heating. The parabolic design ensures that solar radiation effectively focuses on the receiver tube, enhancing thermal efficiency. The ability of the receiver tube to reach high temperatures increases the overall system's efficiency, making PTCs effective in large-scale solar thermal applications that demand consistent and high thermal outputs.

(ii) **Tracking Systems**

To optimize the quantity of solar energy collected, PTCs use tracking units that adjust the orientation of the collector throughout the day. Traditional PTC configurations predominantly employ two principal tracking systems: single-axis and dual-axis.

**Single-Axis Tracking**

The collector rotates about a horizontal axis in single-axis tracking systems, typically aligned in the north–south direction. This configuration tracks the sun's movement across the sky, following its path from east to west throughout the day.

Single-axis trackers modulate the collector's orientation to sustain an optimal angular alignment with the sun, thereby augmenting solar energy capture relative to stationary configurations. These systems are more straightforward and economically advantageous than dual-axis counterparts, potentially enhancing solar energy yield by 20–30% compared to fixed collectors.

However, single-axis tracking systems only adjust for the sun's daily movement and do not account for seasonal changes in the sun's position. Despite this limitation, they significantly improve fixed systems and are widely used in various solar thermal applications.

**Dual-Axis Tracking**

Dual-axis tracking systems provide a more sophisticated solution by allowing the collector to rotate around both horizontal and vertical axes. This system can compensate for diurnal and seasonal variations in the sun's angular displacement, providing optimal alignment with the sun throughout the year.

Dual-axis trackers maintain the collector's orientation to ensure it is always oriented tangentially toward the solar apex, regardless of the time of day or season. This capability significantly augments the efficacy of solar radiation absorption, with potential increases in solar energy collection of 30–40% compared to single-axis systems.

Although dual-axis tracking systems exhibit greater complexity and incur higher costs than single-axis configurations, their ability to maximize solar energy collection makes them an attractive option for applications requiring high levels of energy capture and conversion.

### (iii) Receiver Tubes and Heat Transfer Fluids

Receiver tubes are integral to the PTC design as they are the components that absorb the concentrated solar energy. The working liquid circulating within these conduits assumes a pivotal function in transferring the absorbed heat to storage systems or heat exchangers [1].

#### Receiver Tubes

The receiver tube is engineered to capture the concentrated solar radiation converged by the parabolic reflector. Typically fabricated from high thermal conductivity metals or composites, it is designed to optimize thermal transfer. The tube is coated with a selective material that maximizes absorptance, thereby enhancing solar radiation absorption, while minimizing emissivity to reduce thermal loss via radiation.

#### Design Considerations

The design of the receiver tube involves choosing materials that can withstand high temperatures while maintaining efficient thermal conductivity. The tube's surface coating is critical for maximizing heat absorption and minimizing heat losses. Innovations in materials and coatings have led to improved performance in the modern era, enhancing the efficiency of PTC systems.

#### Heat Transfer Fluids (HTFs)

HTFs are utilized to transfer the thermal energy collected by the tube to other components of the solar thermal system, like heat storage systems or heat exchangers. Common HTFs include.

- **Water**: Water is widely used due to its elevated volumetric heat capacity and availability. It is effective for lower temperature applications but may require additional safety measures to handle high temperatures and pressures.
- **Thermal Oils**: These fluids are used in higher temperature applications due to their ability to operate at elevated temperatures without boiling. They have lower specific heat capacities compared to water, but are effective in transferring heat efficiently.
- **Molten Salts**: These are utilized in concentrated solar power units that require very high temperatures. Molten salts have high specific heat capacities and can store thermal energy for extended periods, making them suitable for systems with large-scale energy storage needs.

The choice of HTF affects the overall thermal effectiveness of the collector. The HTF must efficiently absorb and transfer heat, and its properties must be suitable for the operating temperature range of the collector system.

## 3.4 Technological Evolution

The advancement of Parabolic Solar Trough Collectors began in the 1970s, during a period of heightened interest in alternative energy sources due to global energy crises. Since then, technological advancements have significantly enhanced the efficiency as well as performance of PTC units. Innovations include improvements in reflective materials, advanced tracking systems, and high-performance heat transfer fluids. These advancements have made PTCs more effective and economically viable.

Looking forward, the potential for further improvements in PTC technology remains substantial. Ongoing research aims to enhance the efficiency of reflective coatings, optimize tracking mechanisms, and develop new materials for better heat transfer. Additionally, integrating PTC systems with energy storage solutions and hybrid power systems could further expand their applicability and effectiveness in meeting global energy needs.

**Historical Development of Parabolic Trough Collector Technology**

The development of Parabolic Solar Trough Collectors (PTCs) has evolved significantly from early conceptual designs to advanced systems. In the second century BC, Diocles conceptualized parabolic mirrors focusing solar rays, igniting the path to solar energy. By the late nineteenth century, John Ericsson actualized the first functional PTC, forging a new era in thermal power. This innovation included a flat silvered glass reflector, but commercial success eluded him due to secrecy and his premature death. Frank Shuman's 1912 plant in Egypt utilized metal reflectors and steam engines, advancing PTC technology despite its destruction in WWI. Innovations continued with Lewis Shipman's 1925 design using lenses and reflectors, and C. G. Abbot's 1959 improvements with aluminum reflectors. The 1970s saw significant advancements, including automated tracking mechanisms and covered systems by inventors like Garner. Each iteration has contributed to enhanced performance and cost reduction, shaping modern PTC technology [2]. Figure 3.2 elaborates on the construction of a parabolic solar trough collector by Frank Shuman [3], highlighting the early design of the reflector, receiver tube, and support system for solar energy concentration.

The idea of concentrating solar energy dates to distant past civilizations using simple reflectors, but practical development of Parabolic Solar Trough Collectors (PTCs) began in the mid-twentieth century. The energy crises of the 1970s, driven by rising fossil fuel prices, sparked increased interest in solar thermal technologies, paving the way for the commercialization of PTC systems. Frank Shuman, an early pioneer in solar energy, made notable contributions in the early twentieth century by experimenting with solar concentrators to generate steam. Using parabolic mirrors to focus sunlight onto a central receiver, his work laid important groundwork for future innovations in solar energy. In the 1970s, engineers began developing early PTC prototypes with manually adjustable parabolic reflectors and linear receiver tubes. These early systems were primarily experimental, but they provided valuable

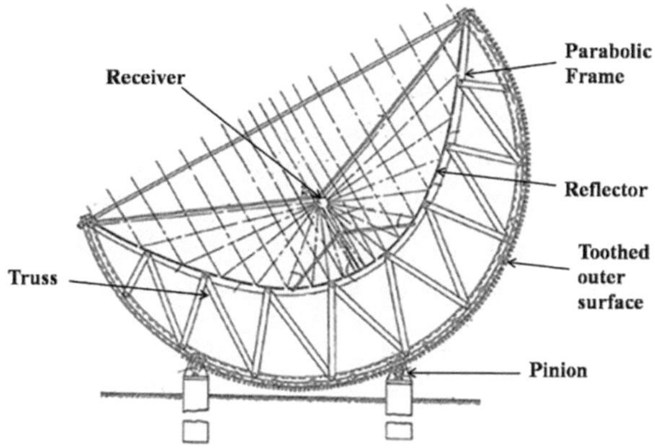


**Fig. 3.2** Frank Shuman's PTC [3, 4]

insights into the principles of solar concentration, which would later evolve into more advanced and efficient solar thermal technologies [2].

## 3.5 Technological Milestones

The progression of PTC technology is marked by several key technological milestones that have significantly enhanced its performance and commercial viability. Each of these milestones represents a major leap forward in the development of PTC systems.

**Introduction of Improved Reflective Materials**

The evolution of reflective materials has significantly improved the performance of Parabolic Solar Trough Collectors (PTCs). Initially, polished metals like stainless steel as well as aluminum were utilized due to their reflective properties. However, these metals had limitations, including reflectivity levels of 80–90% and susceptibility to oxidation and corrosion. This degradation not only reduced their effectiveness but also increased maintenance needs and costs.

A breakthrough occurred in the 1980s with the introduction of silvered polymer films. These films, which feature a thin layer of silver on a polymer substrate, offer reflectivity levels of up to 95–98%. This higher reflectivity improves the concentration of solar radiation on the receiver tube, enhancing system performance. Additionally, the polymer substrate protects the silver coating from environmental damage, extending the lifespan of the reflective surface and reducing maintenance requirements. The flexibility and lightweight nature of these films also facilitated easier handling and installation, promoting innovative designs [2].

Further advancements were made with enhanced glass coatings. Multi-layer dielectric coatings applied to glass substrates achieved reflectivity levels comparable to or exceeding those of silvered polymer films. These coatings maximize sunlight transmission while minimizing losses due to absorption and scattering. The durability of coated glass ensures resistance to UV radiation, moisture, and temperature fluctuations, contributing to lower operational costs and reduced maintenance [2].

Overall, these innovations have substantially grown the efficiency as well as longevity of PTC systems, making them effective and also versatile in various applications. As research progresses, further improvements in reflective materials are expected to advance PTC technology even more.

**Advancements in Tracking Mechanisms**

Accurate sun tracking is crucial for optimizing the execution of Parabolic Solar Trough Collectors (PTCs), vital in concentrated solar power units. Initially, PTCs used manual tracking, requiring frequent adjustments to align the reflectors with the sun. This labor-intensive method often led to inefficiencies and higher operational costs due to challenges in maintaining precise alignment throughout the day.

The 1980s and 1990s saw a major advancement with the introduction of automated tracking systems. Single-axis trackers, which rotate the reflectors along a longitudinal axis, follow the Sun's longitudinal directional movement, reducing manual adjustments and improving energy capture. However, these systems didn't account for the sun's changing altitude [2].

Dual-axis trackers addressed this limitation by enabling movement in both east-to-west and north-to-south directions. This capability allows precise adjustments for the Sun's changing location and altitude all day long and seasons. Dual-axis monitors significantly enhance solar energy capture, improving overall system efficiency.

Automated tracking systems have reduced labor costs and operational complexity, making large-scale solar power plants more viable. They also facilitate scalability, allowing for more reliable and efficient solar installations, crucial for meeting growing energy demands and sustainability goals.

**Development of Efficient Thermal Fluids**

The selection of thermal fluid in Parabolic Solar Trough Collectors (PTCs) is critical for optimizing performance. Early systems used water or basic oils, which were limited in efficiency due to their low boiling points and thermal stability. Water, though cost-effective, restricted the maximum operating temperature, while basic oils lacked the necessary thermal conductivity and stability. In the late 1980s and early 1990s, synthetic oils were introduced, offering improved temperature stability and thermal conductivity. These oils allowed PTC systems to operate at higher temperatures without degradation, increasing efficiency. Further advancements came with molten salts, composed of sodium and potassium nitrates, which provided an even broader temperature range and superior thermal storage capabilities. Molten salts enabled PTC systems to maintain consistent energy production, even during intermittent solar radiation, making them ideal for large-scale power plants and industrial applications. This innovation enhanced the reliability and commercial viability of PTC systems in renewable energy [2].

### Commercialization and Large-Scale Deployment

The commercialization of Parabolic Solar Trough Collector (PTC) technology began in the 1980s, marking a significant shift from theoretical research to practical application. This period saw the establishment of several notable solar power plants in Europe as well as the United States, which demonstrated the feasibility and potential of large-scale solar thermal energy. Among the pioneering projects were Solar Two and Solar One in California. Solar One, initiated in the final years of the 1970s and operational by the beginning of the 1980s, was one of the first large-scale implementations of PTC technology. Located in the Mojave Desert, it used a field of parabolic trough collectors to concentrate solar energy and generate electricity, proving that PTC systems could be effective on a commercial scale. Building on this success, Solar Two began operations in the early 1990s with advanced features such as an expanded collector area, and also an innovative thermal storage unit using molten salts. This allowed for more efficient energy storage and extended operational hours, further validating the technology's potential. The successful operation of these early plants was crucial in establishing PTC technology as a viable and reliable option for grid-scale solar power generation, paving the way for further development and deployment of solar thermal technologies. Their success helped build confidence in solar thermal power, leading to increased investment and interest in the field, and ultimately contributing to the broader adoption and growth of solar energy solutions worldwide [5].

### Evolution in Design and Materials

Over the decades, the design and materials used in Parabolic Solar Trough Collectors (PTCs) have evolved significantly, reflecting advancements in technology and materials science. This evolution has transformed PTC systems from relatively simple designs into sophisticated, high-performance systems capable of efficiently capturing and converting solar energy. Key advancements include improvements in reflective coatings, the integration of computer-controlled tracking systems, and the development of advanced heat transfer fluids.

### Reflective Coatings

The evolution of reflective coatings has a significant impact on enhancing the performance of PTC systems. Early PTC designs utilized polished metal mirrors as reflective surfaces. While these mirrors were effective to some extent, they had limitations in reflectivity and durability. Polished metal surfaces were prone to oxidation and other forms of environmental degradation, which reduced their efficiency over time.

Significant progress was made with the development of silvered polymer films and advanced glass coatings. Silvered polymer films are a major innovation in reflective coatings. These films are coated with a thin layer of silver, which provides excellent reflectivity and is more resistant to environmental damage compared to polished metal. They are also lightweight and flexible, making them easier to handle and install. Additionally, the use of advanced glass coatings has further enhanced the performance of PTC systems. These coatings are designed to endure the most brutal environmental extremes, including searing temperatures and relentless ultraviolet radiation, while maintaining high reflectivity. Modern reflective materials, with their superior reflectivity and durability, have led to significant improvements in the efficiency and longevity of PTC systems [2].

**Computer-Controlled Tracking Systems**

The advent of computer-controlled tracking systems represents another major advancement in PTC technology. Early PTC systems required manual adjustments to maintain the precise alignment of the parabolic reflectors with the sun's trajectory. This manual tracking was labor-intensive, often imprecise, and could lead to suboptimal solar energy collection. The integration of computer-controlled tracking systems has revolutionized the design and functionality of parabolic troughs. These units use a combination of sensors, actuators, and computerized controls to continuously adjust the position of the parabolic reflectors, ensuring they remain optimally positioned with the sun all day. The tracking systems employ algorithms and real-time data from sensors to make precise adjustments, which maximizes solar energy capture. This automation not only enhances the efficiency of solar energy collection but also reduces the need for manual intervention, lowering operational costs and improving the overall reliability of the PTC systems. Advanced tracking technologies have enabled PTC systems to achieve higher levels of performance and efficiency, bolstering their dominance in the inexhaustible energy market [1].

**Advanced Heat-Carrying Fluids**

The evolution of high-performance heat-carrying fluids has been another critical factor in the evolution of PTC technology. Early PTC systems used water and basic oils as heat transfer fluids. While these fluids were functional, they had limitations in terms of operating temperatures and thermal conductivity. Water, for instance, was constrained by its low boiling point, which limited the maximum temperature that the system could achieve. Basic oils, while offering some improvements over water, still had issues with thermal stability and conductivity.

Modern PTC systems benefit from the utilization of advanced heat-carrying liquids, such as synthetic oils and molten salts. Synthetic oils are engineered to function at elevated temperatures than traditional oils, offering better thermal stability and improved thermal conductivity. These properties allow PTC systems to achieve higher operating temperatures and enhance their overall efficiency.

Molten salts represent an even more significant advancement. Composed of a blend of sodium nitrate and also potassium nitrate, molten salts have a broader operating temperature range and excellent thermal storage capabilities. They can store large amounts of thermal energy, which can be utilized to form electricity even when direct sunlight is not present. This ability to capture and discharge thermal energy over extended periods enhances the reliability and consistency of power formation. The use of molten salts has enabled PTC systems to perform more efficiently and be used in a wider range of applications, from power generation to industrial processes [6].

**Table 3.1** Core principles and governing equations

| Sl. No. | | Equations [7, 8] |
|---|---|---|
| 1 | Calculation for the equation of continuity | $\nabla \cdot \left(\rho \vec{V}\right) = 0$ (1) |
| 2 | Calculation for the equation of momentum | $\nabla \rho \vec{V} \vec{V_i} = \frac{-\partial p}{\partial X_i} + \nabla\left(\mu \cdot \nabla \vec{V_i}\right) + S_i$ (2)<br>$V = (V_R, V_\theta, V_Z)$ field velocity<br>$X_i = R, \theta, Z$<br>$S_i$ = Remaining Viscous Terms<br>i = 1, radial direction<br>$S_1 = \rho \vec{V_\theta} \frac{\vec{V_\theta}}{r} - \mu\left[\frac{\vec{V_R}}{r} + \left(\frac{2}{r^2}\right)\left(\frac{\partial \vec{V_\theta}}{\partial \theta}\right)\right]$ (3)<br>i = 2 tangential direction<br>$S_2 = \mu\left[\left(\frac{2}{r^2}\right)\frac{\partial \vec{V_R}}{\partial \theta} - \frac{\vec{V_\theta}}{r^2}\right] - \rho V_R \frac{\vec{V_\theta}}{r}$ (4)<br>i = 3 axial direction<br>$S_3 = 0$ (5) |
| 3 | Calculation for the equation of energy | $\nabla\left(\rho \dot{v} c_p T\right) = \nabla \times (k \nabla T)$ (6) |

**Equations for Energy Transfer and Fluid Dynamics**

Systems utilized fluids flowing through tubes along with inserts, governed by conservation equations: momentum (Navier–Stokes), mass (continuity), and energy (energy equation). Assumptions include incompressible, steady flow with constant properties of nanofluid, and neglect of viscous dissipation, as per Table 3.1. These equations model nanofluids' behavior, ensuring perfect predictions as well as practical application insights.

## 3.6 Parabolic Trough Geometry

The parabolic trough collector is shaped like a parabolic dish that concentrates solar energy onto a tube. The key dimensions are as follows:

Let $W_a$ = Aperture width.

$d$ = The collector's depth.

$f$ = Focal length.

$L$ = Reflector length.

$\theta$ = Angle of the collector's field of view, which can be calculated based on the collector's specific design parameters.

**Aperture Width Calculation**

To determine the width of the aperture in the parabolic trough:

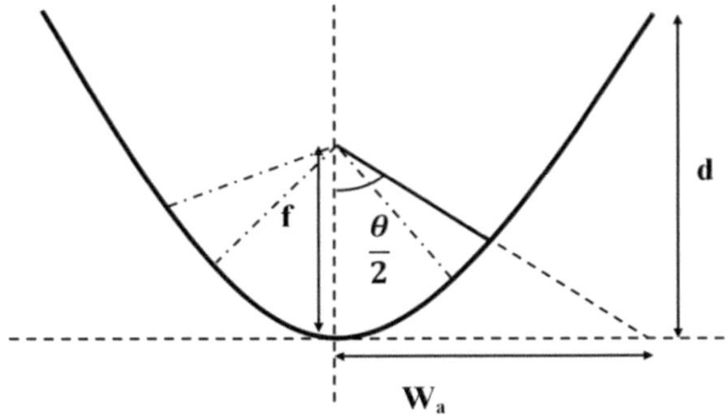


**Fig. 3.3** Structure of a parabola

$$W_a = 2 \cdot f \cdot \tan\left(\frac{\theta}{2}\right) \tag{3.1}$$

### Focal Length Calculation

The focal length $f$ can be derived from the dimensions of the parabolic trough and its geometry. For a parabolic trough, if the collector's depth ($d$) is known:

$$f = \frac{(W_a)^2}{16 \times d} \tag{3.2}$$

The structure of the parabola, as shown in Fig. 3.3, illustrates its geometric characteristics for parabola.

Here,

Parabola: The curve represents the cross-section of the parabolic trough.

Focal Length ($f$): The precise separation between the parabola's vertex and its focal point, where concentrated solar radiation converges onto the receiver.

Width ($W_a$): The horizontal span between the edges of the collector at its aperture.

Depth ($d$): The vertical displacement from the vertex of the parabola to the focal line, where concentrated rays converge (receiver).

The width $W_a$ of the collector is in terms of the focal length $f$ and the angle of the collector's field of view $\theta$. Field of View Angle ($\theta$) is the total angular span that the collector captures. The half-angle is $\frac{\theta}{2}$. At the edge of the collector, a right triangle is formed with

- One leg as $\frac{W_a}{2}$ (half the width of the collector).
- The other leg as the focal length $f$.
- The angle at the focal point is $\frac{\theta}{2}$.

The tangent of the angle $\frac{\theta}{2}$ is given by

$$\tan\left(\frac{\theta}{2}\right) = \frac{\frac{W_a}{2}}{f} \tag{3.3}$$

Rearrange to solve for $W$:

$$\frac{W_a}{2} = f \times \tan\left(\frac{\theta}{2}\right) \tag{3.4}$$

$$W_a = 2 \times f \times \tan\left(\frac{\theta}{2}\right) \tag{3.5}$$

**Focal Length (*F*) from the Collector's Depth (*D*)**

The convergence point in optics defines the focal length ($f$) in terms of the width $W$ and the depth d of the collector.

The equation of a parabola with its vertex at the origin and a horizontal axis of symmetry is $x^2 = 4 \times f \times y$.

**Depth of the Parabola**: The depth d is the vertical distance from the vertex to the focal line, aligned with the axis of symmetry.

Where $x = \frac{W_a}{2}$

$$d = \frac{\left(\frac{W_a}{2}\right)^2}{4 \times f} \tag{3.6}$$

Simplify to solve for $f$:

$$f = \frac{(W_a)^2}{16 \times d} \tag{3.7}$$

## 3.7 Thermal and Optical Functionality of Parabolic Trough Collectors

**Thermal Performance**

PTCs operate on the principle of concentrating solar radiation onto a tube, where a heat-conducting liquid absorbs the thermal energy. Process efficiency depends on intricate variables, conditions, and parameters such as the collector's geometric design, the quality of the reflective surface, and the type of heat-carrying liquid utilized.

**Principles of Heat Collection and Transfer**

Parabolic trough collectors (PTCs) are designed to focus sunlight onto a tube, which collects the heat energy.

### Heat Collection

The heat collection process in a parabolic trough collector involves two main stages.

**Concentration of Solar Radiation**: The parabolic reflector directs the sunlight onto the tube. The amount of solar power concentrated is described by the concentration ratio as follows.

The local concentration ratio is derived from the ratio of areas:

$$C_L = \frac{A_{\text{aperture}}}{A_{\text{receiver}}} \tag{3.8}$$

where $A_{\text{aperture}}$ is the area of the collector's aperture (the area exposed to sunlight), and $A_{\text{receiver}}$ is the area of the receiver tube. This ratio indicates how much the collector focuses solar energy onto the receiver.

**Absorption and Transfer of Heat**: The receiver tube intercepts concentrated solar radiation, subsequently transferring thermal energy to the heat transfer fluid (HTF). The net thermal output ($Q_u$) of the collector is expressed as follows:

Let $m$ = Mass flow rate of the working liquid.
$c_p$ = Specific Heat Capacity of the working liquid.
$T_{\text{in}}$ = Liquid flow inlet temperature.
$T_{\text{out}}$ = Liquid flow outlet temperature.
$A_a$ = Aperture area.
$G_a$ = Solar Direct Beam Irradiance.
$\Delta Q_{\text{tube}}$ = Heat absorbed by the collector tube.
$\rho_r$ = Reflectivity of the collector's mirror.

$$Q_u = m \times c_p \times (T_{\text{out}} - T_{\text{in}}) \tag{3.9}$$

The term $Q_s$ denotes the available solar energy, which is derived from the interaction of the collector's aperture area, the incident direct beam solar irradiance, and the system's concentration ratio. It quantifies the energy captured by the system, factoring in the size of the collector, irradiance intensity, and optical concentration:

$$Q_s = C_L \times A_a \times G_a \tag{3.10}$$

The thermal efficiency is $\eta_{\text{th}}$:

$$\eta_{\text{th}} = \frac{Q_u}{Q_s} \tag{3.11}$$

The local concentration ratio is ascertained through the subsequent mathematical formulation:

$$C_L = \frac{\Delta Q_{\text{tube}}}{\rho_r \times G_a \times A_a} \tag{3.12}$$

Thermal dissipation primarily manifests through radiative losses, as delineated by Eq. (3.13). It is imperative to acknowledge that, owing to the complete evacuation of the glass tube, convective heat transfer between the receiver and the collector is rendered negligible, thereby mitigating significant thermal losses within the system:

Let $D_{co}$ = Collector outer diameter.
$L$ = Collector length.
$h_w$ = Heat transfer coefficient due to wind.
$T_{co}$ = Temperature at the collector outer surface.
$T_a$ = Ambient temperature.
$T_{sky}$ = Sky temperature.

$$Q_{loss} = \frac{A_{ro} \times \sigma \times \left(T_r^4 - T_c^4\right)}{\frac{1}{\varepsilon_r} + \frac{(1-\varepsilon_c)}{\varepsilon_r} \times \left(\frac{A_{ro}}{A_{ci}}\right)} \tag{3.13}$$

$$\varepsilon_r = 0.000327T_r - 0.065971 \tag{3.14}$$

Once more, the heat losses attributed to wind velocity cannot be disregarded; these losses are accounted for through the following relations:

$$Q_{loss} = \pi \times D_{co} \times L \times h_w \times (T_{co} - T_a) + \varepsilon_c \times \pi \times D_{co} \times L \times \sigma \times \left(T_{co}^4 - T_{sky}^4\right) \tag{3.15}$$

The parameters $h_w$ and $T_{sky}$ represent the standard heat transfer coefficient between the outer surface of the glass tube and the surrounding environment (ambient), as well as the sky temperature, respectively, which are incorporated into the analysis as delineated in the subsequent relations:

$$h_w = V_w^{0.58} \times D_{co}^{-0.42} \tag{3.16}$$

$$T_{sky} = 0.0552 \times T_a^{1.5} \tag{3.17}$$

Incorporating thermal losses, the optical efficiency of the collector is determined through the ensuing relationship. This can alternatively be expressed in terms of mirror reflectivity; the intercept factor is contingent upon the angle of incidence of the rays, while $\tau\alpha$ denotes the product of the receiver's absorptance and the cover's transmittance.

Let $\rho_r$ = Reflectivity of the mirror.
$\gamma(\theta)$ = Intercept factor, which accounts for the angle of incidence.
$\tau$ = Transmittance of the cover.
$\alpha$ = Absorptance of the receiver.

$$Q_s \times \eta_{opt} = Q_u + Q_{loss}$$

$$= \rho_r \times \gamma(\theta) \times (\tau\alpha) \tag{3.18}$$

**Heat Transfer Mechanisms**

The thermal exchange within the collector involves conduction and convection processes. In conduction, thermal energy is transferred from the receiver tube's surface to the heat transfer fluid (HTF), with the effectiveness of this process depending on the thermal conductivity of the receiver material and the quality of the interface between the tube and the HTF. In convection, thermal energy is transferred from the receiver tube to the HTF through convective mechanisms. The convective heat transfer coefficient, denoted as $h$, plays a crucial role in determining the effectiveness of this heat transfer, and its calculation is essential for assessing the overall thermal performance of the system. Its calculation is formulated as follows.

Let $D_{\text{ri}}$ = Tube's innermost diameter.
$L$ = Tube's length.
$T_{\text{fm}}$ = Mean fluid temperature.
$\mu$ = Dynamic viscosity.
$k$ = Thermal conductivity.

$$h = \frac{Q_u}{(\pi D_{\text{ri}}) \times L \times (T_r - T_{\text{fm}})}$$

$$T_{\text{fm}} = \frac{(T_{\text{in}} + T_{\text{out}})}{2} \tag{3.19}$$

The Reynolds number is presented as Re:

$$\text{Re} = \frac{4 \cdot m}{\pi \times D_{\text{ri}} \times \mu} \tag{3.20}$$

Under fully developed flow conditions, the Nusselt number is mathematically represented as follows:

$$\text{Nu} = hD/k \tag{3.21}$$

**Hydraulic Analysis**

The pump work is ascertained through the subsequent equation.

Let $\Delta p$: Pressure drop during the flow.
$\rho$: Density of the working fluid.
$V$: Mean velocity.

$$W = \frac{m \times \Delta p}{\rho} \tag{3.22}$$

The friction factor ($f_{\text{th}}$) is calculated by using the relation shown below

$$f = \frac{\Delta p}{\frac{1}{2} \times \rho \times V^2 \times \left(\frac{D_{\text{ri}}}{L}\right)} \tag{3.23}$$

Mean velocity is considered by using the relation below

$$V = \frac{m}{\rho \times \pi \times \left(\frac{D_{\text{ri}}}{2}\right)} \tag{3.24}$$

### Exergy Efficiency

$E_u$ represents the exergy generated while the flow is happening:

$$\begin{aligned} E_u = {} & m \times c_p(T_{\text{out}} - T_{\text{in}}) - m \times c_p T_o \\ & \times \ln\left(\frac{T_{\text{out}}}{T_{\text{in}}}\right) - T_o \times m \times \frac{\Delta p}{T_{\text{fm}} \times \rho_{\text{fm}}} \end{aligned} \tag{3.25}$$

### Exergy Available ($E_s$)

The exergy derived from solar direct beam irradiation is quantified as follows.

$E_s$ represents exergy available from solar direct beam irradiation:

$$E_s = Q_s\left[1 - \frac{4}{3} \times \left[\frac{T_0}{T_{\text{sun}}}\right] + \frac{1}{3} \times \left[\frac{T_o}{T_{\text{sun}}}\right]^4\right] \tag{3.26}$$

Reference temperature ($T_0$) is 298.15 K [9]:

$$T_{\text{sun}} = 5770 \text{ K}$$

The exergy efficiency ($\eta_{\text{ex}}$) is calculated as below

$$\eta_{\text{ex}} = \frac{E_u}{E_s} \tag{3.27}$$

### Overall Efficiency

Overall efficiency is presented by $\eta_{\text{over}}$

$$\eta_{\text{over}} = \frac{\left(Q_u - \frac{W_p}{\eta_{\text{el}}}\right)}{Q_s} \tag{3.28}$$

Electrical efficiency ($\eta_{\text{el}}$) is 32.7% [9].

## Factors Influencing Thermal Efficiency

### Temperature

The efficiency of a solar collector is influenced by several factors, including collector temperature, ambient temperature, solar irradiance, and the incident angle of sunlight. As the collector temperature rises, the amount of heat collected increases, but so do thermal losses, primarily due to enhanced radiation and convective heat transfer. This means that at higher temperatures, the collector's efficiency can decrease due to these increased losses, making it a delicate balance between heat collection and heat loss. Ambient temperature also plays a crucial role in the collector's performance. When ambient temperatures are higher, the thermal gradient between the collector and its surroundings decreases, reducing heat losses and potentially improving efficiency. Solar irradiance, or the amount of solar energy hitting the collector, is directly proportional to the collector's efficiency. Higher levels of solar irradiance lead to more concentrated energy being absorbed, resulting in higher heat output. Additionally, the incident angle of sunlight significantly impacts the collector's ability to absorb energy. The optimal angle varies throughout the day, and collectors equipped with tracking systems can adjust their orientation to ensure they maintain this optimal angle, maximizing solar absorption and overall performance. These combined factors determine the overall effectiveness of the solar collector system.

### Fluid Type

The performance of a solar collector is significantly influenced by the thermophysical characteristics of the heat transfer fluid (HTF), including its thermal conductivity, specific heat capacity, and thermal stability. Fluids with higher thermal conductivity and specific heat capacity are more efficient at transferring heat, as they can absorb and transport greater amounts of thermal energy. Additionally, the thermal stability of the HTF ensures that the fluid maintains its performance over extended periods without degrading, which is crucial for long-term operational efficiency. Another important factor is the kinematic viscosity of the HTF, which affects the volumetric flow rate and the thermal transfer efficiency. Fluids with lower viscosity flow more easily, enhancing the convective heat transfer dynamics and improving overall heat transfer efficiency. However, while low-viscosity fluids typically offer better heat transfer, they may require special materials or design considerations to withstand the high temperatures they might encounter in the system. This could involve selecting appropriate materials that can handle these temperatures without failure or excessive wear. Therefore, the choice of HTF involves balancing factors like thermal conductivity, specific heat capacity, viscosity, and thermal stability to optimize the collector's efficiency and longevity.

### Optical Properties

The efficiency of a solar collector system is significantly influenced by both the reflectivity of the mirrors and the absorptance of the receiver. The reflectivity of the mirrors plays a vital role in concentrating solar energy, as mirrors with higher reflectivity direct more sunlight toward the receiver tube. This increases the amount of solar radiation collected and concentrated on the receiver, which is essential for maximizing the heat transfer. Higher reflectivity mirrors ensure that as much solar energy as possible is directed to the receiver, improving the overall energy efficiency of the system. Similarly, the absorptance of the receiver tube is critical in determining how effectively the captured solar radiation is converted into usable heat. Materials with higher absorptance are more efficient at absorbing solar radiation, leading to increased heat collection. In addition to absorptance, the receiver tube often has low-emissivity coatings that help minimize heat loss through thermal radiation. These coatings reduce the amount of heat radiated away from the receiver, ensuring that more of the collected energy is retained for transfer to the heat transfer fluid. By optimizing both the reflectivity of the mirrors and the absorptance of the receiver, a solar collector system can operate more efficiently, leading to better overall performance and energy conversion.

**Advances in Thermal Performance Over Time**

Recent advancements in solar energy technology have significantly enhanced the performance and efficiency of solar collectors. These innovations focus on improved materials, advanced tracking systems, and more efficient thermal storage solutions. Key developments include better reflectors, selective coatings for receiver tubes, two-axis tracking systems, automated controls, and advanced thermal storage solutions like molten salt and phase change materials.

Here's a summary in Table 3.2 form.

These innovations collectively enhance the solar collector's ability to harness and store solar energy more efficiently, contributing to greater sustainability and energy reliability.

**Heat Transfer Improvements**

Recent advancements in heat transfer fluids (HTFs) and heat transfer technologies have significantly enhanced the performance of solar collector systems. Research into new HTFs with improved thermal and physical properties has led to fluids that can operate at higher temperatures, thus increasing their efficiency in transferring heat. These fluids are designed to withstand harsher operating conditions, which not only extends the lifespan of the system but also improves its overall operational efficiency. The development of such HTFs ensures better heat transfer rates, making the system more effective in capturing and transporting thermal energy. Furthermore, innovations in heat exchanger designs and materials have played a critical role in improving the thermal transfer process from the HTF to the storage medium or power generation systems. Modern heat exchangers have been optimized to reduce thermal losses during the transfer process, ensuring that more of the collected heat is effectively stored or converted into usable energy. These advancements in heat exchanger technology help maintain high efficiency and reduce energy waste, contributing to the overall performance of solar energy systems. By combining enhanced HTFs

**Table 3.2** Advancements in parabolic solar trough collector

| Category | Advancements | Benefits |
|---|---|---|
| Improved materials | *Enhanced reflectors*: advances in reflective coatings and materials | Better reflectivity, increased durability, and sustained high optical performance |
| | *Selective coatings*: new coatings for receiver tubes that enhance absorptance and reduce emissivity | Improved solar radiation absorption and minimized heat losses, leading to better thermal performance |
| Advanced tracking systems | *Two-axis tracking*: adjusts orientation horizontally and vertically to follow the sun's path | Maximizes solar exposure and improves overall collector efficiency |
| | *Automated controls*: real-time adjustments of collector position through sensors and algorithms | Optimizes performance throughout the day and across different seasons |
| Thermal storage solutions | *Molten salt storage*: usage of molten salts as a medium to store heat for extended periods | Enables continuous power generation even when sunlight is unavailable |
| | *Phase change materials (PCMs)*: stores thermal energy by changing phase at specific temperatures | Helps maintain stable temperatures and boosts the efficiency of thermal storage systems |

and improved heat exchangers, solar collectors can achieve better thermal performance, increase energy output, and become more reliable for long-term use in various applications.

## 3.8 Optical Performance of Parabolic Trough Collectors

### Principles of Optical Performance

### Reflectivity and Absorptance

- **Reflectivity ($\rho$)**: Reflectivity measures how much solar radiation is deflected by the collector's mirror. It is defined as

$$\rho = \frac{\text{Reflected Radiation}}{\text{Incident Radiation}}$$

High reflectivity ensures that a greater proportion of solar radiation is focused onto the tube. Modern mirrors use coatings that enhance reflectivity and reduce losses due to scattering and absorption.

- **Absorptance ($\alpha$)**: Absorptance measures how effectively the receiver absorbs solar radiation. It is defined as

$$\alpha = \frac{\text{Absorbed Radiation}}{\text{Incident Radiation}}$$

High absorptance means the receiver captures a larger fraction of the concentrated solar energy. Selective coatings are designed to maximize absorptance while minimizing thermal radiation losses.

**Optical Efficiency ($\eta_{\text{opt}}$)**

The optical efficiency represents how effectively the collector converts incident solar radiation into absorbed heat. It is defined as

$$\eta_{\text{opt}} = \frac{Q_u}{Q_s} \tag{3.29}$$

Thus, rearranging the terms gives

$$Q_u = m \times c_p \times (T_{\text{out}} - T_{\text{in}}) \tag{3.30}$$

$$Q_u = \eta_{\text{opt}} \times Q_s \tag{3.31}$$

**Heat Losses ($Q_{\text{loss}}$)**

The heat losses from the collector occur due to several factors, including radiation and convection. The total heat losses ($Q_{\text{loss}}$) can be expressed as

$$Q_{\text{loss}} = Q_s \times \eta_{\text{opt}} - Q_u \tag{3.32}$$

**Optical Efficiency in Terms of Optical Properties**

The optical efficiency is influenced by several factors, including the reflectivity of the mirrors, the geometric configuration, and the radiative characteristics of the receiver. Specifically

- **Intercept Factor ($\gamma(\theta)$)**: The geometric factor or intercept factor that accounts for the angular orientation of the incident solar radiation. It measures how effectively the reflected sunlight is captured by the receiver.
- **Transmittance ($\tau$)**: The fraction of solar radiation transmitted through the cover or glass tube.

Combining the expressions for $Q_u$, $Q_s$, and $\eta_{\text{opt}}$,

$$Q_u = \eta_{\text{opt}} \times Q_s \tag{3.33}$$

Substitute the expression for $\eta_{\text{opt}}$

$$\eta_{\text{opt}} = \rho_r \times \gamma(\theta) \times (\tau\alpha) \tag{3.34}$$

Thus:

$$Q_u = (\rho_r \times \gamma(\theta) \times (\tau\alpha)) \times Q_s \tag{3.35}$$

Now, the total heat collected ($Q_s \times \eta_{\text{opt}}$) includes both the useful heat output and the heat losses:

$$Q_u + Q_{\text{loss}} = \eta_{\text{opt}} \times Q_s \tag{3.36}$$

Substitute $Q_u$ into this equation:

$$(\rho_r \times \gamma(\theta) \times (\tau\alpha)) \times Q_s + Q_{\text{loss}} = \eta_{\text{opt}} \times Q_s \tag{3.37}$$

This completes the derivation, showing that the available solar energy times the optical efficiency equals the sum of the useful heat collected and the heat losses, which is in line with

$$Q_s \times \eta_{\text{opt}} = Q_u + Q_{\text{loss}} = \rho_r \times \gamma(\theta) \times (\tau\alpha) \tag{3.38}$$

The optical efficiency is contingent upon the orientation of the collector, the integrity of the reflective surfaces, and the thermo-physical characteristics of the receiver.

**Advances in Optical Performance**

Recent advancements in solar collector technology have led to significant improvements in the performance and efficiency of key components. One such advancement is improved reflective coatings, which have enhanced the reflectivity and durability of mirrors. These new coatings increase the concentration of solar energy directed toward the receiver, boosting overall system performance and ensuring that mirrors maintain high reflectivity over time, even under harsh environmental conditions. Similarly, advanced receiver designs have introduced new materials and optimized tube structures that improve the absorptance of solar radiation and enhance thermal performance. These innovations help to reduce thermal losses and improve heat transfer efficiency, ensuring that more of the captured solar energy is effectively converted into usable heat. In addition, tracking technology has seen significant improvements, with more advanced tracking systems that maintain the optimal angle of incidence by following the sun's path throughout the day. By accurately adjusting the collector's orientation, these tracking systems maximize solar exposure, ensuring that the system captures the maximum amount of solar energy. Collectively, these innovations in reflective coatings, receiver designs, and tracking technology work together to increase the efficiency, reliability, and performance of solar energy systems, helping them deliver greater energy output with reduced losses.

## 3.9 Practical Considerations

Parabolic Trough Collectors (PTCs) require periodic maintenance to ensure optimal performance. This includes cleaning the parabolic mirrors to remove dust and debris that can reduce reflectivity and inspecting the tracking systems for proper alignment. The receiver tubes and heat transfer fluid (HTF) systems must also be checked for leaks or damage. Although the initial cost of PTC systems can be significant, they offer substantial long-term benefits in energy conservation and reduced reliance on fossil fuels. Advances in technology have led to cost reductions and enhanced economic viability, while government incentives help make renewable energy options more accessible. Environmentally, PTCs provide a sustainable solution by capturing solar radiation and converting it into thermal energy, thus reducing greenhouse gas emissions and lessening dependence on non-renewable resources.

Recent technological innovations have further improved the performance of PTCs. Enhancements in reflective coatings have increased solar energy absorption efficiency, while advanced materials for receiver tubes have boosted durability and thermal performance. More accurate tracking systems, with improved sensors and control mechanisms, have optimized solar energy capture. Additionally, PTCs are increasingly integrated with thermal energy storage systems and hybrid power plants, enabling reliable energy production even during cloudy weather or at night. Ongoing research and development focus on refining materials, designs, and system performance, aiming to make PTCs more efficient and economically viable for large-scale solar thermal energy generation, with significant environmental and financial advantages.

## 3.10 Analysis of Traditional Parabolic Trough Collector (PTC) Designs: Advantages and Disadvantages

Parabolic Trough Collectors (PTCs) are a prominent technology in solar thermal energy systems. They offer several distinct advantages but also come with notable disadvantages. Here's a concise analysis of their key benefits and drawbacks.

**Advantages of Parabolic Trough Collectors**

Parabolic Trough Collectors (PTCs) are a widely adopted technology for capturing solar energy. They have demonstrated high efficiency and proven reliability, making them a preferred choice for large-scale solar thermal power plants. Table 3.3 shows some key advantages of PTCs.

**Table 3.3** Advantages of Parabolic Trough Concentrator (PTC) Systems

| Advantage | Description | Key benefits |
|---|---|---|
| High efficiency in capturing solar radiation | PTCs use parabolic mirrors to concentrate solar energy onto a receiver tube, significantly increasing solar radiation intensity and enhancing thermal energy collection | Enables high-temperature applications like electricity generation and industrial heating |
| Proven technology with a strong track record | PTCs have been in use since the mid-twentieth century, with a long history of successful implementation in solar thermal power plants worldwide | Demonstrates reliability and effectiveness in large-scale solar projects |
| Scalability and integration | PTCs are adaptable to different installation sizes, from small systems to large solar farms. They can also integrate with thermal energy storage systems for a continuous energy supply | Provides flexibility for various energy demands and enhances energy reliability in hybrid systems |

**Disadvantages of Parabolic Trough Collectors**

Parabolic Trough Collectors (PTCs) are an efficient and reliable technology for solar energy generation, but they come with several drawbacks that must be considered. Table 3.4 is a summary of the challenges associated with PTCs.

## 3.11 Construction of Parabolic Solar Trough Collectors

The construction of Parabolic Trough Collectors (PTCs) is a detailed process that requires careful attention to design parameters, material selection, and manufacturing techniques to ensure the system operates efficiently in harnessing solar energy. The reflectors, which are responsible for concentrating sunlight onto the receiver tubes, are typically made from highly reflective materials such as aluminum or silver-coated glass. Aluminum is often chosen for its lightweight properties and ease of manufacturing, while glass mirrors, although heavier, offer superior durability and reflectivity, but are more fragile. The receiver tubes, which capture the concentrated solar radiation, are made from materials like stainless steel or specialized alloys capable of withstanding the high temperatures and thermal stresses associated with solar energy capture. To further improve efficiency, the receiver tubes are often coated with selective absorbers, which enhance the ability of the tube to absorb sunlight while minimizing thermal losses.

The trough structure, which supports the reflective surface and maintains its parabolic shape, is typically constructed from steel or aluminum. These materials are selected for their strength, durability, and resistance to environmental conditions such

**Table 3.4** Drawbacks of parabolic trough concentrator (PTC) systems

| Drawback | Description | Impact |
|---|---|---|
| High initial cost | The upfront capital cost of PTCs is substantial, covering the manufacturing and installation of mirrors, receiver tubes, tracking systems, and infrastructure | Significant financial barrier, particularly in regions with limited resources |
| Space requirements and land use | PTC systems require large land areas to operate efficiently. The size of the collector field depends on the desired energy output | Limits feasibility in densely populated areas or regions where land is scarce or in conflicts with agriculture or urban development |
| Maintenance and operational challenges | Regular maintenance is needed to address wear and tear on tracking systems, mirrors, and receiver tubes. Mechanical issues and environmental factors increase maintenance costs | Can lead to operational downtime, affecting energy output and system efficiency |
| Heat transfer fluid management | The selection and management of heat transfer fluids (HTFs) are crucial for system performance. Handling HTFs, particularly at high temperatures, require specialized equipment | Necessitates additional safety measures and specialized equipment, increasing operational complexity |

as wind and thermal expansion. The support system also includes tracking mechanisms that enable the PTC to follow the sun's path across the sky, optimizing solar energy capture throughout the day. There are two main types of tracking systems used in PTCs: single-axis and dual-axis tracking. Single-axis tracking systems rotate the collector on one axis, typically oriented from north to south, to follow the sun's east-to-west movement, while dual-axis tracking systems allow for adjustments in both horizontal and vertical axes, enabling the collector to follow the sun more precisely, including adjusting for seasonal variations. Automatic tracking systems are commonly used, as they provide greater accuracy and minimize the need for manual adjustments.

The manufacturing process for PTCs involves several fabrication techniques for both the reflectors and the receivers. Reflectors are typically formed into the desired parabolic shape through rolling for aluminum or molding for glass, followed by the application of reflective coatings to ensure optimal solar reflection. The receiver tubes are manufactured by forming metal tubes and then coating them with selective absorbers to maximize solar absorption and minimize thermal losses. Precision is critical during this stage, as even minor defects can significantly impact the system's efficiency. Once the components are manufactured, assembly begins. The reflector panels are integrated into the trough structure, and the receiver tubes are carefully aligned along the focal line of the parabolic trough. The alignment is crucial to ensuring the maximum amount of solar energy is concentrated onto the receiver tube. The tracking system is then installed and calibrated to ensure it moves the collector along the sun's path accurately.

The installation of PTC systems requires careful site preparation, which includes leveling the ground, setting up foundations, and installing the structural and tracking components. Site-specific considerations, such as soil conditions and wind loads, are factored into the design and construction to ensure long-term stability. Once installed, the system is calibrated, with particular attention given to the alignment of the trough and tracking system to ensure optimal performance.

In practical terms, PTCs are used in large-scale solar thermal power plants to generate electricity. The concentrated solar energy heats a heat transfer fluid (HTF), typically oil or molten salt, which is used to generate steam that drives turbines for electricity generation. PTCs are also commonly employed in industrial applications that require significant thermal energy, such as water desalination, chemical production, and food processing. Furthermore, PTCs can be integrated with thermal energy storage systems, allowing them to store excess heat for later use, ensuring a continuous energy supply even during periods of cloud cover or at night. Their scalability allows for use in both large solar farms and smaller installations, making them a versatile solution for a range of solar energy needs. Despite the complexity of their construction and installation, the long-term benefits of PTCs, including their ability to reduce reliance on fossil fuels, lower energy costs, and provide a sustainable source of energy, make them a crucial component of the transition to renewable energy systems worldwide.

## 3.12 Case Studies

The Solar Energy Generating Systems (SEGS) plants, located in the Mojave Desert in California, USA, represent one of the most successful and pioneering implementations of Parabolic Trough Collector (PTC) technology. The project began in 1984 with the completion of SEGS I, and it has since expanded to include nine plants (SEGS I–IX), collectively generating over 350 megawatts (MW) of electricity. SEGS is one of the largest solar thermal power plants in the world, playing a crucial role in the development and deployment of solar thermal technology. These plants use parabolic trough collectors, which are curved mirrors that focus sunlight onto receiver tubes positioned along the focal point of the troughs. These collectors are arranged in linear arrays and are equipped with single-axis tracking systems that allow them to follow the sun's east-to-west movement, ensuring optimal solar radiation concentration throughout the day.

The receiver tubes are filled with a heat transfer fluid (HTF), usually synthetic oil, which absorbs the concentrated solar energy. The HTF is heated to temperatures of around 390 °C (734 °F) as it flows through the receiver tubes. This heated fluid is then pumped through a series of heat exchangers, where it transfers its thermal energy to water, generating steam. The steam drives a steam turbine coupled to an electrical generator, which converts thermal energy into electricity. The electricity is then fed into the power grid for distribution.

One of the key features of the SEGS plants is their integration of thermal storage systems. These systems store heated HTF in insulated tanks, allowing the plants to continue generating electricity even during periods when sunlight is unavailable, such as at night or on cloudy days. This ability to store thermal energy enhances the reliability and consistency of the power generation process, making SEGS an even more valuable source of renewable energy.

The SEGS project has proven the scalability and effectiveness of PTC technology for large-scale energy production. Over the decades, it has consistently provided significant amounts of renewable electricity, reducing the need for fossil-fuel-based power generation and lowering greenhouse gas emissions. Furthermore, SEGS has contributed to local economic growth, creating jobs in the Mojave Desert region and supporting further investments in solar thermal technologies. The success of SEGS has been instrumental in advancing solar thermal power and setting a benchmark for future projects worldwide [10]. The operational layout of the SEGS solar trough power plant (Fig. 3.4) [10] and the elevated view of the five 30 MWe parabolic trough plants at Kramer Junction, California (Fig. 3.5) [10], offer valuable insights into the structure and functionality of large-scale solar power systems.

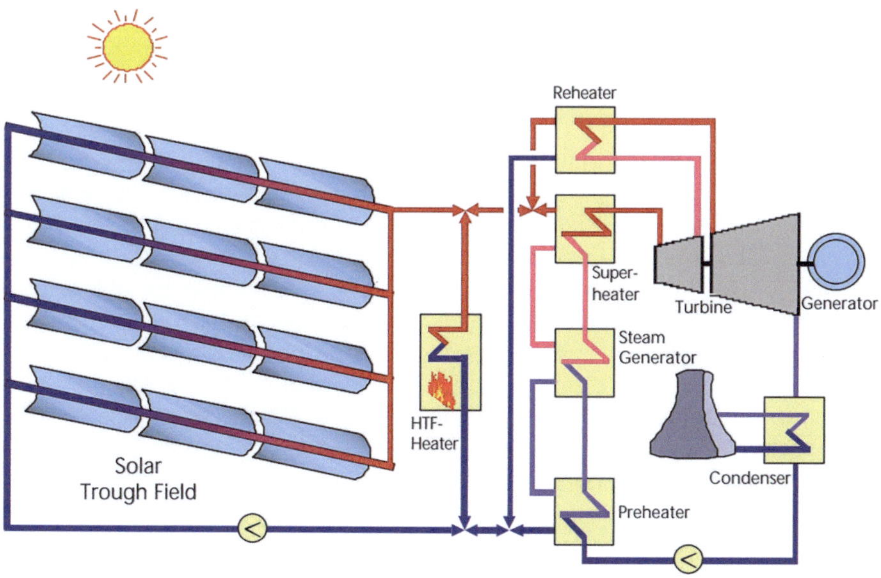


**Fig. 3.4** SEGS solar trough power plant operational diagram [10]

**Fig. 3.5** Aerial view of the five 30 MWe parabolic trough plants at Kramer Junction, California (Courtesy of KJC) [10]

### The Andasol Solar Power Station

The Andasol Solar Power Station, located in Andalusia, Spain, is one of the most advanced applications of parabolic trough technology in Europe. The project is divided into three phases: Andasol 1, Andasol 2, and Andasol 3, with a combined generation capacity of 150 MW. Andasol 1 began operation in 2008, followed by Andasol 2 in 2009 and Andasol 3 in 2011, making it a landmark in solar thermal power generation. The plant uses parabolic trough collectors, which are arranged in linear arrays and designed to concentrate solar radiation onto receiver tubes. The collector system employs a dual-axis tracking mechanism, enabling the mirrors to adjust both horizontally and vertically. This ensures that the solar collectors can capture solar energy more efficiently by following the sun's path across the sky, optimizing energy capture throughout the day and across seasonal changes.

One of the standout features of Andasol is the use of molten salt as the heat transfer fluid (HTF) and thermal energy storage medium. Unlike conventional systems that rely on synthetic oils, molten salt has a high heat retention capacity, which enables it to store thermal energy for extended periods. The molten salt is stored in large, insulated tanks and is used to absorb and retain heat from the solar collectors during the day. This innovative storage solution allows the plant to continue generating electricity even when direct sunlight is unavailable, such as during the night or on cloudy days, making it highly reliable and efficient in maintaining a stable power supply.

In the heat exchangers, the molten salt transfers its stored thermal energy to water, converting it into steam. This steam drives a steam turbine connected to an electrical generator, producing electricity that is fed into the grid. The ability to generate electricity around the clock, regardless of sunlight availability, is a key feature of the Andasol project, enhancing its contribution to grid stability and reliability.

The Andasol Solar Power Station also exemplifies technological innovation, as it integrates advanced parabolic trough systems with molten salt thermal storage technology. This combination enhances the plant's ability to produce consistent, uninterrupted power, even during times of reduced solar radiation. The project has made a significant contribution to Spain's renewable energy goals by reducing its reliance on fossil fuels and minimizing greenhouse gas emissions. Additionally, it has created numerous jobs in the Andalusia region, providing a positive impact on the local economy.

Furthermore, the integration of thermal storage provides grid stability, enabling the Andasol plant to provide power even when solar radiation is intermittent. This capability is critical for supporting the broader integration of renewable energy sources into the grid, reducing dependency on conventional fossil fuel-based power generation. Overall, the Andasol Solar Power Station represents a significant advancement in solar thermal energy technology, providing a cleaner and more reliable energy source while supporting local economies and contributing to environmental sustainability. The operational layout of the 50 MW Andasol 1 solar power plant with its thermal energy storage system is shown in Fig. 3.6, while Fig. 3.7 provides its schematic layout [11].

The Solar Energy Generating Systems (SEGS) plants and the Andasol Solar Power Station are two exemplary case studies that showcase the effectiveness and scalability of Parabolic Trough Collectors (PTCs). The SEGS plants highlight the successful application of PTC technology in large-scale solar power generation, while the Andasol project demonstrates the benefits of integrating advanced thermal storage solutions. Both projects underscore the potential of PTC technology to contribute to sustainable energy production and provide valuable insights for future solar thermal power developments.

## 3.13 Applications of Parabolic Trough Collectors

Parabolic Trough Collectors (PTCs) are a proven and highly effective solar energy technology, widely used to concentrate sunlight and generate high-temperature heat for a range of applications. Over time, PTC systems have evolved from early prototypes to sophisticated solutions in various sectors, including power production, industrial heating, refrigeration, air-conditioning, and even agricultural and residential usage. In power production, PTCs are a critical component of Concentrated Solar Power (CSP) plants, where they focus solar radiation onto receiver tubes filled

**Fig. 3.6** 50 MW solar power plant Andasol 1 with solar thermal energy storage system [11], photo with permission from solar millennium

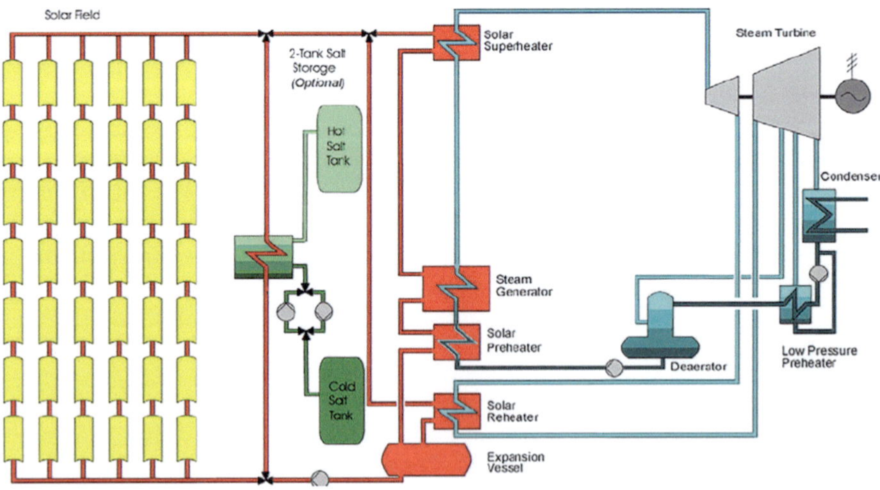


**Fig. 3.7** Schematic of Andasol solar power station [11]

with a heat transfer fluid that drives a steam turbine to generate electricity. These systems are most effective in regions with high direct sunlight but can incorporate thermal energy storage, like molten salts, to ensure reliable energy production even during intermittent sunlight. In industrial processes, PTCs are used to provide high-temperature heat for tasks such as drying, pasteurization, and chemical reactions, significantly reducing reliance on fossil fuels and lowering greenhouse gas emissions. This integration can help industries meet sustainability targets and reduce fuel costs, though it may require initial infrastructure adjustments. In refrigeration and air-conditioning, PTCs are utilized for both direct and indirect solar cooling applications. Direct solar cooling systems use PTCs to power absorption chillers, providing efficient cooling for commercial, industrial, and residential sectors, while indirect systems generate electricity to power conventional cooling systems, reducing grid dependence. Furthermore, PTCs are used in agricultural applications, heating water for irrigation, livestock, and greenhouse heating, helping to maintain optimal conditions for crops and animals. In residential settings, PTC systems can be incorporated into hot water systems, offering a sustainable and cost-effective alternative to conventional energy sources for everyday household needs. The versatility of PTC technology makes it a promising solution for a wide array of energy needs, contributing to a more sustainable and energy-efficient future.

**Challenges and Considerations**

Despite the numerous benefits of Parabolic Trough Collectors (PTCs), some challenges need to be addressed. The initial capital investment can be significant, and integrating PTC technology into existing infrastructure may require careful planning and additional resources. The efficiency of these systems is also affected by geographical and climatic conditions, which necessitate thorough site assessment and design optimization. However, advancements in materials, tracking systems, and heat transfer fluids are helping to improve PTC performance. Ongoing research aims to enhance efficiency and broaden the applications of PTCs. Beyond power generation, PTCs are also valuable for industrial heating, refrigeration, air-conditioning, and residential applications. These advancements are crucial for reducing fossil fuel reliance and promoting sustainability, making PTC systems increasingly vital in the global shift toward renewable energy.

**Future Trends and Innovations in Parabolic Trough Collectors**

Parabolic Trough Collectors (PTCs) have significantly progressed since their inception, evolving into a key technology in solar energy implementations. As the field of solar thermal energy continues to grow, several future trends and innovations are shaping the development of PTC technology. These include emerging technologies such as hybrid systems and advances in materials and efficiency, as well as ongoing research and potential breakthroughs. Understanding these trends is crucial for anticipating the next steps in PTC technology and its role in the global energy landscape.

## Emerging Technologies

### Hybrid Systems Combining PTC with Other Renewable Technologies

A highly promising development in PTC technology is the integration of hybrid systems that combine PTCs with other renewable energy technologies. These hybrid systems aim to improve overall energy efficiency as well as reliability by leveraging the strengths of different technologies.

For example, hybrid PTC systems that incorporate photovoltaic (PV) panels represent an innovative approach to maximizing energy production. In such systems, PV panels are mounted alongside or integrated into PTC installations. While the PTCs focus on generating thermal energy, the PV panels convert sunlight directly into electricity. This combination allows for the simultaneous production of both thermal and electrical energy, making the most of available solar resources and providing a more stable and versatile energy output.

Another hybrid approach involves combining PTC systems with biomass or geothermal energy sources. In this configuration, PTCs provide thermal energy during sunny periods, while biomass or geothermal systems supply additional heat when solar energy is insufficient. This hybrid strategy enhances the reliability and continuity of energy supply, making it suitable for applications where a consistent thermal output is essential.

### Advances in Materials and Efficiency

Advancements in materials and efficiency are fundamental to the future of Parabolic Trough Concentrator (PTC) technology, driving improvements in performance, durability, and cost-effectiveness. One of the primary areas of innovation is in reflective coatings, which directly affect the optical efficiency of PTC systems. Traditional reflective surfaces, like polished metal mirrors, face limitations related to reflectivity and long-term durability, often degrading due to environmental exposure. Recent developments in reflective coatings, including silvered polymer films and advanced glass coatings, have significantly improved performance. These modern materials offer higher reflectivity, greater resistance to environmental degradation, and better long-term performance, which ultimately enhances solar energy capture and improves overall system efficiency.

Another critical area of advancement lies in heat transfer fluids (HTFs), which play a vital role in transferring heat from the solar collector to the power conversion system. Traditional HTFs, such as thermal oils, have limitations in terms of thermal stability and efficiency. Newer fluids, such as synthetic oils with improved thermal stability and molten salts with enhanced thermal storage capacity, are now being integrated into PTC systems. These advanced fluids allow PTC systems to operate at higher temperatures, enhancing thermal efficiency and energy storage capabilities. Ongoing research into more effective HTFs focuses on increasing temperature tolerance, improving thermal conductivity, and reducing costs, which will further boost the overall efficiency of PTC systems.

In addition to coatings and HTFs, advances in structural materials are also making a significant impact on PTC technology. Lightweight, high-strength composite materials are being introduced to reduce the weight and cost of PTC systems, all while maintaining structural integrity and performance. These materials help optimize system designs, making PTC systems more cost-effective and efficient without sacrificing durability. As a result, the combination of improved reflective coatings, innovative heat transfer fluids, and advanced structural materials is transforming PTC technology, making it more competitive with other renewable energy technologies. These advancements are crucial for increasing the scalability and sustainability of solar energy production.

## 3.14 Research and Development

Parabolic Solar Trough Collectors (PTCs) are a proven technology in solar thermal energy production. They concentrate solar energy using parabolic mirrors to heat a heat transfer fluid, which can be utilized for electricity generation or industrial heating. Despite their established utilization, ongoing research and development (R&D) are critical to enhancing their performance, efficiency, and overall competitiveness in the inexhaustible energy sector. As the global demand for clean energy increases, advancing PTC technology remains a priority, and several key research areas have emerged to address current limitations and unlock new possibilities.

### *3.14.1 Thermal Storage Systems*

One of the major challenges faced by Parabolic Trough Concentrator (PTC) systems is their dependence on direct sunlight, which can be intermittent due to factors such as weather conditions or the time of day. To address this issue, thermal storage systems have become a key focus of research, as they can store excess heat during periods of high solar intensity for later use when sunlight is low. Phase Change Materials (PCMs) are being explored as a promising solution, as they store thermal energy by changing phases between solid and liquid at specific temperatures. Researchers are investigating new materials with higher thermal storage densities and better thermal stability, such as organic, inorganic, and hybrid PCMs, to withstand the high temperatures typical of PTC systems. Additionally, advanced molten salts, which are commonly composed of sodium nitrate and potassium nitrate, are widely used in thermal energy storage systems due to their exceptional heat retention properties and thermal stability at high temperatures. Ongoing research is focused on developing improved formulations of molten salts that enhance thermal conductivity, reduce costs, and increase stability, including the incorporation of additives to lower the melting point and boost heat transfer efficiency. These innovations in thermal storage are crucial for ensuring a continuous energy supply from PTC systems, even during cloudy weather or at night. By providing backup power when sunlight is unavailable, effective thermal storage solutions will enhance the reliability, grid stability, and economic viability of solar thermal energy, making PTC systems more practical and competitive in the renewable energy market.

### 3.14.2 Tracking Systems

PTC systems depend on precise solar tracking to maximize energy capture, and their performance is directly influenced by the accuracy with which they follow the sun's trajectory throughout the day. Research in tracking systems aims to enhance precision, increase efficiency, and improve cost-effectiveness. Innovations in sensor technology are central to this progress, focusing on developing cost-effective, high-precision sensors that provide real-time data about the sun's position. By combining GPS technology with inertial measurement units (IMUs), tracking systems can adjust dynamically to changing conditions, ensuring better alignment with the sun. Furthermore, advancements in control algorithms help improve the responsiveness and accuracy of the tracking systems. Artificial intelligence (AI) and machine learning (ML) are increasingly integrated into tracking systems to optimize performance. AI can analyze historical weather data and solar radiation forecasts, allowing for real-time adjustments that enhance tracking accuracy. By learning from past performance, AI systems can minimize errors caused by weather variations or mechanical failures, ensuring that the system remains aligned with the sun for optimal energy capture. These technologies enable PTC systems to better adapt to environmental conditions, reduce operational costs, and maximize efficiency. With improvements in sensors, AI-driven optimization, and control systems, PTC solar tracking is becoming more precise, reliable, and cost-effective, helping to improve the overall performance and economic viability of solar thermal energy systems.

### 3.14.3 Integration with Smart Grids

Another key area of focus in solar thermal energy research is integrating PTC systems with smart grids, which use advanced communication and control technologies to optimize energy distribution and overall management. Research efforts are directed at creating seamless integration between PTC systems and these smart grids to enhance the efficiency of energy production, storage, and consumption. One significant development is the creation of advanced communication systems, which involve new communication protocols that enable real-time data exchange between PTC systems and smart grids. This integration ensures that energy production, storage, and system status are synchronized with grid demands. Additionally, innovations in Energy Management Systems (EMS) are helping to optimize the production, storage, and distribution of energy from PTC systems. These systems use sophisticated algorithms to balance supply and demand, reduce energy losses, and improve efficiency. By integrating PTC systems with smart grids, researchers aim to enhance the distribution of renewable energy, improving grid stability and enabling more effective management of solar power, ultimately leading to a more reliable and efficient energy network.

## 3.15 Potential Breakthroughs

Beyond ongoing research, several potential breakthroughs are poised to impact PTC systems significantly. Developing next-generation materials, such as novel reflective coatings, advanced heat transfer fluids, and lightweight, durable structural materials, could significantly enhance PTC systems' efficiency, durability, and cost-effectiveness. For example, nanomaterials and advanced coatings are being explored to improve the reflectivity of parabolic mirrors, allowing for greater solar energy capture. Research into high-efficiency concentrators, including advanced optical designs and multi-junction solar cells, could lead to more efficient solar concentration and energy conversion, reducing optical losses and making PTC systems more compact and cost-effective. Additionally, integrating PTC technology with alternative sustainable energy sources, like hydrogen production or geothermal systems, is gaining attention as a promising approach to creating hybrid systems. These systems could boost the efficiency and versatility of renewable energy solutions by providing more reliable and integrated energy sources. As PTC technology evolves through breakthroughs in materials, concentrators, and hybrid systems, it will become increasingly competitive and effective in meeting global energy demands. The ongoing advancements in thermal storage, tracking systems, smart grid integration, and materials science will further improve the efficiency and scalability of PTC systems, allowing them to play a pivotal role in the transition to a sustainable, renewable energy future.

## References

1. Nsengiyumva W, Chen SG, Hu L, Chen X (2018) Recent advancements and challenges in solar tracking systems (STS): a review. Renew Sustain Energy Rev 81:250–279
2. Gharat PV, Bhalekar SS, Dalvi VH, Panse SV, Deshmukh SP, Joshi JB (2021) Chronological development of innovations in reflector systems of parabolic trough solar collector (PTC): a review. Renew Sustain Energy Rev 145:111002
3. Shuman F, Boys CV (1912) Sun boiler. US Patent, p 1240890
4. Suman S, Khan MK, Pathak M (2015) Performance enhancement of solar collectors: a review. Renew Sustain Energy Rev 49:192–210
5. Fredriksson J, Eickhoff M, Giese L, Herzog M (2021) A comparison and evaluation of innovative parabolic trough collector concepts for large-scale application. Sol Energy 215:266–310

6. Krishna Y, Faizal M, Saidur R, Ng KC, Aslfattahi N (2020) State-of-the-art heat transfer fluids for parabolic trough collector. Int J Heat Mass Transf 152:119541
7. Chakraborty O, Das B, Gupta R (2022) Performance of parabolic trough collector having rotating receiver: effect of elliptical insert and hybrid nanofluid. Int J Energy Res 46:15254–15275. https://doi.org/10.1002/er.8225
8. Chakraborty O, Roy S, Das B, Gupta R (2022) Computational analyses of parabolic trough solar collector in the presence of helical coil-insert. Int J Environ Sci Technol 137:87. https://doi.org/10.1007/s13762-021-03891-1
9. Bellos E, Daniil I, Tzivanidis C (2018) Multiple cylindrical inserts for parabolic trough solar collector. Appl Therm Eng 143:80–89
10. Michael A, Quaschning V. Solar thermal power the seamless solar link to the conventional power world. Renew Energy World · 2000, 15:184–191.
11. Rainer-Aringhoff R, Geyer M, Herrmann U, Kistner R, Navab P, Osuna R (2002) AndaSol: 50MW solar plants with 9 h storage for Southern Spain. In: Proceedings of the 1th SolarPACES international symposium on concentrated solar power and chemical energy technologies at: Zurich

# Chapter 4
# Thermal Performance of Parabolic Solar Trough Collectors: Effect of Receiver Tube Geometry

In this chapter, the discussion narrows to the effect of receiver tube geometry on the thermal performance of PTCs. It presents a comparative analysis of differently shaped tubes based on energy and exergy perspectives. Drawing from contemporary literature, it evaluates pressure drops, heat transfer characteristics, and design trade-offs. Each geometry is analyzed for its specific benefits and shortcomings in real-world performance. This focused approach enables readers to understand how minor design alterations can significantly impact thermal efficiency.

## 4.1 Early Designs

The earliest receiver tubes were predominantly cylindrical, chosen for their simplicity and cost-effectiveness. These designs were straightforward to manufacture and effective in absorbing solar radiation, though they had limitations in maximizing thermal efficiency. The primary materials used during this period included copper and stainless steel, known for their good thermal conductivity and durability. However, these materials also had drawbacks, such as high cost and relatively high thermal losses due to radiation.

During the same period, flat-plate receivers emerged as an alternative to cylindrical designs. By increasing the surface area exposed to solar radiation, flat-plate receivers aimed to enhance heat absorption. This design modification improved thermal efficiency by providing a larger area for solar energy collection. Advances in materials, such as the introduction of selective coatings and aluminum, further boosted the performance of flat-plate receivers by improving absorptance and reducing radiative heat losses.

 
O. Chakraborty and B. Das, *Parabolic Solar Trough Collector*, Mechanical Engineering Series, https://doi.org/10.1007/978-3-032-11229-3_4

**Advances in Geometry and Materials**

The 1990s saw the introduction of elliptical receiver tubes, which were designed to better align with the focal line of parabolic mirrors. This geometry allowed for improved absorption of concentrated solar energy, enhancing overall thermal efficiency. Elliptical tubes addressed some limitations of cylindrical designs by better matching the curvature of the reflected solar rays, thereby capturing more solar energy.

During this period, there was a shift toward more advanced materials and coatings. High-performance ceramics and specialized coatings were developed to improve heat retention and reduce thermal losses. These innovations not only enhanced the efficiency of receiver tubes but also contributed to their durability and resistance to environmental factors [1].

The early 2000s marked a significant evolution in solar thermal collector technology, driven by advancements in receiver designs and manufacturing techniques. The introduction of finned and multi-tube receivers notably improved heat transfer rates and distribution uniformity, leading to enhanced system efficiency. Advances in precision machining, computer-controlled welding, and fabrication techniques allowed for the production of intricate geometries with high accuracy, boosting collector performance.

**Recent Developments and Trends**

Recent research has continued to explore hybrid designs that combine various geometries to optimize performance. Innovations in materials, such as advanced composites and nanocoatings, are also being investigated to further enhance thermal efficiency and reduce costs. The focus has shifted toward balancing performance improvements with cost-effectiveness and system durability.

In summary, the historical development of receiver tube designs in PTCs reflects a continuous effort to enhance thermal performance through geometric optimization and material innovation. From simple cylindrical tubes to complex multi-tube and finned arrangements, the evolution of receiver tube designs demonstrates the progress made in solar thermal technology and its growing sophistication.

Parabolic trough collectors (PTCs) are a widely used technology in solar thermal systems. They consist of a parabolic-shaped mirror that concentrates sunlight onto a receiver tube running along the focal line. The receiver tube absorbs the concentrated solar energy and transfers it to a heat transfer fluid, which is then used for various applications such as generating steam for electricity or heating water. The performance of PTCs is highly dependent on the design of the receiver tube. This analysis explores how different receiver tube shapes impact the thermal efficiency and heat transfer rates of PTCs.

## 4.2 Performance of Differently Shaped Receiver Tubes

### Cylindrical Receiver Tubes

Cylindrical receiver tubes, a conventional design in parabolic trough collectors (PTCs), play an integral role in solar thermal systems by capturing concentrated solar radiation reflected by the parabolic mirror and transferring it to a heat transfer fluid. However, the cylindrical shape does not align precisely with the focal line of the parabolic mirror, resulting in inefficiencies in solar energy absorption. This misalignment causes a portion of the concentrated sunlight to miss the receiver tube, thereby reducing its energy absorption capacity. As a result, cylindrical tubes generally exhibit lower thermal efficiency compared to advanced receiver designs that are better optimized for solar ray capture. In addition, heat losses, especially when insulation or coatings are not optimal, further hinder the system's overall performance. While enhancements such as improved insulation and coatings can mitigate some of these drawbacks, the fundamental geometric limitations of the cylindrical design persist. The relatively modest heat transfer rates in cylindrical tubes are attributed to the limited surface area available for heat exchange, as the basic cylindrical form offers less surface area compared to advanced geometries like finned or corrugated tubes. Moreover, the heat transfer coefficient, which governs the rate at which heat is transferred from the tube to the fluid, is limited by the tube's geometry, contributing to lower heat transfer efficiency. Exergy analysis underscores the inefficiencies of cylindrical tubes, as the misalignment and heat losses increase exergy destruction, which refers to the loss of the potential useful work of energy. This results in a less efficient conversion of solar energy into thermal energy. While cylindrical tubes remain cost-effective and widely used, their performance is ultimately constrained by their basic geometric design, as depicted in Fig. 4.1.

### Elliptical Receiver Tubes

Elliptical receiver tubes are engineered to align more precisely with the focal line of the parabolic mirror, improving the concentration of solar radiation onto the tube and enhancing solar energy absorption. In contrast to cylindrical tubes, the elliptical shape reduces the angle of incidence between the incoming solar rays and the tube's surface, facilitating better capture of concentrated sunlight. This alignment results in significant improvements in thermal efficiency, as research demonstrates enhanced performance due to more effective energy capture. The optimized geometry also promotes better heat transfer by reducing the temperature gradient between the receiver tube and the heat transfer fluid, resulting in a more efficient heat exchange process. Additionally, the reduced misalignment helps minimize heat losses, further boosting the

**Fig. 4.1** Cylindrical tube

**Fig. 4.2** Elliptical tube

system's overall performance. Exergy analysis reveals that elliptical tubes exhibit moderate exergy destruction, reflecting their improved efficiency in comparison to cylindrical tubes. The lower level of irreversibility in elliptical designs makes them a more efficient choice for solar thermal systems. The precision alignment of the elliptical tube ensures better capture of solar energy, absorbing a larger portion of incident radiation and yielding higher thermal efficiency. The superior heat transfer performance, driven by the optimized alignment, reduces the temperature difference between the tube and the fluid, thereby enhancing heat transfer rates. While elliptical tubes provide notable performance improvements, the complexity of their design and manufacturing can raise production costs. Despite these challenges, the superior solar energy capture and heat transfer capabilities make elliptical receiver tubes a valuable advancement over cylindrical designs. They offer substantial efficiency gains for solar thermal systems, making them an attractive option for high-performance applications. Elliptical receiver tubes are illustrated in Fig. 4.2.

**Converging–Diverging Receiver Tubes**

Converging–diverging receiver tubes feature a tapered geometry that narrows and then widens, optimizing fluid flow and enhancing heat absorption by improving heat transfer dynamics and reducing thermal gradients. This innovative shape significantly boosts thermal efficiency by minimizing thermal stress and improving heat distribution. The tapered design promotes more uniform fluid flow, leading to more consistent heat absorption and a more effective conversion of solar radiation into thermal energy. Studies have shown that this design leads to substantial improvements in thermal efficiency, as it facilitates enhanced fluid mixing and greater contact with the tube surface, thus increasing the rate of heat transfer. The reduction in thermal gradients further contributes to higher efficiency in heat transfer, making the converging–diverging design particularly beneficial for high-performance solar thermal systems. Exergy analysis reveals that this design minimizes exergy destruction by optimizing heat absorption and reducing thermal gradients, resulting in lower irreversibility and more efficient energy conversion. The improved heat transfer and reduced thermal stress contribute to a more effective conversion of solar energy into thermal energy, ultimately enhancing system performance. Also known as nozzle-shaped or DeLaval tubes, converging–diverging receiver tubes are specifically designed to optimize solar energy capture and heat transfer in solar thermal systems. The converging section of the tube focuses and concentrates solar radiation onto a smaller area, improving alignment with concentrated sunlight and boosting absorption efficiency. The diverging section then distributes the absorbed heat more effectively, reducing thermal losses and facilitating better heat transfer to the heat transfer fluid (HTF). This specialized geometry enhances thermal efficiency, especially in high-concentration systems like parabolic trough collectors and solar power plants. Despite the increased

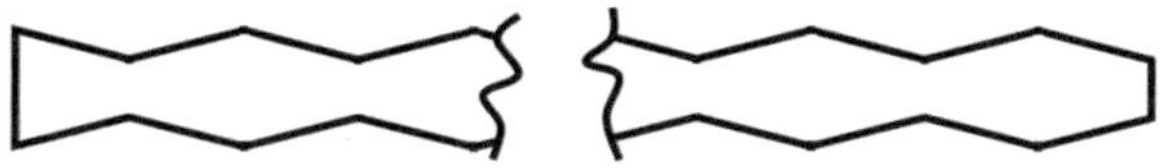

**Fig. 4.3** Converging–diverging receiver tube

manufacturing and installation complexity, as well as higher maintenance requirements, the efficiency gains make converging–diverging tubes a valuable choice for high-performance solar thermal systems. These tubes are shown in Fig. 4.3.

**Convex Corrugated Receiver Tubes**

Convex corrugated receiver tubes are designed with a convex, corrugated surface that increases the contact area between the fluid and the tube, significantly improving heat transfer and minimizing thermal strain on the solar collector. The convex shape, with its ridges and valleys, enhances the surface area exposed to solar radiation, thereby optimizing the absorption of concentrated sunlight. This increased surface area enables more efficient heat capture and minimizes heat losses, directly boosting thermal efficiency. Additionally, the corrugated design facilitates better fluid mixing within the receiver tube, which promotes more uniform heat transfer from the tube to the heat transfer fluid. This optimization of heat distribution helps reduce thermal gradients and enhances the overall performance of the solar thermal system. Research indicates that convex corrugated tubes can substantially improve thermal performance by enhancing both heat absorption and transfer efficiency. The increase in surface area, coupled with the improved fluid dynamics, results in higher heat transfer rates, leading to better thermal conversion and reduced energy losses. Furthermore, the design helps lower exergy destruction, as the enhanced fluid interaction and improved heat transfer contribute to a more efficient use of solar energy. Despite these advantages, the design complexity of convex corrugated receiver tubes increases manufacturing and maintenance costs. However, the improved performance, particularly in terms of energy capture and heat management, justifies these additional expenses, especially in high-performance solar thermal systems. This design is especially beneficial in applications where maximizing solar energy capture and improving thermal efficiency are paramount. Overall, convex corrugated receiver tubes offer a highly effective solution for advanced parabolic trough collectors aiming for higher efficiency and superior performance in solar thermal energy systems. These tubes are depicted in Fig. 4.4.

**Sinusoidal and Wavy Receiver Tubes**

Sinusoidal and wavy receiver tubes are advanced designs utilized in solar thermal systems, featuring intricate patterns on their inner surfaces that optimize fluid

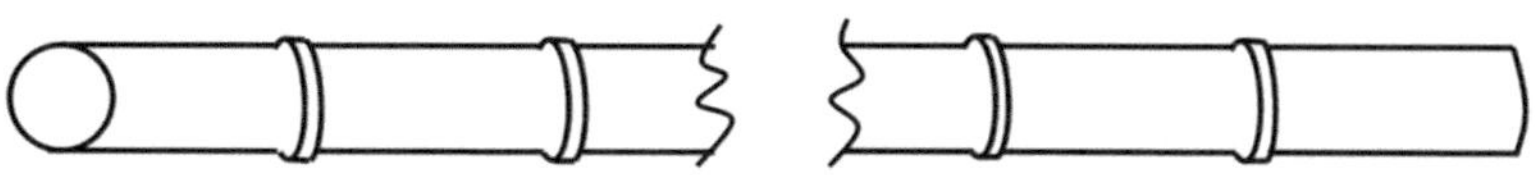

**Fig. 4.4** Convex corrugated receiver tube

**Fig. 4.5** Sinusoidal/wavy receiver tubes

dynamics and heat transfer. These wavy patterns reduce temperature fluctuations and improve heat distribution within the tube, significantly enhancing thermal efficiency. By promoting more uniform fluid flow, these designs minimize temperature gradients and optimize heat absorption, leading to superior system performance. Studies have demonstrated that sinusoidal tubes, in particular, achieve higher Nusselt numbers, indicating improved heat transfer characteristics. This enhanced fluid dynamics, facilitated by the wavy surface, results in higher heat transfer rates and reduced pressure losses, ultimately boosting the overall efficiency of the system. Additionally, the optimized flow dynamics and heat distribution contribute to minimal exergy destruction, ensuring that solar energy is utilized more effectively. The reduction in exergy losses is a key factor that makes sinusoidal and wavy receiver tubes highly efficient for solar thermal applications. The sinusoidal design increases the surface area exposed to concentrated solar radiation, improving solar absorption by reducing the angle of incidence of the solar rays. This shape also minimizes thermal gradients within the tube and promotes better fluid mixing, which further enhances heat transfer. While sinusoidal tubes offer significant advantages in terms of energy capture and heat management, their design complexity increases both production and maintenance costs. Despite these challenges, the substantial performance improvements in thermal efficiency and solar energy utilization make them invaluable for high-performance solar thermal systems. Particularly beneficial for parabolic trough collectors, sinusoidal and wavy receiver tubes provide a sophisticated solution for maximizing solar energy capture. These tubes are shown in Fig. 4.5.

**Flat-Plate Receivers**

Flat-plate receivers are a prevalent design in solar thermal systems, characterized by a flat absorber surface directly exposed to solar radiation. Typically employed in flat-plate solar collectors, these receivers are often coated with a selective coating to enhance solar absorption while minimizing thermal losses. The efficiency of flat-plate receivers is heavily dependent on the quality of the absorber coating and insulation, with moderate to high thermal efficiency being achievable under optimal conditions. These receivers perform well under low to moderate solar radiation, but their efficiency diminishes under high radiation levels, as overheating and increased convective losses become more pronounced. The heat transfer rates in flat-plate receivers are influenced by the design of the absorber plate, including its surface area and the configuration of the heat transfer fluid. Although the large, flat surface area facilitates reasonable heat transfer rates, convective and radiative losses can significantly reduce performance. In terms of exergy, flat-plate receivers exhibit moderate exergy destruction. The simplicity of the design allows for effective heat absorption, but the unavoidable losses through convection and radiation result in a reduction of overall exergy efficiency. This exergy destruction is intrinsically tied to the efficiency of

**Fig. 4.6** Flat-plate receiver tube

heat capture and transfer within the system. Unlike cylindrical or elliptical designs, flat-plate receivers do not concentrate solar rays but rely on a broad surface area for solar energy capture. While this design proves particularly effective in scenarios where solar concentration is not critical, the lack of a concentration mechanism limits their performance in systems designed for high solar flux. Despite these limitations, the cost-effectiveness and simplicity of flat-plate receivers make them an appealing choice for many solar thermal applications. A flat-plate receiver tube is depicted in Fig. 4.6.

**Finned Tube Receivers**

Finned tube receivers are engineered with tubes incorporating extended surfaces or fins to enhance the surface area for heat transfer. This design is particularly utilized in high-performance solar thermal systems, where the fins are crucial in improving the heat transfer between the fluid inside the tubes and the absorbed solar radiation. The extended surface area provided by the fins increases the convective heat transfer coefficient, thereby reducing thermal resistance and significantly improving thermal efficiency. As a result, finned tube receivers generally exhibit superior thermal efficiency, allowing for more effective capture and conversion of solar energy into thermal energy. The fins also optimize the heat transfer dynamics by facilitating more uniform heat distribution, which leads to enhanced heat absorption and transfer to the heat transfer fluid. Consequently, these receivers offer higher heat transfer rates than simpler, traditional designs. From an exergy standpoint, the increased surface area and improved heat transfer characteristics reduce exergy destruction. By minimizing irreversibilities in the system, the fins enhance the overall efficiency of the heat transfer process, making finned tube receivers a highly effective choice for solar thermal applications. The primary advantage of these receivers lies in their ability to improve fluid mixing and mitigate thermal gradients within the receiver, promoting more even heat distribution and higher outlet temperatures for the heat transfer fluid. However, adding fins complicates the design and manufacturing process, leading to higher production costs. Moreover, the larger surface area may demand additional space and maintenance efforts. Despite these challenges, finned tube receivers are particularly beneficial in high-performance solar thermal systems, prioritizing efficient heat management. A finned receiver tube is shown in Fig. 4.7.

**Multi-tube Receivers**

Multi-tube receivers consist of multiple parallel tubes arranged within a solar collector, significantly enhancing the overall surface area for solar absorption. This design enables the collector to capture a larger volume of solar energy and distribute it more uniformly, which leads to a marked improvement in system performance. The increased surface area and better heat distribution contribute directly to high thermal efficiency, allowing the system to capture more solar energy and transfer

**Fig. 4.7** Finned receiver tube

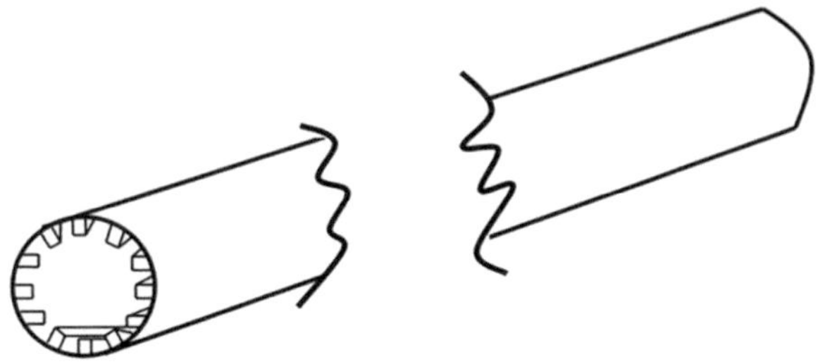

it efficiently to the heat transfer fluid. The use of multiple tubes amplifies heat transfer rates by increasing the total surface area available for heat exchange, thereby reducing thermal gradients and improving the distribution of solar energy across the tubes. This results in more effective heat transfer and overall performance. Multi-tube receivers generally exhibit lower exergy destruction compared to single-tube designs, as the increased surface area and optimized heat distribution minimize system irreversibilities. Consequently, multi-tube receivers are more efficient for converting solar energy into thermal energy and improving solar thermal system performance. These multi-tube arrangements facilitate enhanced fluid mixing and reduced thermal gradients, further boosting heat transfer rates. However, the complexity of designing and installing multi-tube systems is greater than that of single-tube designs. The higher number of components and interconnections leads to increased manufacturing and installation costs and potentially higher maintenance requirements. Despite these drawbacks, multi-tube systems provide significant advantages in improved heat transfer and energy capture, making them ideal for high-performance solar thermal applications where maximizing efficiency is crucial. The flexibility inherent in multi-tube designs allows for customization based on specific operational needs and conditions. Multi-tube arrangements are depicted in Fig. 4.8.

**Fig. 4.8** Multi-tube arrangements

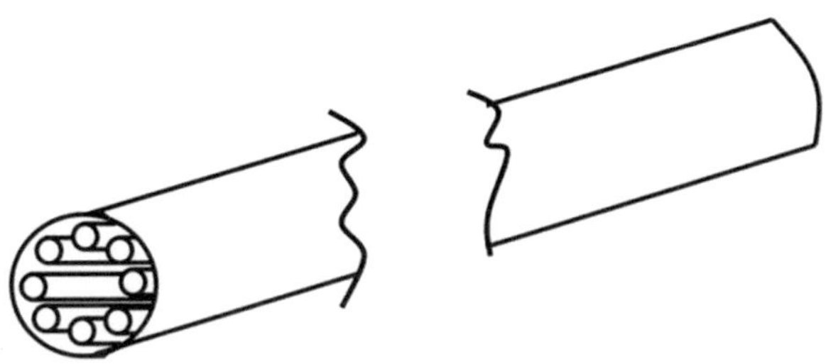

## 4.3 Mathematical Relations for Parabolic Trough Collector Receiver Tubes

To understand the energetic performance of parabolic trough collectors (PTCs) with different receiver tube shapes, it's crucial to delving into the detailed mathematical relations. These relations help quantify heat transfer, thermal efficiency, and overall performance. Below, we break down the mathematical relations for various tube geometries:

**Heat Transfer Rate (*Q*)**

The heat transfer rate for a cylindrical receiver tube is given by

$$Q = U \times A \times \Delta T \tag{4.1}$$

Here:

$U$ is the overall heat transfer coefficient (W/m$^2$ K).

$A$ is the surface area of the tube (m$^2$).

$\Delta T$ is the temperature difference between the receiver tube and the heat transfer fluid (K).

The overall heat transfer coefficient $U$ combines several factors, including the thermal conductivity of the tube material and convective heat transfer coefficients:

$$U = \frac{1}{R_{\text{total}}} = \frac{1}{R_{\text{conductive}} + R_{\text{convective}}} \tag{4.2}$$

$$R_{\text{conductive}} = \frac{d_{\text{tube}}}{k_{\text{tube}} \times A_{\text{tube}}} \tag{4.3}$$

$$R_{\text{convective}} = \frac{1}{h_{\text{fluid}} \times A_{\text{outer}}} \tag{4.4}$$

Here:

$R_{\text{total}}$ = Total Thermal Resistance.

$R_{\text{conductive}}$ = Conductive Resistance.

$R_{\text{convective}}$ = Convective Resistance.

$A_{\text{tube}}$ = The surface area of the tube (m$^2$).

$d_{\text{tube}}$ = is the thickness of the tube.

$k_{\text{tube}}$ = is the thermal conductivity of the tube material (W/m K).

$h_{\text{fluid}}$ = is the convective heat transfer coefficient of the heat transfer fluid (W/m$^2$ K).

**Exergy Analysis**

The solar exergy absorbed by the cylindrical tube can be expressed as

$$E_{\text{in}} = \eta_{\text{abs}} \times A \times G \tag{4.5}$$

The useful exergy delivered to the heat transfer fluid is

$$E_{\text{out}} = \eta_{\text{thermal}} \times Q \tag{4.6}$$

Heat losses due to radiation, convection, and conduction are significant:

$$E_{\text{lost}} = U \times A \times (T_{\text{tube}} - T_{\text{ambient}}) \tag{4.7}$$

For any thermal system, the exergy of a component (or receiver tube) can be calculated as

$$E_{\text{ex}} = E_{\text{in}} - E_{\text{out}} - E_{\text{lost}} \tag{4.8}$$

Here:

$\eta_{\text{abs}}$ = Absorptance of the tube.

$\eta_{\text{thermal}}$ = Thermal efficiency of the tube.

$G$ = Solar irradiance (W/m$^2$).

$Q$ = Heat delivered (W).

$T_{\text{tube}}$ = Temperature of the receiver tube.

$T_{\text{ambient}}$ = Ambient temperature.

$E_{\text{ex}}$ = Exergy of the system.

$E_{\text{in}}$ = Incoming energy (solar energy absorbed).

$E_{\text{out}}$ = Useful energy output (thermal energy delivered to the working fluid).

$E_{\text{lost}}$ = Energy losses (including heat losses to the environment).

**Surface Area ($A$)**

**Cylindrical Receiver Tubes**

For a cylindrical tube, the surface area $A$ is

$$A = \pi \times D \times L \tag{4.9}$$

Here:

$D$ is the diameter of the tube.

$L$ is the length of the tube.

**Elliptical Receiver Tubes**

For an elliptical receiver tube [2]:

The surface area of an elliptical tube $A_{\text{ellipse}}$ can be approximated as

$$A_{\text{ellipse}} = \pi \times a \times b \tag{4.10}$$

Here:

$a$—is the semi-major axis of the ellipse.

$b$—is the semi-minor axis of the ellipse.

**Converging–Diverging Receiver Tubes**

$A_{\text{effective}}$ is the effective heat transfer area, which varies along the length of the tube due to its shape [3]:

$$A_{\text{effective}} = \pi \times \left(\frac{(D_{\min} + D_{\max}}{2}\right) \times L \tag{4.11}$$

Here

$D_{\min}$—is the minimum diameter.

$D_{\max}$—is the maximum diameter.

**Convex Corrugated Receiver Tubes**

$A_{\text{corrugated}}$ is the increased surface area due to the corrugations. It is typically larger than that of a smooth tube [4]:

$$A_{\text{corrugated}} = \pi \times D \times L \times (1 + \text{Correction Factor}) \tag{4.12}$$

$D$ is the diameter of the tube.

The correction factor depends on the amplitude and frequency of the corrugations.

**Sinusoidal and Wavy Receiver Tubes**

$A_{\text{wavy}}$ is the increased surface area due to the sinusoidal or wavy patterns [5, 6]:

$$A_{\text{wavy}} = \pi \times D \times L \times (1 + \text{Wavy Factor}) \tag{4.13}$$

$D$ is the diameter of the tube $(m)$,

Where the wavy factor is derived from the amplitude and wavelength of the wave patterns.

**Flat-Plate Receivers**

For flat-plate receivers, the surface area is given by

$$A = W \times L \tag{4.14}$$

Here:

$W$—is the width of the plate.

$L$—is the length of the plate.

**Finned Receiver Tubes**

The surface area for finned receiver tubes is

$$A = \pi \times D \times L \times (1 + \text{Fin Factor}) \tag{4.15}$$

Here:

$D$ is the diameter of the tube.

$L$ is the length of the tube.

- The Fin Factor is a term that accounts for the additional surface area provided by the fins, which is typically a percentage increase over the basic cylindrical tube area.

**Multi-tube Arrangements**

For multi-tube arrangements, the surface area is

$$A = n \times \pi \times D \times L \tag{4.16}$$

Here:

$n$—is the number of tubes.

$D$—is the diameter of each tube.

$L$—is the length of each tube.

## 4.4 Comparative Analysis of Receiver Tube Designs

Each shape offers distinct advantages and trade-offs when comparing different receiver tube designs. Cylindrical tubes are simple and cost-effective, but may have lower efficiency due to geometric misalignment with concentrated solar rays. Elliptical tubes improve efficiency by providing better alignment with the focal line, enhancing energy absorption and heat transfer. Flat-plate receivers are straightforward and affordable but may not perform well in high-concentration systems. Finned receivers and multi-tube arrangements offer advanced performance by increasing surface area and improving heat transfer, though they come with increased complexity and costs. The choice of receiver tube shape depends on specific application requirements, including efficiency, cost, and design complexity. Understanding these characteristics helps select the most appropriate receiver design for optimizing solar thermal system performance and achieving desired energy outcomes. Each design presents unique benefits that can be leveraged based on the specific goals and constraints of the solar thermal system. A comparative analysis of differently designed receiver tubes is presented in Table 4.1.

## 4.5 Advantages and Disadvantages of Various Geometries

In solar thermal systems, selecting the right receiver tube geometry is essential for optimizing performance, efficiency, and cost. Each geometry has unique attributes that affect its operation in different contexts. The following table provides a comprehensive comparison of cylindrical tubes, elliptical tubes, flat-plate receivers, finned tubes, and multi-tube arrangements, highlighting their advantages and disadvantages. The advantages and disadvantages of various receiver tube designs are outlined in Table 4.2.

## 4.6 Pressure Drop Analysis in Solar Thermal Receiver Tubes

In solar thermal systems, the design and efficiency of receiver tubes are crucial factors that influence the overall performance. One critical aspect of receiver tube performance is pressure drop. Pressure drop is the reduction in pressure that occurs as a fluid flows through a tube or duct due to friction and other resistances. Managing pressure drop is essential as it affects the energy required to pump the liquid, thus influencing the operational costs and efficiency of the system.

This detailed analysis explores the basic concepts of pressure drop, mathematical models and equations used to calculate it, case studies of different receiver tube shapes, and a comparative analysis of their pressure drop characteristics.

In the context of solar thermal systems, pressure drop is significant because

i. Higher pressure drops require more energy to pump the fluid through the system. This increases operational costs and can reduce the system's overall efficiency.
ii. Pressure drop affects the design and sizing of pumps and piping, impacting both installation and operational costs.

**Table 4.1** Comparative analysis of differently designed receiver tube [3–10]

| Design | Description | Advantages | Disadvantages | Best For |
|---|---|---|---|---|
| Cylindrical tubes | Simple hollow cylinders are positioned along the focal line of the parabolic mirror | • Cost-effective<br>• Simple design<br>• Easy manufacturing and integration | • Lower efficiency due to misalignment with concentrated solar rays<br>• Higher thermal losses | Cost-sensitive applications with moderate performance requirements |
| Elliptical tubes | Tubes with an elliptical shape for improved alignment with concentrated solar flux | • Enhanced absorption efficiency<br>• Improved heat transfer<br>• Better alignment with solar flux | • More complex and costly to design and manufacture<br>• Requires precise alignment | Systems requiring higher efficiency and better thermal performance, with a higher budget |
| Flat-plate receivers | Broad, flat surfaces that absorb solar energy without concentrating it | • Simple and cost-effective<br>• Easy to manufacture<br>• Suitable for low-concentration systems | • Lower efficiency compared to concentrated designs<br>• Less effective for high-concentration applications | Applications where cost is prioritized and high concentration is not critical |
| Finned receivers | Tubes with extended surfaces (fins) to enhance heat transfer and reduce thermal gradients | • Improved heat transfer<br>• More uniform heat distribution<br>• Enhanced fluid mixing | • Increased design complexity<br>• Higher production and maintenance costs | High-performance systems where efficient heat management is essential |
| Multi-tube arrangements | Multiple tubes are arranged in parallel or other configurations to increase surface area and heat transfer | • Increased energy capture<br>• Better heat distribution<br>• Improved fluid mixing | • Higher complexity<br>• Increased manufacturing, installation, and maintenance costs | Large-scale systems requiring extensive energy capture and efficient heat transfer |
| Converging–diverging tubes | Nozzle-shaped tubes with a converging section to focus solar radiation and a diverging section for heat distribution | • Optimized solar energy capture<br>• Reduced thermal losses<br>• Enhanced heat transfer | • Complex design and manufacturing<br>• Higher costs<br>• Requires precise integration | High-performance systems with a focus on maximizing energy capture and minimizing losses |

(continued)

**Table 4.1** (continued)

| Design | Description | Advantages | Disadvantages | Best For |
|---|---|---|---|---|
| Convex corrugated tubes | Tubes with a convex, corrugated surface to increase surface area and improve absorption | • Increased solar energy absorption<br>• Better heat transfer<br>• Enhanced thermal gradient management | • Complex and costly to design and manufacture<br>• More intricate maintenance | Advanced systems aiming for higher efficiency and performance with higher costs |
| Sinusoidal tubes | Tubes with a sinusoidal, wavy surface profile to enhance solar energy capture and heat distribution | • Improved absorption efficiency<br>• More uniform heat distribution<br>• Better fluid mixing | • Higher manufacturing and design complexity<br>• Increased installation and maintenance needs | High-performance systems where optimizing energy capture and thermal efficiency is critical |

iii. Excessive pressure drop can lead to reduced flow rates, which can negatively affect the heat transfer performance of the system.

**Factors Affecting Pressure Drop**

i. Fluid Velocity: As the velocity of the fluid increases, the frictional forces between the fluid and the tube walls increase, leading to a higher pressure drop. Higher velocities generally result in a higher pressure drop.
ii. Tube Geometry: The shape and size of the tube impact the flow characteristics. Different geometries, such as cylindrical or elliptical tubes, affect how the fluid interacts with the tube walls and, consequently, the pressure drop.
iii. Surface Roughness: The roughness of the tube's inner surface contributes to resistance against the fluid flow. Rougher surfaces increase friction and pressure drop.
iv. Length of the Tube: The longer the tube, the greater the resistance to fluid flow. Pressure drop increases linearly with its length.
v. Flow Regime: The flow regime (laminar or turbulent) affects the pressure drop. Laminar flow, characterized by smooth and orderly fluid motion, generally has a lower pressure drop than turbulent flow, where the fluid movement is chaotic and irregular.

**Pressure Drop Analysis in Various Receiver Tube Geometries**

Analyzing pressure drop in solar thermal systems involves understanding how different geometries affect fluid flow and resistance. This section provides detailed derivations of pressure drop calculations for several receiver tube geometries: cylindrical tubes, elliptical tubes, flat-plate receivers, finned tubes, and multi-tube arrangements.

**Table 4.2** Advantages and disadvantages [3–10]

| Receiver tube geometry | Advantages | Disadvantages |
|---|---|---|
| Cylindrical tubes | • *Simplicity*: Straightforward design; easy to manufacture<br>• *Manufacturability*: Cost-effective production<br>• *Durability*: Generally robust and resistant to mechanical stress<br>• *Ease of Installation*: Simple and quick setup | • *Potentially Lower Heat Transfer Efficiency*: May not optimally align with concentrated solar rays<br>• *Limited Absorption Area*: Smaller effective surface area compared to some other designs<br>• *Thermal Losses*: Greater heat loss due to less effective insulation and design |
| Elliptical tubes | • *Improved Heat Transfer Characteristics*: Better alignment with solar flux<br>• *Higher Thermal Efficiency*: Enhanced energy capture and transfer<br>• *Reduced Reflection Losses*: More effective at capturing concentrated light<br>• *Optimized Absorption*: More consistent heat distribution | • *Complex Manufacturing*: More intricate design and production process<br>• *Higher Cost*: Increased manufacturing expenses<br>• *Maintenance Challenges*: Difficult to clean due to shape<br>• *Potential Alignment Issues*: Precision required for optimal performance |
| Flat-plate receivers | • *Enhanced Thermal Performance*: Broad surface area captures more solar energy<br>• *Easier Cleaning*: A Simple surface facilitates maintenance<br>• *Lower Cost*: Generally cheaper to produce and install<br>• *Versatile Application*: Effective in a range of solar thermal systems | • *Higher Pressure Drops*: Increased resistance to fluid flow<br>• *Potential for Heat Losses*: Less effective in high-concentration systems<br>• *Lower Efficiency in Direct Concentration Systems*: Not ideal for systems with high solar flux concentration<br>• *Larger Footprint*: Requires more space due to broad surface area |
| Finned tubes | • *Increased Surface Area*: Larger area for heat transfer<br>• *Improved Heat Transfer*: Better fluid mixing and heat exchange<br>• *Enhanced Efficiency*: Reduced thermal gradients within the receiver<br>• *Reduced Heat Losses*: Better heat retention due to larger surface area | • *Increased Pressure Drop*: Higher resistance affecting system efficiency<br>• *Complexity in Design*: More complex and potentially costly<br>• *Maintenance Issues*: Fins can accumulate debris and require more frequent cleaning<br>• *Space Requirements*: May need more space due to extended surfaces |

(continued)

**Table 4.2** (continued)

| Receiver tube geometry | Advantages | Disadvantages |
|---|---|---|
| Multi-tube arrangements | • *High Efficiency*: Greater surface area and improved energy capture<br>• *Excellent Heat Transfer*: Better distribution and fluid mixing<br>• *Flexibility*: Configurable to meet specific system requirements<br>• *Reduced Thermal Gradients*: Better heat distribution among tubes | • *Increased Pressure Drop*: Higher system resistance due to multiple tubes<br>• *Design Complexity*: More components and higher installation costs<br>• *Maintenance Requirements*: Higher maintenance due to the increased number of tubes and connections<br>• *Higher Space Consumption*: Requires more space for installation compared to single-tube designs |
| Converging–diverging tubes | • *Optimized Solar Energy Capture*: Focuses and concentrates solar radiation more effectively<br>• *Reduced Thermal Losses*: Improved distribution of absorbed heat<br>• *Enhanced Heat Transfer*: Superior alignment and heat management<br>• *Higher Efficiency*: Improved energy conversion compared to simpler designs | • *Complex Design and Manufacturing*: More intricate and costly to produce<br>• *Higher Installation Costs*: Requires precise alignment and additional space<br>• *Maintenance Challenges*: Intricate shape may complicate cleaning and maintenance<br>• *Integration Issues*: May need precise system integration |

(continued)

**Table 4.2** (continued)

| Receiver tube geometry | Advantages | Disadvantages |
|---|---|---|
| Convex corrugated tubes | • *Increased Absorption Area*: More surface area exposed to solar radiation<br>• *Enhanced Heat Transfer*: Improved heat management and fluid mixing<br>• *Better Absorption Efficiency*: Effective at capturing concentrated solar flux<br>• *Improved Thermal Gradient Management*: Reduced heat losses and more efficient thermal conversion | • *Complex and Costly Manufacturing*: Difficult to design and produce<br>• *Higher Maintenance Requirements*: Corrugated surface can be challenging to clean<br>• *Higher Initial Costs*: More expensive due to complex design<br>• *Integration Complexity*: May require more space and careful system integration |
| Sinusoidal tubes | • *Improved Absorption Efficiency*: Better alignment with concentrated solar flux<br>• *Enhanced Heat Distribution*: More uniform heat transfer and reduced thermal gradients<br>• *Increased Surface Area*: Greater exposure to solar radiation<br>• *Effective Fluid Mixing*: Reduces localized hot spots | • *Higher Design and Manufacturing Complexity*: More intricate and costly to produce<br>• *Increased Installation Costs*: Requires precise fitting and alignment<br>• *Maintenance Challenges*: A Wavy surface may complicate cleaning<br>• *Higher Space Requirements*: Potentially larger footprint due to wavy geometry |

## Cylindrical Tubes

### Darcy–Weisbach Equation

The Darcy–Weisbach equation is used to calculate the pressure drop due to friction along a tube. This equation applies to laminar and turbulent flow regimes and is derived from the fundamental principles of fluid mechanics.

The flow is smooth in laminar flow (Reynolds number Re < 2000), and the pressure drop is primarily due to viscous effects. For laminar flow in a cylindrical pipe, the Hagen–Poiseuille equation is used [11–14]:

$$\Delta P = \frac{128\mu L Q}{\pi D^4} \tag{4.17}$$

Here:

$\Delta P$ = Pressure drop

$\mu$ = Dynamic viscosity

$L$ = Length of the tube

$Q$ = Volumetric flow rate

$D$ = Diameter of the tube.

To relate this to the Darcy–Weisbach equation, note that

$$\Delta P = f \times \frac{L}{D} \times \frac{\rho v^2}{2} \tag{4.18}$$

where $f$ is the Darcy friction factor. For laminar flow, the friction factor $f$ is

$$f = \frac{64}{\mathrm{Re}} \tag{4.19}$$

For turbulent flow (Reynolds number Re > 4000), the flow is chaotic, and both viscous and inertial effects influence the pressure drop. The Darcy–Weisbach equation:

$$\Delta P = f \times \frac{L}{D} \times \frac{\rho v^2}{2} \tag{4.20}$$

where $f$ must be determined using empirical correlations. The Colebrook–White equation is commonly used:

$$\frac{1}{\sqrt{f}} = -2\log_{10}\left(\frac{\epsilon/D}{3.7} + \frac{2.51}{\mathrm{Re}\sqrt{f}}\right) \tag{4.21}$$

where the Reynolds number Re is defined as

$$\mathrm{Re} = \frac{\rho v D}{\mu}$$

where

$\rho$ = Density of the fluid

$v$ = Velocity of the fluid.

**Hazen–Williams Equation (for Water)**

For water flow, the Hazen–Williams equation simplifies the pressure drop calculation:

$$\Delta P = 10.67 \frac{L}{C^{1.852} D^{4.865}} Q^{1.852} \tag{4.22}$$

where
$C$ = Hazen–Williams roughness coefficient.

**Elliptical Tubes**

**Generalized Pressure Drop Calculation**

Elliptical tubes have a more complex cross-section, requiring specialized equations to estimate pressure drop [11–14].

The pressure drop in an elliptical tube can be approximated using the following formula:

$$\Delta p = \frac{8 \mu L Q}{\pi^2 r_1^2 r_2^2} \left( \frac{1}{r_1} + \frac{1}{r_2} \right) \tag{4.23}$$

Here:

$r_1$ = Semi-major axis of the ellipse

$r_2$ = Semi-minor axis of the ellipse

$L$ = Length of the tube

$Q$ = Flow rate

$\mu$ = Dynamic viscosity.

**Flat-Plate Receivers**

Flat-plate receivers generally do not involve pressure drop calculations in the same way as cylindrical or elliptical tubes. Instead, they focus on heat absorption and thermal performance. However, if flow through channels or ducts in flat-plate receivers is analyzed, the pressure drop can be calculated similarly to that of rectangular ducts.

For flow through a rectangular channel [11–14]:

$$\Delta P = f \times \frac{L}{D} \times \frac{\rho v^2}{2} \tag{4.24}$$

where $D_h$ is the hydraulic diameter, defined as

$$D_h = \frac{2 \text{ width} \times \text{hight}}{\text{width} + \text{hight}} \tag{4.25}$$

### Finned Tubes

Finned tubes have extended surfaces to enhance heat transfer. The increased surface area leads to higher pressure drops due to additional resistance.

The pressure drop for finned tubes can be approximated using [11–14]

$$\Delta p = \frac{8\mu LQ}{\pi^2 r_1^2 r_2^2}\left(\frac{1}{r_1} + \frac{1}{r_2}\right) \times \text{correction factor} \tag{4.26}$$

The correction factor accounts for the increased resistance due to fins. It is determined empirically or through detailed flow simulations considering the fin geometry and spacing.

### Multi-tube Arrangements

Multi-tube arrangements involve multiple parallel or series tubes to increase the effective surface area and improve heat transfer.

For a multi-tube system with n tubes in parallel [11–14]:

$$\Delta P = \frac{1}{n} \times (\text{Pressure droppettube}) \tag{4.27}$$

### Converging–Diverging Tubes

The pressure drop can be complex due to varying cross-sectional areas in a converging–diverging tube (often resembling a nozzle shape). The pressure drop in such tubes can be approximated by integrating the pressure drop over the varying cross-sections. For practical purposes, it is often calculated using [11–14]

$$\Delta P = \int_{A_1}^{A_2} \frac{f \times \frac{L}{D} \times \frac{\rho v^2}{2}}{A} \mathrm{d}A \tag{4.28}$$

where $A$ is the cross-sectional area, and $A_1$ and $A_2$ are the inlet and outlet areas, respectively. The friction factor f can be obtained using empirical correlations for turbulent flow or specific formulas for laminar flow.

### Convex Corrugated/Sinusoidal/Wavy Tubes

Convex corrugated tubes have a wavy surface that increases turbulence and pressure drop. The pressure drop in these tubes is typically higher than in smooth tubes due to increased surface area and turbulence. The pressure drop can be approximated as [11–14]

$$\Delta P = f \times \frac{L}{D} \times \frac{\rho v^2}{2} \tag{4.29}$$

where $D_h$ is the hydraulic diameter, which for corrugated tubes can be estimated as

$$D_h = \frac{4 \times \text{FlowArea}}{\text{WettedPerimeter}} \tag{4.30}$$

For corrugated tubes, the wetted perimeter and flow area are derived from the geometry of the corrugations, which complicates the calculation. Often, empirical correlations are used to estimate the friction factor f.

**Notes**

i. **Friction Factor ($f$)**: For laminar flow, $f = \frac{64}{\mathrm{Re}}$ for turbulent flow, $f$ is typically determined using empirical correlations such as the Colebrook–White equation or other correlations specific to the tube's geometry.
ii. **Hydraulic Diameter ($D_h$)**: For non-circular tubes, the hydraulic diameter is a useful parameter to describe the effective cross-sectional area and perimeter for calculating pressure drop.
iii. **Empirical Data**: For complex geometries like corrugated, sinusoidal, and wavy tubes, empirical data and correlations derived from experiments or detailed simulations are often used to obtain accurate pressure drop values.

## 4.7 Case Studies of Different Receiver Tube Shapes

### Cylindrical Tubes

Several factors critically influence pressure drop characteristics in fluid flow within tubes. The pressure drop in cylindrical tubes is directly proportional to the tube's length, as the extended surface area in contact with the fluid increases resistance, which in turn causes frictional losses. Consequently, doubling the length of the tube results in approximately doubling the pressure drop for a given fluid velocity. Fluid velocity also plays a pivotal role in determining the pressure drop, as higher velocities correspond to increased frictional forces exerted on the tube walls, thus elevating the pressure drop. This relationship is typically nonlinear, with even modest increases in velocity leading to substantial rises in pressure drop. Cylindrical tubes, commonly utilized in solar thermal systems, exhibit moderate pressure drops, which make them optimal for use in applications such as parabolic trough solar collectors. These tubes effectively balance system performance with ease of integration due to their moderate pressure drop and simple manufacturing process. Their advantages lie in their well-understood pressure drop characteristics, supported by straightforward models and equations such as the Darcy–Weisbach or Hazen–Williams equations, making them accessible for precise calculation and efficient integration into solar thermal systems.

**Non-cylindrical Designs**

**Elliptical tubes** in solar thermal systems present significant challenges regarding pressure drops compared to their cylindrical counterparts, primarily due to the fluid dynamics disruptions caused by their non-circular cross-section. The elliptical geometry introduces increased friction and turbulence, significantly contrasting the smoother flow found in cylindrical designs. These disruptions lead to a rise in flow separation, boundary layer effects, drag, and turbulence, culminating in an elevated pressure drop. In Concentrated Solar Power (CSP) systems, this higher pressure drop compels the use of more powerful pumps to sustain fluid flow, thereby increasing operational costs. Case studies in CSP systems have illustrated that elliptical tubes contribute to higher frictional losses, necessitating more pumping energy and thus elevating costs. Consequently, choosing elliptical tubes requires meticulous design consideration, weighing potential benefits in heat transfer efficiency and alignment with solar flux against the added costs arising from higher pressure drops. A comprehensive cost–benefit analysis becomes essential, ensuring that the advantages in heat absorption or design efficiency outweigh the additional energy consumption and operational expenditures. Additionally, system designers must account for these higher pressure drops, potentially necessitating the incorporation of more robust pumps and optimized fluid flow management to mitigate the losses.

In contrast, flat-plate receivers are characterized by lower pressure drops, attributable to their broad, flat surface area. This configuration reduces resistance to fluid flow compared to tubular designs, with the larger cross-sectional area promoting a more uniform fluid distribution, thereby minimizing frictional losses. A typical example of such a system is residential solar thermal heating systems, where flat-plate collectors demonstrate minimal pressure drops, contributing to reduced operational costs and improved efficiency. However, flat-plate designs are less effective in applications requiring high solar concentration, such as parabolic troughs or dish collectors, where more focused solar flux is critical. These receivers are best suited for low-to-moderate temperature applications where ease of maintenance, cost-effectiveness, and low-pressure drops are paramount. However, they may not deliver optimal performance in high-concentration systems.

Multi-tube receivers offer a hybrid approach to balancing heat transfer efficiency and pressure drop. These receivers incorporate several parallel tubes within a single unit, distributing the fluid flow across multiple paths and reducing the pressure drop per tube. The configuration of these tubes—whether staggered or aligned—affects the overall pressure drop, and fluid velocity must be carefully controlled to optimize performance. Multi-tube systems, commonly employed in high-concentration systems like parabolic trough collectors, strike a balance by providing large heat transfer areas without excessively increasing pressure drop. A thoughtful design in these systems ensures fluid flows efficiently through the tubes with minimal turbulence.

Conversely, converging–diverging tubes, often used in nozzle designs, present unique pressure drop characteristics. The converging section accelerates the fluid, while the diverging section decelerates it, leading to complex flow dynamics and changes in pressure drop. These designs can enhance heat transfer by increasing flow velocity, but they require precise engineering to avoid excessive pressure losses. Similarly, convex corrugated tubes induce significant turbulence and friction due to their wavy surfaces, resulting in higher pressure drops, although they offer better heat transfer efficiency. This trade-off must be carefully evaluated to ensure that the benefits of improved thermal performance justify the increased operational costs.

Lastly, **sinusoidal and wavy tubes** enhance heat transfer by improving fluid mixing, which results in higher pressure drops due to their complex geometries. Both types of tubes disrupt the flow, leading to increased friction and turbulence, thus raising the pressure drop. These designs necessitate more powerful pumps and sophisticated control systems to effectively manage the increased energy consumption. In systems such as solar thermal collectors, the design must carefully balance the enhanced heat transfer with the higher operational costs associated with managing the increased pressure drop. Similarly, finned tubes, which increase heat transfer surface area, contribute to higher friction and turbulence, leading to increased pressure drops. This requires careful consideration in system design to ensure the enhanced thermal performance justifies the added cost of larger pumps and fluid flow optimization.

In conclusion, whether elliptical, flat-plate, multi-tube, or any other specialized form, each tube design involves a trade-off between pressure drop and heat transfer efficiency. The impact on system performance must be meticulously analyzed to ensure that the increased costs from higher pressure drops are offset by the advantages of heat transfer, ultimately ensuring that the system remains efficient and cost-effective.

**Comparative Analysis**

Table 4.3 presents a comparative analysis based on pressure drop. This table highlights the performance differences among various receiver tube designs under different operating conditions.

## 4.8 Implications for System Design and Operation

1. **Cylindrical Tubes**

**Design Considerations**: Suitable for systems where cost and simplicity are key. Ensure that pumping power is sufficient for the expected pressure drop.

**Operational Impact**: The moderate pressure drop means the energy requirements for pumping are manageable, making cylindrical tubes a reliable choice for many solar thermal applications.

**Table 4.3** Comparative analysis based on pressure drop [15–23]

| Receiver tube geometry | Pressure drop characteristics | Typical values (Pa/m) | Factors influencing values |
|---|---|---|---|
| Cylindrical tubes | Moderate pressure drop; straightforward calculations | 10–50 | Tube length, fluid velocity, and tube diameter |
| Elliptical tubes | Higher pressure drop due to complex shape, increased friction, and turbulence | 30–100 | Tube dimensions, aspect ratio, and fluid velocity |
| Flat-plate receivers | Lower pressure drop due to broad surface area; minimal resistance to fluid flow | 5–15 | Flow rate, channel geometry, surface area |
| Finned tubes | Increased pressure drop due to additional surface area from fins; higher resistance and turbulence | 20–80 | Density and geometry of fins, flow rate, and fin arrangement |
| Multi-tube arrangements | Higher pressure drops are due to multiple tubes; configuration and arrangement affect resistance | 50–150 | Number of tubes, tube arrangement, fluid velocity, and tube spacing |
| Converging–diverging tubes | Complex pressure drop varies with cross-sectional area changes, fluid velocity, and turbulence | 40–120 | Geometry of converging/diverging sections, fluid velocity |
| Convex corrugated tubes | Increased pressure drop due to wavy surface, additional friction, and turbulence | 30–100 | Corrugation pattern, tube dimensions, and fluid velocity |
| Sinusoidal tubes | Increased pressure drop due to periodic wave patterns causes turbulence and flow disruption | 30–90 | Wave amplitude, frequency, and fluid velocity |
| Wavy tubes | Higher pressure drops are due to irregular surfaces, increased turbulence, and friction | 40–100 | Wave pattern, tube dimensions, and fluid velocity |

2. **Elliptical Tubes**

**Design Considerations**: Useful for applications where improved heat transfer efficiency is needed, despite higher pressure drop. Careful design and optimization are required to manage the increased pressure drop.

**Operational Impact**: Increased pumping power results in higher operational costs. The benefits of enhanced heat transfer efficiency must be weighed against these costs.

3. **Flat-Plate Receivers**

**Design Considerations**: They are ideal for systems where a low pressure drop is advantageous, such as residential solar thermal heating. However, they may not be suitable for high-concentration systems.

**Operational Impact**: Lower energy requirements for pumping result in reduced operational costs. The broader surface area can be advantageous for capturing solar energy in less concentrated systems.

4. **Finned Tubes**

**Design Considerations**: This is best for high-performance systems where enhanced heat transfer is critical. The increased pressure drop must be accounted for in the system design.

**Operational Impact**: Due to increased pressure drop, pumping requires more energy. The trade-off is improved heat transfer efficiency and potentially better overall system performance.

5. **Multi-tube Arrangements**

**Design Considerations**: This method is effective for maximizing heat transfer and energy capture. However, it requires careful planning to manage the increased pressure drop and system complexity.

**Operational Impact**: Increased pressure drop leads to higher pumping power requirements. In many high-performance systems, the improved performance and efficiency justify the higher operational costs.

6. **Converging–Diverging Tubes**

**Design Considerations**: Converging–diverging tubes, with their variable cross-sectional area, present complex design challenges. Careful attention is needed to balance fluid acceleration and deceleration while managing the associated pressure drop. Optimization of these tubes involves precise control of flow dynamics to handle the complex pressure drop behavior effectively.

**Operational Impact**: The pressure drop for converging–diverging tubes ranges from 40 to 120 Pa/m. This requires careful system design to ensure that the increased pressure drop does not outweigh the benefits of controlled fluid flow. The design must account for these dynamics to avoid inefficiencies and maintain operational effectiveness.

7. **Convex Corrugated Tubes**

**Design Considerations**: Convex corrugated tubes feature a wavy surface that enhances heat transfer but introduces additional friction and turbulence. Design considerations should focus on managing the higher pressure drop while leveraging the improved surface area for heat transfer.

**Operational Impact**: With a pressure drop ranging 30–100 Pa/m, convex corrugated tubes demand higher pumping power. The increased resistance must be balanced with the advantages of enhanced heat transfer, requiring a careful evaluation of overall system performance and energy costs.

8. **Sinusoidal Tubes**

**Design Considerations**: Sinusoidal tubes, with their periodic wave patterns, are designed to enhance mixing and heat transfer. However, their complex geometry leads to additional turbulence and pressure drop. System design must ensure that the performance benefits of improved mixing justify the higher energy costs for pumping.

**Operational Impact**: The pressure drop for sinusoidal tubes ranges 30–90 Pa/m. This results in increased energy requirements for pumping. Designers must assess whether the improved heat transfer efficiency compensates for the additional operational costs and energy consumption.

9. **Wavy Tubes**

**Design Considerations**: Wavy tubes enhance turbulence and heat transfer through their irregular surface. The design should address the higher pressure drop by maximizing the benefits of enhanced heat transfer while managing the increased resistance and pumping power requirements.

**Operational Impact**: With a pressure drop typically ranging 40–100 Pa/m, wavy tubes require more energy for pumping. The trade-off between the improved heat transfer performance and the higher operational costs must be carefully considered to ensure system efficiency.

## 4.9 Studies for Different Receiver Tube Geometries in Parabolic Solar Trough Collectors

Researchers have recently explored various modifications to traditional tubular absorber tubes in Parabolic Trough Collectors (PTCs) to improve their thermal performance and heat transfer capabilities. These modifications aim to enhance the efficiency of PTCs, especially under non-uniform flux conditions, while also reducing issues such as thermal stress and pressure losses. Several studies have presented different geometrical changes to the collector tubes, each offering unique thermal performance and fluid dynamics benefits.

Bellos et al. [3] conducted a detailed numerical investigation to evaluate the effect of a converging–diverging shaped tube on the thermal behavior of a PTC. The study focused on collector parameters, including efficiency, heat loss coefficient, collector temperature, and pressure losses, under varying flux conditions. The results demonstrated that modifying the absorber tube's geometry improved the collector's overall efficiency by ~4.55%. The study highlighted the importance of optimizing the collector tube shape to enhance heat transfer, especially in the context of non-uniform flux, which is a common challenge in solar thermal applications. The converging–diverging tube shape provided better heat absorption and distribution, resulting in improved overall performance.

In a similar vein, Fuqiang et al. [4] investigated the impact of using a convex corrugated collector tube on the performance of a PTC. Their study, which utilized numerical simulations, showed that the convex corrugated design significantly improved the thermal performance factor, with a maximum increase of 148%. This modification also helped in achieving a more uniform heat distribution along the collector surface, which in turn resulted in a 26.8% reduction in the Von Mises thermal strain. The reduction in thermal strain was particularly notable because it suggested that the modified tube design could help mitigate the mechanical stresses typically associated with temperature gradients in conventional absorber tubes. This improvement in uniform heat distribution not only enhanced thermal performance but also contributed to the structural integrity of the collector, ensuring that the modified design would be more durable under operational conditions.

Bitam et al. [5] explored a novel sinusoidal collector tube design through numerical simulations. Their study found that the sinusoidal tube design helped reduce the maximum temperature difference within the tube to 35 K, indicating a more even temperature distribution. This uniform distribution of heat significantly reduced thermal stresses, which are known to impact the long-term performance and durability of the absorber tube. Furthermore, the study revealed that the modified collector tube resulted in a notable increase in both the Nusselt number and the friction factor, with enhancements of 63 and 40.8%, respectively. The Performance Evaluation Criterion (PEC), which is a ratio of the heat transfer enhancement to the increase in pressure drop, was observed to be 1.35. This value indicated a clear advantage of using the sinusoidal collector tube over the traditional plain collector tube, as it enhanced heat transfer without incurring excessive pressure losses. The results of this study provided strong evidence that modifying the tube geometry with a sinusoidal design could significantly improve the efficiency of PTCs, particularly in terms of heat transfer and fluid dynamics.

Building on these findings, Yang and Ordonez [19] numerically evaluated a symmetric wavy collector tube, considering varying amplitudes and wave numbers in their design. Their study found that the first law efficiency of the collector significantly improved at larger mass flow rates, which was attributed to better energy transport within the wavy tube. The maximum PEC observed in this case was 1.25, which demonstrated a marked performance improvement compared to conventional tube designs. However, the study also highlighted an important limitation of the wavy tube design. Specifically, when the amplitude exceeded 0.006 m, the PEC dropped below unity due to the increased pressure drop in the collector tube. This finding suggested that while wavy tubes could enhance heat transfer, they also posed challenges related to fluid flow and pressure losses. The researchers recommended using smaller amplitudes and wave numbers to balance the benefits of enhanced heat transfer with the drawbacks of higher pressure losses. This insight emphasized the need for a careful optimization of the tube geometry to achieve the best performance while minimizing operational drawbacks such as excessive pressure drop.

In another study, Khelkar et al. [20] investigated a sinusoidal inner surface collector tube using Therminol VP-1 and water as heat transfer fluids. The study found that the modified collector tube provided significantly higher PEC values when using Therminol VP-1 compared to water. Specifically, the PEC for Therminol VP-1 was 8.09, whereas for water, it was 1.03. This disparity can be attributed to the different thermal properties and heat transfer characteristics of the two fluids. Therminol VP-1, being a synthetic oil, has a higher specific heat capacity and thermal stability at elevated temperatures, making it more suitable for high-temperature applications in solar thermal systems. The results from this study reinforced the idea that the choice of heat transfer fluid, in combination with modified absorber tube designs, could have a profound impact on the overall performance of PTCs. The sinusoidal design proved to be particularly effective in enhancing heat transfer when paired with Therminol VP-1, suggesting that further optimization of fluid and tube geometry combinations could lead to even better performance in solar thermal systems.

Munusamy et al. [24] conducted an experimental study on a copper receiver tube with dimples for water heating. The experiments showed that the dimpled tube design produced the best thermal results when the flow rate was set at 2.5 kg/min. At this flow rate, the overall thermal efficiency of the collector increased by as much as 31.85%. The dimples on the copper tube surface served to disrupt the boundary layer of the fluid, improving the heat transfer between the fluid and the tube wall. This increase in heat transfer efficiency led to higher thermal performance and lower temperature gradients within the tube, further enhancing the system's efficiency. The experimental findings supported the idea that surface modifications, such as dimples, could offer significant improvements in thermal efficiency for water-based systems in solar thermal applications.

Chakraborty and Nath [2] found that hybrid nanofluids with a 1% volume fraction in distilled water and semi-elliptical tubes with half-star-shaped fins enhanced thermal efficiency by 23.08%, while improving the Nusselt number by 43% compared to traditional circular tubes. Similarly, Liu et al. [21] investigated supercritical $CO_2$ with a novel elliptical twisted tube, showing a 107% improvement in heat transfer over conventional tubes. Jebasingh and Herbert [22] demonstrated that elliptical absorber tubes, though leading to a higher pressure drop, offered more uniform heat distribution, improving collector efficiency. Alshehery et al. [23] introduced a triangular prism turbulator (TPT) in flat tubes, yielding a 28% higher Performance Enhancement Coefficient (PEC) and up to a 525% increase in heat transfer. These studies highlight how modifying the absorber tube shape and employing hybrid nanofluids or turbulators can significantly enhance the thermal performance of PTCs.

In conclusion, a range of studies has demonstrated that modifying the geometry of the tubular absorber tube in Parabolic Trough Collectors can substantially improve heat transfer, efficiency, and overall performance. Whether through converging–diverging shapes, convex corrugated surfaces, sinusoidal geometries, or wavy structures, each modification offers unique advantages in terms of heat distribution, pressure losses, and fluid dynamics. However, these modifications must be carefully optimized to balance heat transfer improvements with potential drawbacks, such as increased pressure drops and thermal stress. Furthermore, the choice of heat transfer fluid plays a crucial role in the performance of modified collector tubes. As the research in this field continues, even more advanced and optimized designs will likely emerge, further enhancing the performance and sustainability of solar thermal systems. Table 4.4 shows more studies related to the thermal performance of PTCs and their advancements in absorber tube designs, based on the topic.

**Table 4.4** Summary of studies on thermal performance enhancement in parabolic trough collectors

| Study | Modification | Key findings |
|---|---|---|
| Bellos et al. [3] | 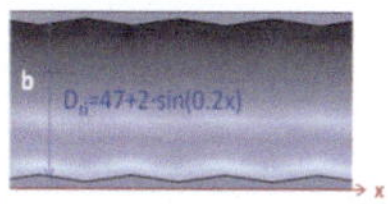 | Overall efficiency improved by 4.55%, enhancing heat absorption, distribution, and reducing thermal stress under non-uniform heat flux conditions |
| Khudheyer [18] | 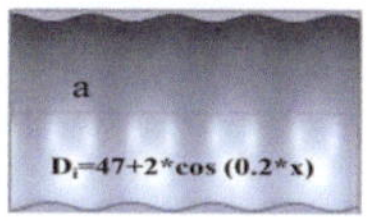 | Optimized absorber tube designs, such as convergent-divergent tubes, improved thermal efficiency by 4.75%, enhancing heat transfer performance significantly |
| Fuqiang et al. [4] | 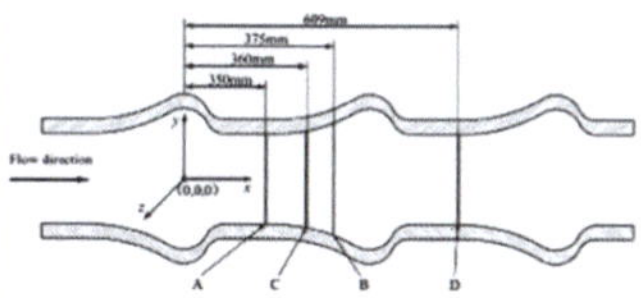 | Thermal performance factor increased by 148; 26.8% reduction in thermal strain, improving heat distribution, durability, and reducing mechanical stresses |
| Bitam et al. [5] | 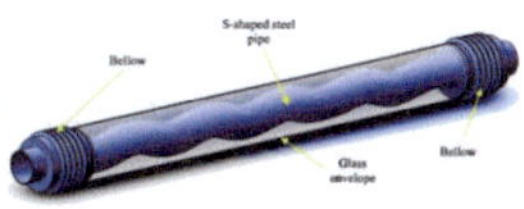 | Nusselt number increased by 63%, friction factor by 40.8%; Performance Evaluation Criterion = 1.35, reduced temperature difference (35 K), better heat transfer |
| Yang and Ordonez [19] |  | Performance Evaluation Criterion = 1.25 at larger mass flow rates, efficiency improves, but pressure drop increases with amplitude >0.006 m |
| Khelkar et al. [20] | 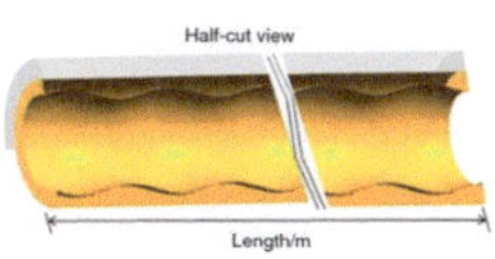 | Performance Evaluation Criterion for Therminol VP-1 = 8.09, for water = 1.03; higher PEC with Therminol VP-1 due to superior thermal properties |
| Chakraborty and Nath [2] | 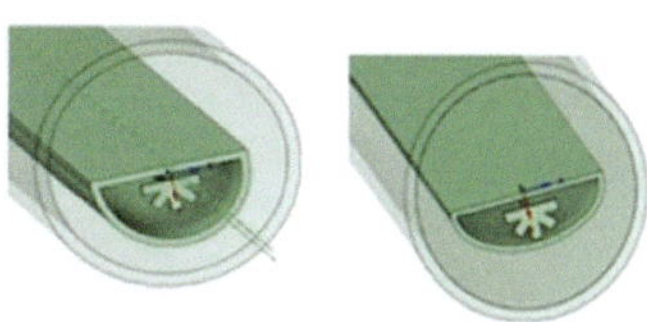 | Thermal efficiency improved by 23.08%; Nusselt number increased by 43%; hybrid nanofluids and modified tube design enhanced heat transfer performance |

(continued)

**Table 4.4** (continued)

| Study | Modification | Key findings |
|---|---|---|
| Liu et al. [21] | 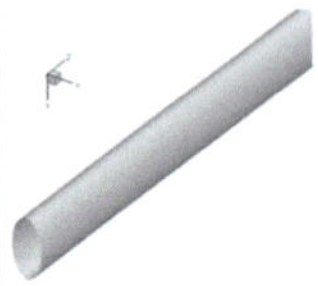 | Heat transfer improved by 107% compared to conventional tubes; substantial enhancement due to supercritical $CO_2$ and novel elliptical twisted design |
| Jebasingh and Herbert [22] | 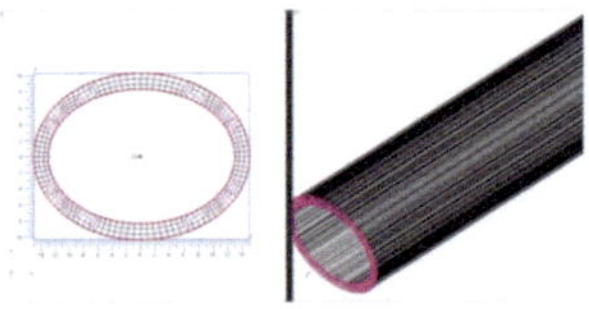 | More uniform heat distribution, improved collector efficiency despite higher pressure drop; enhanced performance with better heat distribution but increased pressure drop |
| Alshehery et al. [23] |  | Performance Evaluation Criteria (PEC) increased by 28%, heat transfer improved by 525%; a significant enhancement with Thermal Plate Technology (TPT), boosting overall performance |
| Chakraborty and Nath [25] | 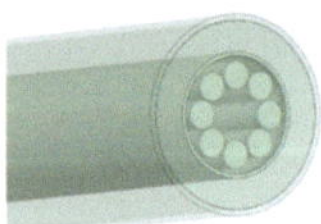 | Enhanced Parabolic Trough Collector (PTC) thermal performance using spinning tubes and ternary nanofluids. Thermal efficiency improved by 16.34%, heat transfer 97.90%, pump work increased |
| Chakraborty et al. [17] |  | Helical absorber tube improves heat transfer, efficiency, and thermal performance. Helical tube increases heat transfer, efficiency; outlet temperature reaches 392 K |
| Shi et al. [26] | 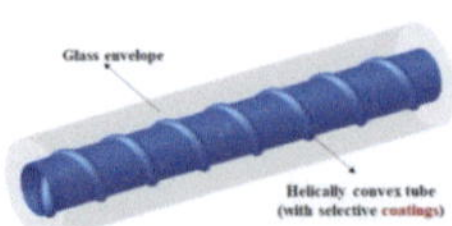  | Helically convex absorber tube improves thermal properties by 34.0%. Helically convex absorber tube improves thermal properties, with optimized parameters |
| Zheng et al. [27] | 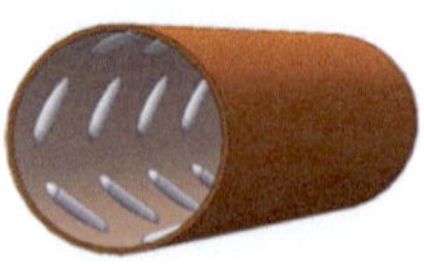 | The ribbed tube increases heat transfer 1.8–3.6 times, optimizing flow. Ribbed tube enhances heat transfer, friction; optimal configuration improves performance |

(continued)

**Table 4.4** (continued)

| Study | Modification | Key findings |
|---|---|---|
| Huang et al. [28] | 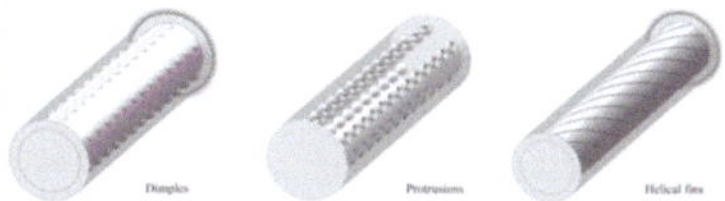 | Dimples significantly improve thermal performance, outperforming fins and protrusions. Dimples enhance heat transfer; deeper, narrower, and more dimples improve performance |
| Xiangtao et al. [29] | 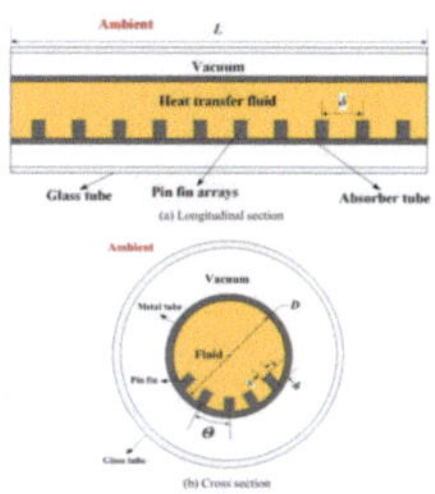 | Thermal performance improved with pin–fin arrays, increasing heat transfer by 12%. Pin–fin arrays enhance heat transfer, increasing the Nusselt number by 9% |
| Bellos et al. [30] | 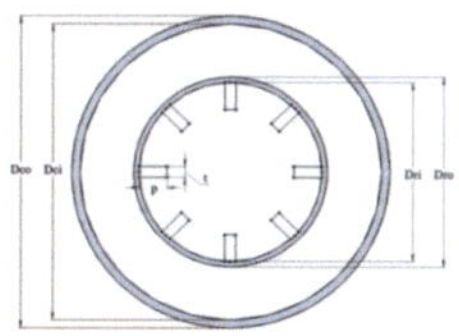 | Higher fin length improves thermal efficiency, Nusselt number, and enhancement. Optimal fin design: 20 mm length, 4 mm thickness enhances thermal performance |
| Bellos et al. [31] |  | Three fins in the lower absorber enhance thermal efficiency by 0.51%. The optimum configuration is three fins in the lower part of the absorber |
| Zhao et al. [10] | 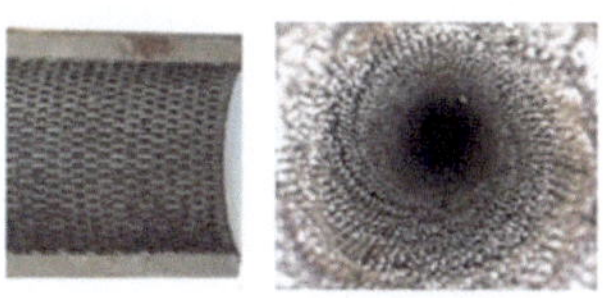 (a) | Thermal performance enhanced by 10.4–14.5% with internally pin–fin tubes. Internally pin–fin tubes improve energy efficiency, exergy, and temperature rise |
| Limboonruang et al. [32] |  (a) Smooth tube  (b) Finned tube | Thermal performance enhanced by 38% with externally finned tubes at 1.0 LPM. Externally finned tubes increase thermal efficiency by 38%, outperforming smooth tubes |

(continued)

**Table 4.4** (continued)

| Study | Modification | Key findings |
|---|---|---|
| Mahmoud et al. [33] | 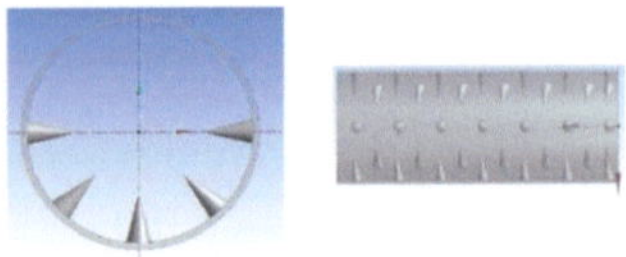 | Conical pin-fins and increased fin height improve thermal performance. Fin height increase leads to a 56.2% enhancement in thermal performance |
| Hamada et al. [34] | 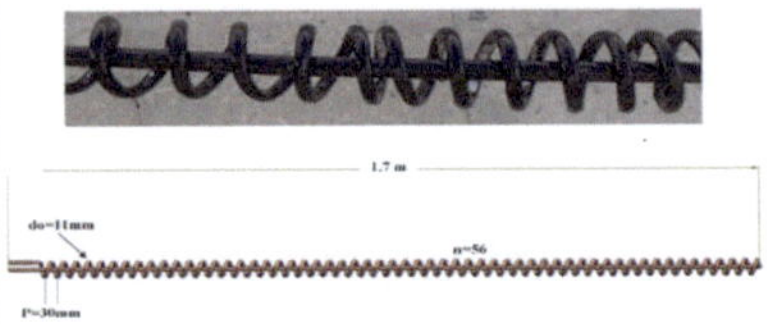 | The helical tube increased thermal efficiency by 83.5% compared to the reference system. Helical tube, nanofluids, and thermal storage improve efficiency by 145.9% |
| Peng et al. [35] | 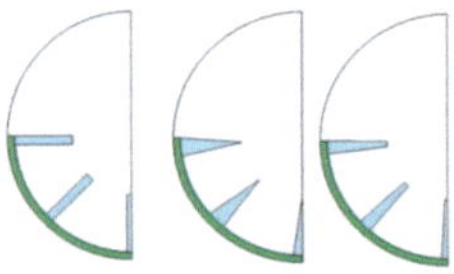 | Thermal performance enhanced by 256.3–838.7% in semi-annular and fin-shaped metal foam. Semi-annular and fin-shaped metal foam insertion enhances thermal performance |
| Javidan et al. [36] | 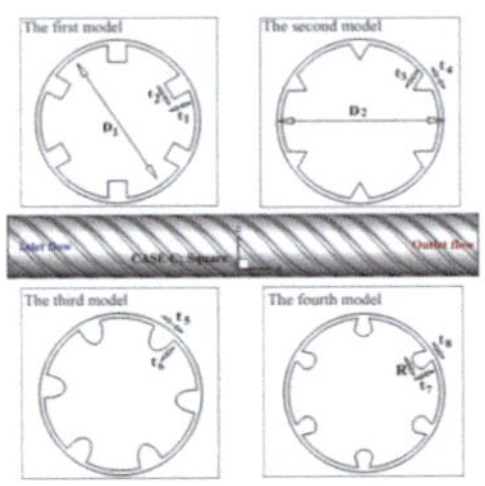  | Combined square-circle fin and Ag–MgO hybrid nanofluid improves thermal efficiency |
| Kursun [37] | 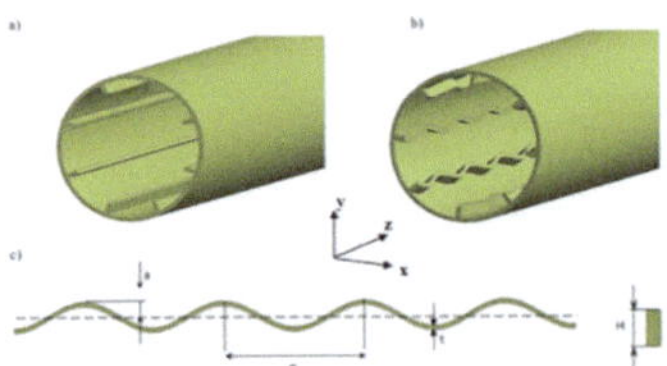 | Sinusoidal lateral surface improves heat transfer with 78% Nusselt number enhancement |
| Kaood et al. [38] | 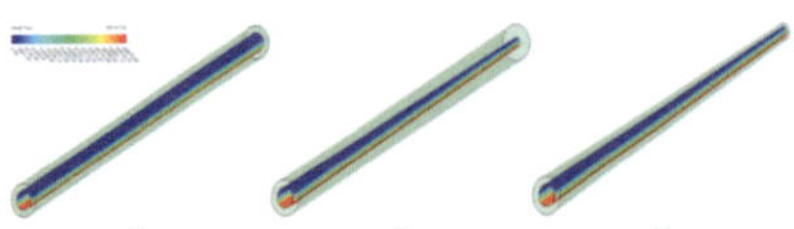 | Conical receiver design improves thermal efficiency by 7–65% |

(continued)

**Table 4.4** (continued)

| Study | Modification | Key findings |
|---|---|---|
| Byiringiro et al. [39] | 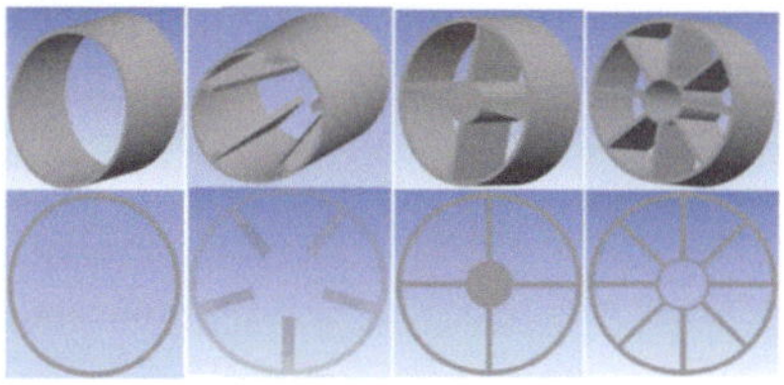 | Hybrid nanofluid improves thermal efficiency, Nusselt number, and reduces temperature. Configuration 3 and the hybrid nanofluid improved thermal performance and efficiency |
| Loni et al. [7] | 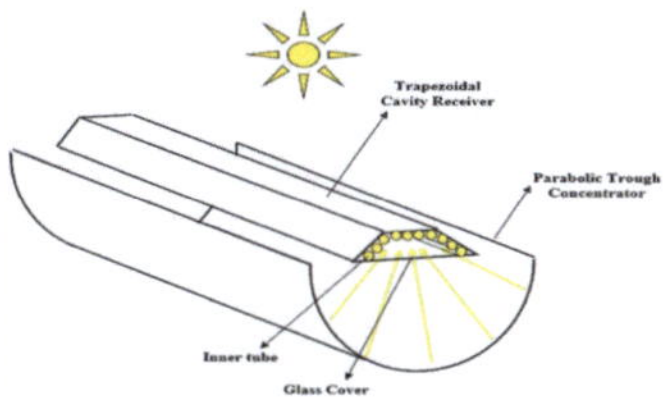  | Cubic cavity receivers perform best; trapezoidal least effective; optimal exergy at 0.075% MXene nanofluid |
| Xiao et al. [8] | 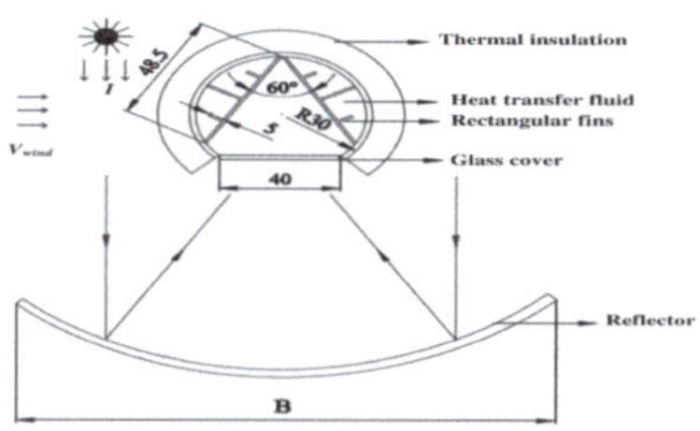  | V-cavity absorber with rectangular fins improves heat transfer, yielding higher outlet temperatures, reduced heat losses, and enhanced thermal management |
| Nazir et al. [40] | 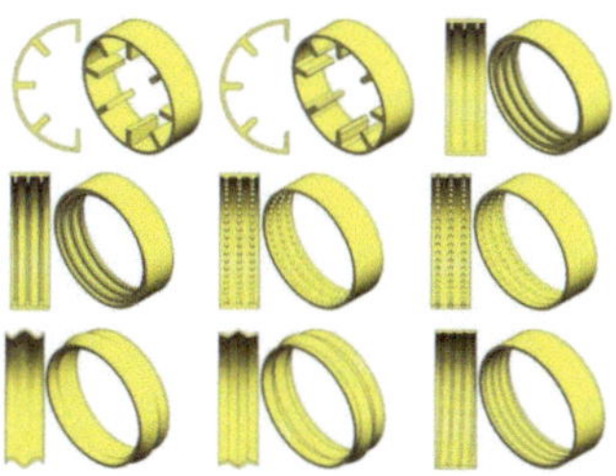 | Various fins and wavy channels enhance thermal performance with minimal pressure drop |
| Cheng et al. [41] | 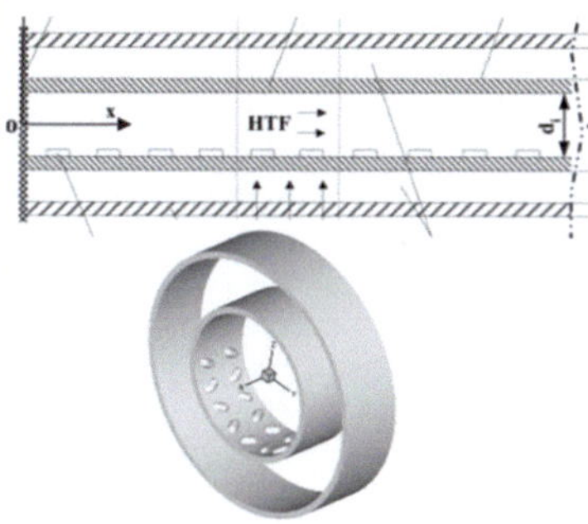  | Unilateral multi-longitudinal vortex generators improve heat transfer, outperforming a smooth absorber tube |

(continued)

**Table 4.4** (continued)

| Study | Modification | Key findings |
|---|---|---|
| Shi et al. [26] | 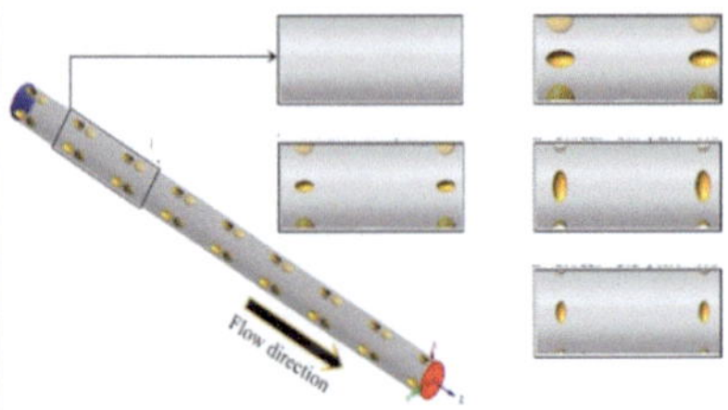 | The addition of 1% $TiO_2$ and 2% $Al_2O_3$ in Syltherm 800, along with elliptical fins (*a*/*b* = 5/3), results in a 38% higher Nusselt number and an 80% thermal efficiency, significantly enhancing heat transfer performance |
| Munusamy et al. [24] | 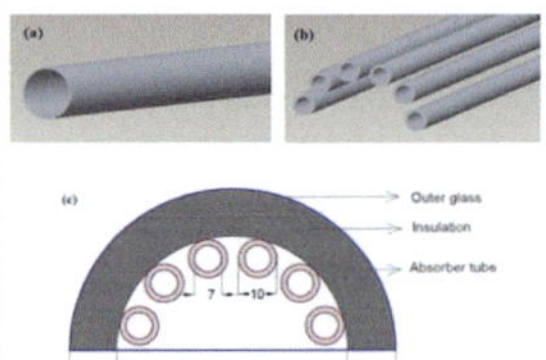 | The semicircular absorber with double inlets improves thermal and exergy efficiencies by 43.9 and 33.6%, respectively. It increases the heat transfer coefficient by 28.5% and enhances heat transfer, benefiting line-focused solar collector designs |
| Vengadesan et al. [42] | 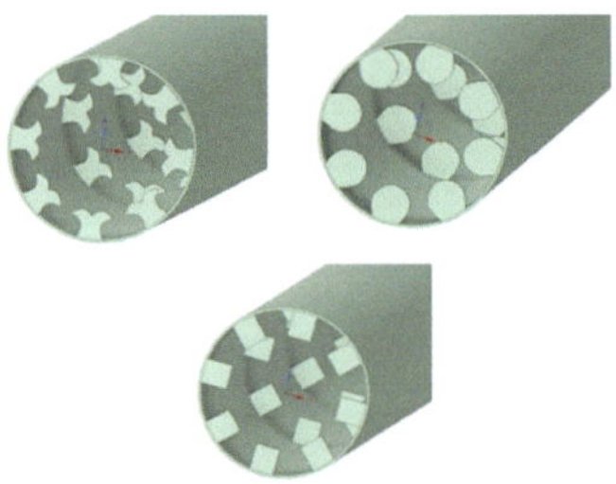 | The study explores the use of ternary nanoparticles in ionic liquids and finned PTC tubes. Results show up to 55.45% improvement in heat transfer, with 22.50% higher thermal efficiency for rotating tubes, optimizing design and performance |

## References


1. Kalogirou SA (2012) A detailed thermal model of a parabolic trough collector receiver. Energy 48:298–306
2. Chakraborty O, Nath S (2024) Thermal performance simulation of parabolic trough collector with ternary nanofluid flows and half-star-shaped fins inserts in different receivers. J Brazil Soc Mech Sci Eng 46:114
3. Bellos E, Tzivanidis C, Antonopoulos KA, Gkinis G (2016) Thermal enhancement of solar parabolic trough collectors by using nanofluids and converging-diverging absorber tube. Renew Energy 94:213–222
4. Fuqiang W, Zhexiang T, Xiangtao G, Jianyu T, Huaizhi H, Bingxi L (2016) Heat transfer performance enhancement and thermal strain restrain of tube collector for parabolic trough solar collector by using asymmetric outward convex corrugated tube. Energy 114:275–292
5. Bitam EW, Demagh Y, Hachicha AA, Benmoussa H, Kabar Y (2018) Numerical investigation of a novel sinusoidal tube collector for parabolic trough technology. Appl Energy 218:494–510
6. Zhu X, Zhu L, Zhao J (2017) Wavy-tape insert designed for managing highly concentrated solar energy on absorber tube of parabolic trough receiver. Energy 141:1146–1155

7. Loni R, Sharifzadeh M (2024) Performance comparison of a solar parabolic trough concentrator using different shapes of linear cavity receiver. Case Stud Therm Eng 60:104603
8. Xiao X, Zhang P, Shao DD, Li M (2014) Experimental and numerical heat transfer analysis of a V-cavity absorber for linear parabolic trough solar collector. Energy Conver Manag 86:49–59
9. Bortolato M, Dugaria S, Col DD (2016) Experimental study of a parabolic trough solar collector with flat bar-and-plate absorber during direct steam generation. Energy 116:1039–1050
10. Zhao Z, Bai F, Zhang X, Wang Z (2020) Experimental study of pin finned receiver tubes for a parabolic trough solar air collector. Sol Energy 207:91–102
11. White FM (2016) Fluid mechanics, 8th edn. McGraw-Hill Education
12. Munson BR, Rothmayer AP, Okiishi TH (2013) Fundamentals of fluid mechanics, 7th edn. Wiley
13. Cengel YA, Cimbala JM (2018) Fluid mechanics: fundamentals and applications, 3rd edn. McGraw-Hill Education
14. Colebrook CF (1939) Turbulent flow in pipes with particular reference to the transition between laminar and turbulent flow. J Instit Civil Eng 11(1):133–156
15. Chakraborty O, Roy S, Das B, Gupta R (2022) Computational analyses of parabolic trough solar collector in the presence of helical coil-insert. Int J Environ Sci Technol 87:3891. https://doi.org/10.1007/s13762-021-03891-1
16. Chakraborty O, Das B, Gupta R (2023) Thermal performance evaluation of parabolic trough collector having different inserts and working with hybrid nanofluid. Energy Environ 35:3010–3037. https://doi.org/10.1177/0958305X231156407
17. Chakraborty O, Roy S, Das B, Gupta R (2021) Effects of helical absorber tube on the energy and exergy analysis of parabolic solar trough collector: a computational analysis. Sustain Energy Technol Assess 44:101083
18. Khudheyer AF (2020) Numerical investigation and exergetic analysis of convergent-divergent absorber tube in concentrated solar trough. J Green Eng 10:10
19. Yang S, Ordonez JC (2019) 3D thermal-hydraulic analysis of a symmetric wavy parabolic trough absorber pipe. Energy 189:116320
20. Khelkar AB, Debnath BK, Debnath K (2020) Use of sinusoidal surface profile in the absorber tube of a parabolic trough solar collector to enhance its thermal performance. J Therm Anal Calorim 141(6):2589–2597
21. Liu Y, Donga Y, Lib T, Zhang CZ (2022) Performance analysis and comparison of different corrugated structures and a novel alternative elliptical twisted tube in supercritical $CO_2$ tower solar receivers. Renew Energy 199:1523–1533
22. Jebasingh VK, Herbert GMJ (2015) Numerical simulation of elliptical absorber tube in parabolic trough collector for better heat transfer. Indian J Sci Technol 8(24):5422
23. Alshehery S, Bouzidi M, Mdallal QA, Alkathiri T, Altohamy AA, Kolsi L (2025) Boosting the thermal efficiency of a flat-tube parabolic trough solar collector with the introduction of innovative slashed triangular prism turbulators. Case Stud Therm Eng 67:105849
24. Munusamy A, Barik D, Sharma P, Medhi BJ, Bora BJ (2023) Performance analysis of parabolic type solar water heater by using copper-dimpled tube with aluminum coating. Environ Sci Pollut Res 542:1152
25. Chakraborty O, Nath S (2025) Performance analysis of a parabolic solar trough collector with multiple revolving tubes for ternary nanofluid and different base fluids. J Therm Anal Calorim 21:5254
26. Shi X, Zhao X, Wang F, Cheng Z, Dong Y, Xu J (2022) Improving overall heat transfer performance of parabolic trough solar receiver by helically convex absorber tube. Appl Therm Eng 213:118690
27. Zheng N, Liu W, Liu Z, Liu P, Shan F (2015) A numerical study on heat transfer enhancement and the flow structure in a heat exchanger tube with discrete double inclined ribs. Appl Therm Eng 90:232–241
28. Huang Z, Yu GL, Li ZY, Tao WQ (2015) Numerical study on heat transfer enhancement in a receiver tube of parabolic trough solar collector with dimples, protrusions and helical fins. Energy Proced 69:1306–1316

29. Xiangtao G, Fuqiang W, Haiyan W, Jianyu T, Qingzhi L, Huaizhi H (2017) Heat transfer enhancement analysis of tube receiver for parabolic trough solar collector with pin fin arrays inserting. Sol Energy 144:185–202
30. Bellos E, Tzivanidis C, Tsimpoukis D (2017) Multi-criteria evaluation of parabolic trough collector with internally finned absorbers. Appl Energy 205:540–561
31. Bellos E, Tzivanidis C, Tsimpoukis D (2018) Optimum number of internal fins in parabolic trough collectors. Appl Therm Eng 137:669–677
32. Limboonruang T, Oyinlola M, Harmanto D, Phunapai N (2025) Optimizing the configuration of external fins of solar receiver tubes for the solar parabolic trough collector. Appl Therm Eng 263:125317
33. Mahmoud MS, Abbas AS, Khudheyer AF (2020) Solar parabolic trough collector tube heat transfer analysis with internal conical pin fins. J Green Eng 10(10):5431
34. Hamada MA, Abdo MR, Khalil H, Al-Sood MMA, Sharshir SW (2025) Investigating the performance of solar parabolic concentrator using helical tubes, nanoparticles, and thermal storage materials. Process Saf Environ Protect 195:106777
35. Peng H, Li M, Hu F, Feng SS (2021) Performance analysis of absorber tube in parabolic trough solar collector inserted with semi-annular and fin shape metal foam hybrid structure. Case Stud Therm Eng 26:101112
36. Javidan M, Gorji-Bandpy M, Al-Araji A (2023) Investigation and simulation of parabolic trough collector with the presence of hybrid nanofluid in the finned receiver tube. Theor Appl Mech Lett 13(6):100465
37. Kursun B (2019) Thermal performance assessment of internal longitudinal fins with sinusoidal lateral surfaces in parabolic trough receiver tubes. Renew Energy 140:816–827
38. Kaood A, Ismail OA, Al-Tohamy AH (2024) Hydrothermal performance assessment of a parabolic trough with proposed conical solar receiver. Renew Energy 222:119939
39. Byiringiro J, Chaanaoui M, Halimi M (2024) Heat transfer enhancement of a parabolic trough solar collector using innovative receiver configurations combined with a hybrid nanofluid: CFD analysis. Renew Energy 233:121169
40. Nazir MS, Shahsavar A, Afrand M, Arıcı M, Nizetic S, Ma Z, Oztop HF (2021) A comprehensive review of parabolic trough solar collectors equipped with turbulators and numerical evaluation of hydrothermal performance of a novel model. Sustain Energy Technol Assess 45:101103
41. Cheng ZD, He YL, Cui FQ (2012) Numerical study of heat transfer enhancement by unilateral longitudinal vortex generators inside parabolic trough solar receivers. Int J Heat Mass Transf 55(21–22):5631–5641
42. Vengadesan E, Rumaney ARI, Mitra R, Harichandan S, Senthil R (2022) Heat transfer enhancement of a parabolic trough solar collector using a semicircular multitube absorber. Renew Energy 196:111–124

# Chapter 5
# Performance Enhancements in Parabolic Solar Trough Collectors via Insert Diversity

This chapter explores the impact of incorporating various types of inserts—both steady and spinning—within the receiver tubes of Parabolic Solar Trough Collectors (PTCs) to improve thermal performance. Through a detailed literature review, it examines how these internal modifications enhance turbulence, heat transfer rates, and overall thermal efficiency without altering the external geometry of the collector. Steady inserts like twisted tapes and wire coils boost the heat transfer coefficient by disrupting laminar flow, while spinning inserts generate swirling motion that further intensifies fluid mixing. The chapter also addresses the trade-offs, including mechanical complexities and pressure drops introduced by these designs. By highlighting both the benefits and challenges, this chapter underscores the significance of insert diversity in advancing PTC technology and contributing to the development of more efficient and cost-effective solar thermal systems.

## 5.1 Inserts in Absorber Tubes

Inserts in Parabolic Solar Trough Collectors improve thermal efficiency by enhancing heat transfer between the heat transfer fluid (HTF) and the collector surface. These inserts promote turbulence in the fluid flow, disrupting the thermal boundary layer and facilitating better heat exchange. Common steady inserts include twisted tapes, wire coils, helical inserts, and ribbed inserts. Twisted tapes are strips of material twisted and placed inside the tube, causing the fluid to swirl and mix, which increases turbulence. Wire coils are helically wound inside the tube to increase surface area and promote mixing. Helical inserts induce a spiral flow in the HTF, further enhancing turbulence and convective heat transfer. Ribbed inserts, with raised surfaces, disrupt the boundary layer of the fluid, improving the overall heat transfer rate.

The main benefit of these inserts is the significant improvement in heat transfer efficiency, leading to better solar collector performance. They also contribute to

 
O. Chakraborty and B. Das, *Parabolic Solar Trough Collector*, Mechanical Engineering Series, https://doi.org/10.1007/978-3-032-11229-3_5

operational stability because they do not have moving parts, making them of low-maintenance and cost-effective. Furthermore, these inserts allow for greater energy absorption, which is particularly important in maximizing the efficiency of solar thermal systems. However, despite these advantages, there are challenges associated with their use. Inserts typically cause an increase in pressure drop, meaning that the fluid faces greater resistance while flowing through the system. This requires more powerful pumps to circulate the fluid, increasing energy consumption. Additionally, inserts may face durability issues, especially in high-temperature environments, where wear and tear can reduce efficiency over time. Optimizing the balance between heat transfer enhancement and pressure loss is essential for achieving the most efficient and cost-effective solar thermal system.

**Twisted Tapes**

Twisted tapes are a specialized type of steady insert employed in Parabolic Solar Trough Collector (PTC) tubes to significantly optimize thermal changes by improving heat transfer efficiency. These inserts consist of thin strips of material twisted into a helical shape and placed inside the absorber tubes. Their primary function is to induce a swirling motion in the heat-carrying liquid as it flows through the tube, effectively promoting turbulence. This increased turbulence enhances the mixing of fluid layers, leading to a more homogeneous thermal distribution over the surface of the absorber and markedly improving the heat transfer coefficient.

**Advantages**

i. **Enhanced Heat Transfer**: Twisted tapes' ability to disrupt laminar flow patterns is among their foremost notable benefits. Twisted tapes induce turbulence by creating a swirling motion within the HTF, accelerating heat exchange. This results in significantly improved thermal performance, allowing PTCs to capture and convert a larger portion of solar energy into usable thermal energy. Studies have shown that using twisted tapes can lead to heat transfer coefficients that are substantially higher than those in systems without inserts.
ii. **Cost-Effectiveness**: Twisted tapes are relatively straightforward to manufacture and install, making them an economically viable solution for enhancing heat transfer. Their simplicity allows for quick integration into existing systems without the need for extensive modifications, which can be particularly advantageous for operators looking to upgrade the efficiency of their solar thermal systems without incurring high costs.
iii. **Temperature Uniformity**: The swirling motion induced by twisted tapes helps maintain a more homogeneous temperature spread along the absorber surface. This reduces thermal stratification, where heat may concentrate unevenly, leading to inefficiencies. Improved temperature uniformity means that the entire surface area of the absorber can work more effectively to capture solar energy, further enhancing overall system efficiency.

iv. **Flexibility in Design**: Twisted tapes can be easily adapted to fit various tube diameters and configurations, providing significant design flexibility. This adaptability allows engineers to customize the inserts based on specific operational conditions, such as varying flow rates and heat transfer fluid properties, optimizing performance across different applications.

**Disadvantages**

i. **Increased Pressure Drop**: One of the primary challenges associated with twisted tapes is that they can lead to higher pressure drops within the system. The introduction of these inserts disrupts the flow, requiring more energy to maintain adequate fluid circulation. This can necessitate using more powerful pumps, which may offset some of the operational cost savings achieved through enhanced heat transfer.
ii. **Potential for Fouling**: The surfaces created by twisted tapes can provide sites for accumulating deposits from the HTF, leading to fouling over time. This buildup can negatively impact heat transfer performance and require regular maintenance and cleaning to ensure optimal system efficiency. Managing fouling is essential for long-term operation, particularly in systems where the HTF is susceptible to contamination.
iii. **Limited Effectiveness at Low Flow Rates**: While twisted tapes effectively promote turbulence at higher flow rates, their benefits may diminish significantly at low flow rates. In such conditions, the swirling action may not be sufficient to overcome the laminar flow characteristics, which can reduce the overall impact on heat transfer efficiency.
iv. **Material Considerations**: The materials used for twisted tapes must be carefully selected to withstand the thermal and chemical stresses imposed by the HTF. Choosing inappropriate materials may degrade and fail the inserts, compromising the system's performance and necessitating replacements or repairs. Twisted tapes are depicted in Fig. 5.1.

**Mathematical Calculation for Twisted Tape Inserts**

The utilization of twisted-shaped tape inserts in absorber tubes can significantly enhance the effectiveness of Parabolic Solar Trough Collectors (PTCs) in heat dissipation. To quantify this enhancement, various mathematical calculations based on fluid dynamics and heat transfer principles can be used. Below are key calculations and equations typically used to evaluate the functionality of twisted-shaped tape inserts.

The heat transfer coefficient for turbulent flow with twisted tape inserts can be derived from the Nusselt number (Nu):

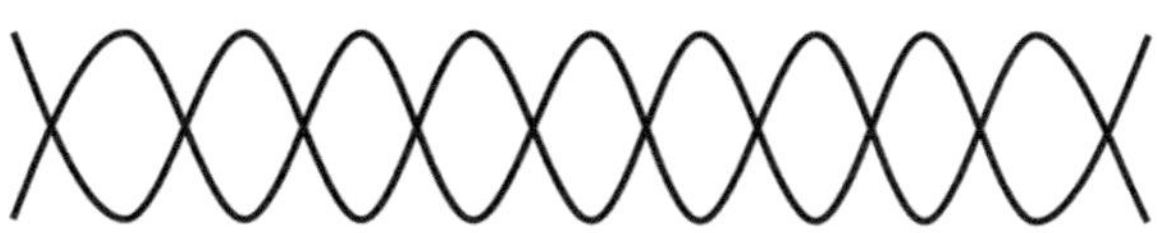

**Fig. 5.1** Twisted tapes

$$\mathrm{Nu} = \frac{hD}{k} \quad (5.1)$$

where

$h =$ convective heat transfer coefficient (W/m$^2$ K)

$D =$ diameter of the tube (m)

$k =$ thermal conductivity of the fluid (W/m K).

For twisted tape inserts, empirical correlations can be employed to calculate the Nusselt number. A common correlation is [1]

$$\mathrm{Nu} = 0.27 \times \mathrm{Re}^{0.6} \times \overset{0.33}{\mathrm{Pr}} \quad (5.2)$$

where

Re = Reynolds number

Pr = Prandtl number.

The Reynolds number is given by

$$\mathrm{Re} = \frac{\rho v D}{\mu} \quad (5.3)$$

where

$\rho =$ density of the liquid (kg/m$^3$)

$v =$ average velocity of the liquid (m/s)

$D =$ tube diameter value (m)

$\mu =$ dynamic viscosity of the liquid (Pa s).

The Prandtl number can be calculated using

$$\mathrm{Pr} = \frac{C_p \mu}{k} \quad (5.4)$$

where

$c_p =$ liquid's specific heat capacity (J/kg K)

$\mu =$ liquid's dynamic viscosity (Pa s)

$k =$ liquid's thermal conductivity (W/m K).

The cumulative heat transfer rate can be calculated using

$$Q = hA(T_s - T_\infty) \quad (5.5)$$

where

$Q =$ heat transfer rate (W)

$h =$ convective heat transfer coefficient (W/m$^2$ K)

$A =$ heat transfer area (m$^2$)

$T_s =$ temperature at the surface of tube (K)

$T_\infty =$ temperature of the fluid (K).

The heat transfer area for a tube can be calculated as

$$A = \pi DL \tag{5.6}$$

where

$L =$ length of the tube.

The total heat transfer coefficient can be determined by considering both convective and conductive resistances:

$$\frac{1}{U} = \frac{1}{h} + \frac{L}{k} + \frac{1}{h_r} \tag{5.7}$$

where

$U =$ overall heat transfer coefficient (W/m$^2$ K)

$h_r =$ external heat transfer coefficient (W/m$^2$ K) (if applicable)

$L =$ thickness of the tube wall.

**Geometrical Calculations for Twisted Tape Inserts**

Geometrical calculations are essential for designing and analyzing twisted strip inserts in tubes of Parabolic Solar Trough Collectors (PTCs). These calculations focus on dimensions, spacing, and orientation to optimize thermal performance. Below are key geometrical aspects and calculations related to twisted strip inserts.

The twisted tape insert's thickness impacts the flow dynamics and heat transfer characteristics. The thickness ($t$) can be expressed as a fraction of the tube diameter ($D$):

$$t = k \times D \tag{5.8}$$

where

$t =$ thickness of the twisted tape

$k =$ thickness ratio (typically between 0.05 and 0.1 [1])

$D =$ diameter of the tube.

The effective length of the twisted tape should cover a significant portion of the receiver tube to improve heat transfer. A common practice is to use [1]

$$L_{\text{tape}} = 0.7 \times L_{\text{tube}} \tag{5.9}$$

where

$L_{\text{tape}} =$ the twisted tape's length insert

$L_{\text{tube}} =$ the absorber tube's total length.

The twist ratio specifies the number of twists per unit of tape length. This is crucial for ensuring adequate turbulence while minimizing pressure loss. The twist ratio (TR) can be calculated as

$$\text{TR} = \frac{N_{\text{twists}}}{L_{\text{tape}}} \tag{5.10}$$

where

$N_{\text{twists}}$ = number of twists in the insert

$L_{\text{tape}}$ = length of the twisted tape insert.

The introduction of twisted tape inserts enhances the effective surface area for heat transfer. The extra surface area ($A_{\text{extra}}$) can be approximated using

$$A_{\text{extra}} = A_{\text{tube}} \times (\text{surface are a enhancement factor}) \tag{5.11}$$

where the surface area enhancement factor depends on the twist configuration and is typically determined through empirical correlations or experimental data.

The volume of the twisted tape can be calculated using

$$V_{\text{tape}} = A_{\text{tape}} \times L_{\text{tape}} \tag{5.12}$$

where

$$A_{\text{tape}} = \pi \times (D_{\text{tube}} + t) \times t \tag{5.13}$$

$D_{\text{tube}}$ = tube's outermost diameter.

$t$ = Thickness of the twisted tape.

The spacing between each twist of the tape is crucial for ensuring effective turbulence. This can be calculated using

$$S = \frac{L_{\text{tape}}}{N_{\text{twists}}} \tag{5.14}$$

where

$S$ = spacing between twists.

**Wire Coils**

Wire coils are another type of steady insert utilized in the tubes of Parabolic Solar Trough Collectors (PTCs) to boost thermal effectiveness by improving heat transfer efficiency. These inserts consist of loops of wire arranged in a coil shape and placed within the tubes to interact with the heat-carrying liquid flowing through it. The primary purpose of wire coils is to induce more turbulence in the liquid, thereby enhancing the overall heat flow process.

**Advantages**

i. **Enhanced Heat Transfer**: Wire coils promote turbulence in the HTF by disrupting the laminar flow patterns. This increased turbulence leads to improved fluid layer mixing, significantly enhancing the heat transfer coefficient. As a result, wire coils can help capture more thermal energy from the sun, increasing the overall efficiency of the solar thermal units.
ii. **Increased Surface Area**: The design of wire coils inherently increases the surface area in contact with the HTF. This expanded contact area facilitates

greater thermal interaction between the fluid and the absorber tube, further improving thermal performance. By providing more surface for heat transfer, wire coils can effectively enhance the energy capture capabilities of PTCs.

iii. **Flexibility in Design**: Wire coils can be manufactured in various sizes and configurations, making them adaptable to different absorber tube diameters and operational requirements. This design flexibility allows optimization based on specific system conditions, such as varying flow rates and fluid properties, thereby enhancing overall performance.

iv. **Cost-Effective Solution**: Wire coils are relatively straightforward to produce and install like twisted tapes. Their simplicity allows for easy integration into existing systems, making them a cost-effective option for operators seeking to improve the efficiency of their solar thermal installations without incurring significant additional expenses.

### Disadvantages

i. **Increased Pressure Drop**: One of the primary challenges of using wire coils is that they can lead to increased pressure drops within the system. The coils disrupt the fluid flow, requiring additional energy to maintain circulation. This can necessitate using more powerful pumps, potentially increasing operational costs and offsetting some of the benefits gained from improved heat transfer.

ii. **Fouling Potential**: Similar to twisted tapes, wire coils can create surfaces that may accumulate deposits from the HTF over time. This fouling can negatively impact heat transfer efficiency and may require regular maintenance to clean the coils, ensuring optimal performance. Managing fouling is crucial to maintaining long-term operational efficiency.

iii. **Limited Effectiveness at Low Flow Rates**: While wire coils effectively promote turbulence at higher flow rates, their performance may diminish at low flow rates. Under these conditions, the benefits of increased turbulence may not be fully realized, potentially reducing the overall impact on heat transfer efficiency.

iv. **Material Considerations**: The choice of materials for wire coils is vital, as they must withstand the thermal and chemical stresses imposed by the HTF. Inadequate material selection can lead to degradation over time, which may necessitate replacements and impact the system's overall reliability. Wire coils are shown in Fig. 5.2.

### Coil Specifications

**Wire Diameter (*d*):** The diameter of the wire must balance strength and heat transfer efficiency.

**Fig. 5.2** Wire coils

**Coil Diameter (*D*):** This is the diameter of the coil itself. It should be sized to fit within the collector design without obstructing the sunlight or affecting the trough's focus.

The number of turns can be calculated based on the collector length and coil pitch:

$$N = \frac{L}{P} \tag{5.15}$$

where $L$ denotes the length of the collector, and $P$ is the pitch (the distance between adjacent turns).

The pitch affects the heat dissipation area and fluid dynamics. It can be determined based on.

**Fluid velocity:** Higher velocities may require tighter coils.

**Heat transfer requirements:** A tighter pitch increases surface area but may restrict fluid flow.

Determine the total heat exchange area using

$$A = \pi DN \tag{5.16}$$

This area is critical for determining the heat transfer rate.

Assuming the wire wraps around the parabola at a radius $R$:

$$R = \frac{D}{2} \tag{5.17}$$

The helix angle can be calculated from the geometry:

$$\tan(\theta) = \frac{P}{2\pi R} \tag{5.18}$$

The volume of the coil can be approximated by

$$V = A \times \text{length of coil} \tag{5.19}$$

The length of the coil can be approximated by

$$\text{Length of coil} = N \times \text{circumference} = N \times 2\pi R \tag{5.20}$$

Thus:

$$V = A \times N \times 2\pi R \tag{5.21}$$

**Essential Design Factors for Effective Solar Trough Collector Systems**

Designing Parabolic Solar Trough Collectors with inserts like wire coils requires careful attention to several key factors to ensure optimal performance. The inlet and outlet design are crucial to preventing stagnation zones and ensure smooth fluid flow through the coils, essential for efficient heat transfer. The choice of fluid type, such as water or oil, directly impacts the system's design due to variations in thermal properties, viscosity, and flow behavior. Thermal expansion of the wire coils must be considered to prevent stress on the system as the material expands and contracts during operation. The system design should allow for movement without compromising its structural integrity. Material selection is equally essential; materials like copper and stainless steel, known for their excellent thermal conductivity and corrosion resistance, are ideal for maintaining system durability and efficiency under high temperatures. A strong support structure is also needed to secure the coils while allowing for easy maintenance and inspection. Finally, installation considerations should prioritize ease of integration, accessibility for maintenance, and seamless alignment with the overall collector system. Addressing these considerations ensures the PTC operates efficiently, remains durable, and maximizes the capture and conversion of solar energy.

**Other Fixed Geometries**

In addition to wire coils and twisted tapes, various other fixed geometries are employed as inserts in Parabolic Solar Trough Collectors (PTCs) tubes to enhance thermal efficiency. These designs promote turbulence, increase heat transfer, and optimize the interaction between the heat-carrying liquid and the absorber surface. Common examples include helical inserts, ribbed structures, perforated plates, baffle inserts, fin designs, conical inserts, and more creative geometries like flower, star, ball, and elliptical ball inserts. **Ribbed structures** consist of ribs that disrupt liquid flow, converting laminar flow into turbulent flow, thus increasing mixing and improving the heat transfer coefficient, and maintaining uniform temperature distribution. **Perforated plates** create localized disturbances in the fluid flow through holes, promoting turbulence and enhancing heat transfer. **Baffle inserts** act as barriers to redirect fluid flow, further boosting turbulence and maintaining temperature uniformity. **Fin designs** increase surface area for heat transfer and generate turbulence in the fluid, improving thermal exchange efficiency. **Conical inserts** accelerate the fluid, promoting turbulence for better heat transfer. **Flower-shaped inserts** create multiple flow paths, improving mixing and thermal exchange, while **star-shaped inserts** induce swirling motion to enhance thermal performance. **Ball-shaped inserts** create chaotic flow patterns for better heat transfer, and **elliptical ball inserts**, with their elongated shape, enhance surface area and flow dynamics, making them optimal for high-performance systems requiring efficient thermal transfer.

**Advantages of Other Fixed Geometries**

i. **Improved Heat Transfer**: All these fixed geometries are designed to enhance turbulence, resulting in significantly improved thermal conductivity profiles in comparison to conventional tubular designs. This enhancement is crucial for maximizing the efficiency of solar thermal systems.

ii. **Customization**: The diverse range of fixed geometries allows for tailored solutions adapted to specific operational requirements, such as varying flow rates and HTF properties. This customization helps optimize performance based on each installation's unique conditions.

iii. **Durability**: Fixed geometries generally have fewer moving parts than dynamic inserts, enhancing system reliability and reducing maintenance needs. Their robustness contributes to the long-term operational efficiency of solar thermal systems.

**Disadvantages of Other Fixed Geometries**

i. **Pressure Drop**: Similar to other inserts, these fixed geometries can increase pressure drops within the system. This indicates a higher energy level may be needed to maintain liquid circulation, potentially growing operational costs.

ii. **Fouling Risk**: Fixed geometries can create surfaces that may accumulate deposits from the HTF over time, leading to fouling. Regular maintenance may be necessary to clean these surfaces and ensure optimal performance.

iii. **Design Complexity**: While customization is an advantage, the design and integration of fixed geometries can add complexity to the system. Engineers must carefully balance the benefits of enhanced heat transfer with the potential drawbacks, such as increased pressure losses.

**Spinning Inserts**

Spinning inserts represent an innovative solution for enhancing heat transfer in thermal systems, particularly in applications like solar thermal collectors. These devices significantly improve the efficiency of heat exchange processes by promoting turbulence and optimizing fluid dynamics. In this detailed examination, the design and mechanics of spinning inserts, along with their benefits and challenges, will be delved into.

**Design and Mechanics**

**Design**

Spinning inserts are dynamic devices used in absorber tubes of solar thermal systems to enhance heat transfer. Unlike fixed geometries, these inserts rotate, promoting turbulence and better fluid mixing. They can be designed in various shapes, such as cylindrical or conical, and may include fins or vanes to facilitate fluid rotation. The design process requires careful consideration of fluid dynamics, material selection, and mechanical functionality. Materials must withstand thermal and pressure extremes while resisting corrosion from aggressive heat transfer fluids (HTFs). Additionally, the insert configuration must minimize resistance to fluid flow while maximizing rotational speed and turbulence to optimize heat transfer efficiency.

**Mechanics**

Spinning inserts enhance heat transfer in absorber tubes by leveraging centrifugal force and fluid dynamics. As the heat transfer fluid (HTF) encounters the rotating insert, it experiences a radial outward force, pushing the fluid against the tube walls. This disrupts laminar flow, turning it turbulent, which increases mixing and the heat transfer coefficient, improving thermal exchange. The rotational speed of the insert can be adjusted to optimize performance for different operational conditions, such as varying flow rates and fluid properties. Engineers can fine-tune these parameters to balance heat transfer efficiency and energy consumption, maximizing overall system performance.

**Advantages of Spinning Inserts**

i. **Enhanced Heat Transfer**

Spinning inserts promote turbulence, increasing the heat transfer coefficient and also improving thermal exchange efficiency. This results in higher energy capture, reduced system size, and cost savings in solar thermal applications.

ii. **Reduced Fouling**

Continuous motion keeps particles suspended in the fluid, reducing deposits on heat transfer surfaces. This minimizes fouling, lowers maintenance needs, and improves long-term performance.

iii. **Adaptability**

Spinning inserts are customizable for varying flow rates, fluid properties, and thermal loads. Their flexibility makes them suitable for diverse conditions, ensuring consistent efficiency.

iv. **Improved Temperature Uniformity**

Enhanced mixing leads to more even temperature distribution, preventing hotspots and extending system lifespan. Consistent thermal performance also reduces the risk of thermal stress and failure.

**Disadvantages of Spinning Inserts**

i. **Mechanical Complexity**

The need for bearings, drive mechanisms, and power sources adds mechanical complexity. This increases the risk of failure, requires more maintenance, and may raise installation costs and lead times.

ii. **Cost**

Spinning inserts are more expensive to manufacture and integrate than fixed designs due to their intricate components. While their benefits may justify the cost, budget constraints could limit their use in some projects.

iii. **Energy Consumption**

Maintaining rotational motion requires energy, which must be factored into the system's overall efficiency. The system's performance may suffer if energy consumption outweighs the heat transfer gains.

iv. **Design Limitations**

Effectiveness can be reduced in low-flow or high-viscosity conditions. To achieve optimal performance, spinning insert designs must be optimized for specific fluid characteristics, requiring careful analysis.

Spinning inserts offer significant benefits in improving heat transfer, reducing fouling, and enhancing system adaptability and uniformity, making them especially useful in solar thermal systems. However, the mechanical complexity, high costs, energy consumption, and design limitations must be carefully managed. With thoughtful integration and maintenance, spinning inserts can significantly improve thermal system efficiency, advancing the transition to a more resilient and environmentally responsible energy paradigm.

## 5.2 Literature Review on Enhanced Designs

### Overview of Recent Studies on Steady Inserts

Parabolic Trough Collectors (PTCs) are among the most extensively employed technologies in solar thermal energy systems. They transform solar radiation into thermal energy for diverse implementations, encompassing energy production, thermal management, and industrial operations. These collectors consist of parabolic-shaped mirrors that concentrate solar irradiance incident upon a receiver conduit aligned along the focal axis of the parabolic reflector. A heat-carrying liquid circulates inside this tube, absorbing the focused solar radiation and transferring it to a thermal storage system or a power generation unit.

Despite their efficiency in harnessing solar energy, traditional PTC designs often encounter challenges related to heat transfer efficiency. The flow of HTF within the receiver tubes tends to be laminar, particularly at lower flow rates, which results in inadequate mixing and reduced heat exchange rates. This phenomenon can lead to significant thermal losses and overall decreased system performance. Researchers and engineers have been exploring various design modifications and enhancements to mitigate these issues. Among these, the integration of twisted tape and coil inserts has emerged as a promising approach to improve thermal performance in PTCs.

### Theoretical Background

### Heat Transfer Mechanisms in PTCs

The efficiency of heat transfer in PTCs is primarily influenced by the convective heat transfer coefficient, which is affected by the flow regime of the HTF. The heat transfer coefficient is relatively low in laminar flow due to the formation of a thermal boundary layer that inhibits heat transfer between the fluid and the tube surface. As flow velocity increases, a transition to turbulent flow occurs, characterized by enhanced mixing and a more homogeneous thermal gradient throughout the fluid medium, leading to higher heat transfer rates.

The Nusselt number (Nu) is a non-dimensional quantity that quantifies convective heat transfer compared to conductive heat conduction and is crucial for evaluating the performance of heat exchangers like PTCs. An increased Nusselt number signifies improved heat transfer efficiency. Traditional smooth tubes exhibit lower Nusselt numbers, necessitating design innovations to promote turbulence and enhance thermal performance.

**Twisted Tape Inserts**

Twisted tape inserts represent a highly efficacious method for augmenting thermal transfer efficiency in parabolic trough collectors (PTCs). These inserts consist of a strip of tape twisted into a spiral shape and inserted into the receiver tube. The design induces secondary flow patterns that disrupt laminar flow, promoting turbulence and enhancing the convective heat transfer coefficient. When a twisted tape is introduced into the tube, it creates a helical flow pattern that improves mixing. This helical flow increases the effective thermal exchange surface area and minimizes the thickness of the thermal flux boundary, leading to more effective heat exchange. The secondary flows generated by the tape also help to distribute temperature evenly, reducing hotspots that could lead to thermal stresses in the tube material. Research has consistently shown that twisted tape inserts can significantly improve thermal performance in PTCs. For instance, Yılmaz [2] reported an impressive 183% increase in heat transfer performance when using wire coil inserts compared to conventional collector designs. This improvement was attributed to the higher Nusselt numbers observed in the presence of the inserts, which facilitate enhanced thermal coupling between the heat transfer fluid (HTF) and the absorber surface.

Eiamsa-ard et al. [3] further validated these findings by demonstrating that a twin-twist insert resulted in minimal discrepancies between experimental and theoretical values for friction factors and Nusselt numbers, with differences of only $\pm 6$ and $\pm 10\%$, respectively. These results underscore the practical applicability of twisted tape inserts in real-world systems.

**Coil Inserts**

Coil inserts represent another innovative approach to enhancing thermal performance in PTCs. These inserts consist of coiled wire or helical structures placed within the receiver tube, creating a continuous spiral pathway for the HTF. Like twisted tape, coil inserts promote turbulence, enhancing heat transfer rates. Coil inserts function by forcing the HTF to follow a helical path, which increases the flow velocity at the tube wall and generates secondary flow patterns. This mechanism enhances mixing within the fluid, reducing the magnitude of the thermal diffusion layer and improving heat exchange with the tube surface. Coil inserts can also improve the effective heat dissipation surface, further boosting thermal performance. Research into coil inserts has demonstrated their effectiveness in improving thermal efficiency in PTCs. For instance, Gunes et al. [4] conducted experiments using coiled wire inserts with equilateral triangular cross-sectional areas, achieving an overall efficiency increase of 36.5% under optimal conditions. This study showed that varying parameters, such as the wire thickness and pitch ratio, significantly impacted heat transfer characteristics, emphasizing the adaptability of coil inserts to different operational conditions. Table 5.1 shows some more studies based on twisted tapes/coil inserts.

**Comparative Studies**

Many studies have compared the performance of PTCs in the presence and absence of twisted tape or coil inserts, providing a comprehensive understanding of their impact on thermal efficiency. Bhuiya et al. [38] empirically investigated using porous twisted inserts for air heating in PTCs. The experimental results revealed that the friction factor, thermal performance coefficient, and Nusselt number for tubes incorporating perforated twisted tape inserts exhibited enhancements ranging 28–59, 110–360, and 110–340%, respectively, relative to conventional tube collectors. Such dramatic improvements underscore the efficacy of these enhancements in optimizing PTC performance. Similarly, Jaramillo et al. [20] utilized computational simulations to investigate the impact of twisted tape on PTCs using thermal oil as the HTF. The outcomes showed significant enhancements in the Nusselt number, friction factor, and thermal efficiency, which were evaluated relative to a baseline empty collector tube configuration, further validating the benefits of using inserts.

**Challenges and Considerations**

While the integration of helical tape and coiled insert enhancements offers substantial improvements in thermal performance, several challenges and considerations must be addressed.

**Increased Pressure Drop**

One of the primary challenges associated with the integration of inserts is the escalation in pressure loss across the collector tube. Introducing twisted tape or coil structures disrupts the fluid flow, leading to higher friction losses. This phenomenon can necessitate a more robust pumping system to maintain adequate flow rates, potentially increasing operational costs. Research by Ghadirijafarbeigloo et al. [32] indicated that the implementation of inserts often leads to a substantial increase in pressure drops, particularly at higher Reynolds numbers.

**Fouling and Maintenance**

Another concern related to fixed geometries such as twisted tape and coil inserts is the potential for fouling. The increased surface area and complex flow paths can lead to the accumulation of deposits from the HTF, adversely affecting heat transfer performance over time. Regular maintenance may be required to clean these surfaces and ensure optimal performance, adding to operational complexities and costs.

**Table 5.1** Studies based on twisted tapes

| Authors | Geometry used | Principle finding |
|---|---|---|
| Prasad et al. [5] | 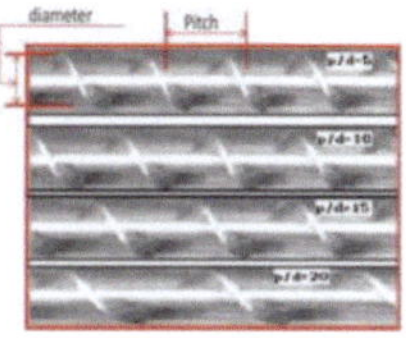 | The Nusselt number (Nu) for the entire pipe exhibits a 32.91% enhancement, with Nu increasing by a factor of 1.38 for a nanofluid at 0.03% concentration, utilizing helical inserts with a twist ratio (TR) of 5, respectively |
| Prasad and Gupta [6] | 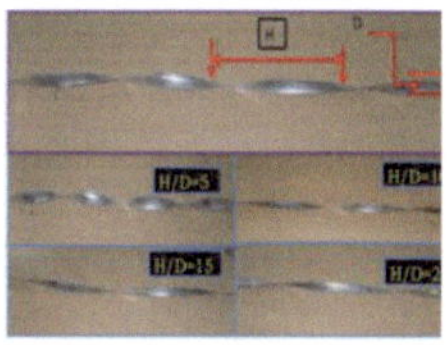 | The Nusselt number (Nu) for the entire pipe with a 0.03% nanofluid concentration demonstrates a 31.28% improvement over water ($H_2O$), while the friction factor ($f$) is augmented by a factor of 1.23 relative to $H_2O$, within helical inserts at a twist ratio (TR) of 5, respectively |
| Sekhar et al. [7] | 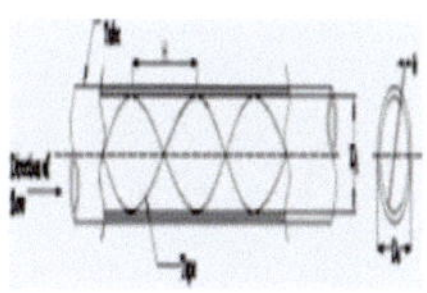 | The enhancement in convective heat transfer coefficient within the smooth tube, facilitated by the utilization of nanofluid, is optimized at 8–12% relative to water ($H_2O$) in a smooth tube configuration. A nanofluid with a 0.5% particle concentration yields the highest friction factor ($f$) in comparison to $H_2O$ |

(continued)

**Table 5.1** (continued)

| Authors | Geometry used | Principle finding |
|---|---|---|
| Azmi et al. [8] | 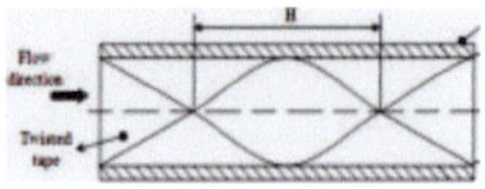 | At a 3% volume fraction and a twist ratio (TR) of 5, the $SiO_2/H_2O$ and $TiO_2/H_2O$ nanofluids exhibit superior performance, surpassing water by 27.9 and 11.4%, respectively |
| Eiamsa-ard et al. [9] | 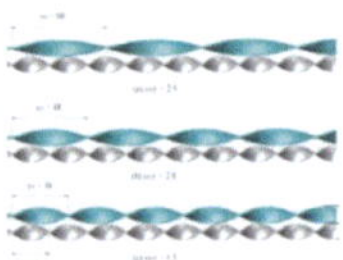 | The simultaneous application of overlapped double twisted tape (TT) with a twist ratio (TR) of 1.5 and a volume fraction of 0.21% resulted in a 9.9–11.2% enhancement in convective heat transfer coefficient (h) and a thermal efficiency increase of up to 4.5%, respectively, compared to single overlapped double TT |
| Aliabadi and Eskandari [10] | 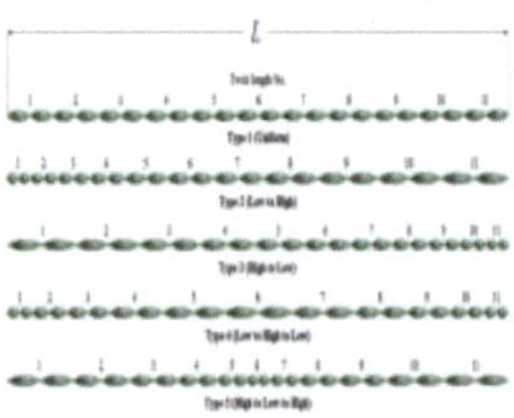 | This approach maximizes the overall performance enhancement ratio by up to 45%, while the simultaneous utilization of $Cu–H_2O$ nanofluid with a 0.3 wt% concentration, combined with twisted tape (TT) inserts spanning the minimum to maximum twist length, elevates the overall improvement ratio by as much as 87% |

(continued)

**Table 5.1** (continued)

| Authors | Geometry used | Principle finding |
|---|---|---|
| Maddah et al. [11] | 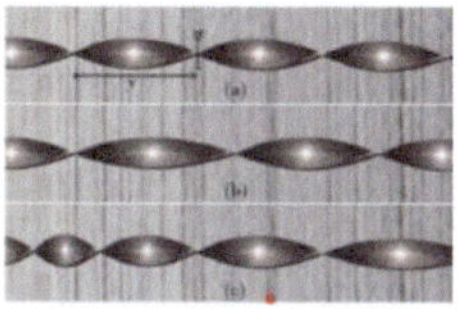 | The application of a minimum geometric progression twist in conjunction with nanofluids results in a simultaneous enhancement of convective heat transfer coefficient ($h$) and friction factor ($f$) by 12–52 and 5–28%, respectively, when compared to conventional tubes incorporating standard twisted tape (TT) and nanofluids |
| Naik et al. [12] |  | The convective heat transfer coefficient (h) was observed to increase by up to 27.95% in the smooth tube with a 0.5% CuO nanofluid concentration. Furthermore, when a twisted tape (TT) insert with a twist ratio (TR) of 5 was employed, $h_c$ was further enhanced by 76.06% relative to the base fluid at the selected Reynolds number (Re) |

(continued)

**Table 5.1** (continued)

| Authors | Geometry used | Principle finding |
|---|---|---|
| Naik et al. [13] | 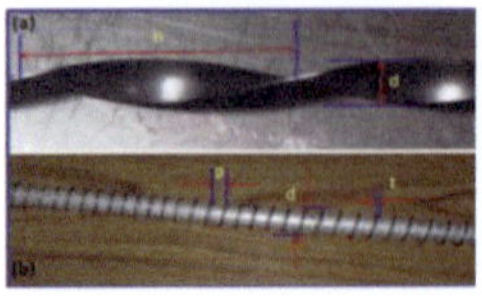 | The Nusselt number (Nu) for a 0.3% nanoparticle concentration in a tube without inserts increased by 17.62%, while the same concentration in a tube with a secondary twisted tape (TT) insert yielded a 31.88% enhancement at a Reynolds number (Re) of 2000, relative to water ($H_2O$) |
| Wongcharee and Eiamsa-ard [14] | 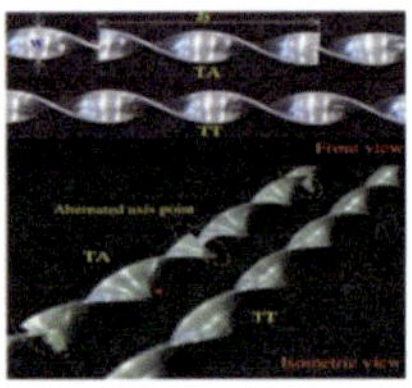 | The simultaneous application of twisted tape (TT) with an alternate axis configuration and nanofluids resulted in a Nusselt number (Nu) enhancement of up to 13.8 times that of the smooth tube. Furthermore, the peak thermal performance efficiency ($\eta$) of 5.53 was achieved with the significant use of $CuO/H_2O$ nanofluid at a 0.7% volume fraction |

(continued)

**Table 5.1** (continued)

| Authors | Geometry used | Principle finding |
|---|---|---|
| Wongcharee and Eiamsa-ard [15] | 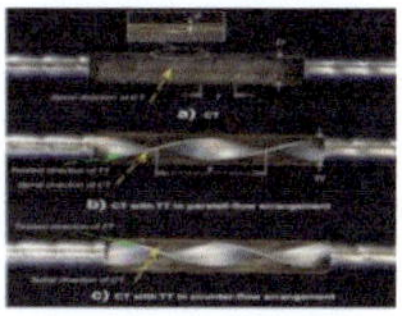 | The maximum thermal performance efficiency ($\eta$) of 1.57 was attained with the utilization of $CuO/H_2O$ nanofluid at a 0.7% volume fraction, within a ribbed tube configuration augmented by twisted tape (TT) with a twist ratio (TR) of 2.7, at a Reynolds number (Re) of 6200 |
| Zheng et al. [16] | 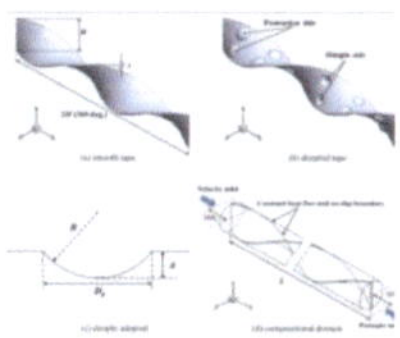 | A substantial reduction in the conventional entropy production rate is accompanied by a marginal increase in the standard entropy production rate, resulting in an overall increase of 29.10% in the total entropy production rate compared to the baseline tube configuration |

(continued)

**Table 5.1** (continued)

| Authors | Geometry used | Principle finding |
|---|---|---|
| Suresh et al. [17] | 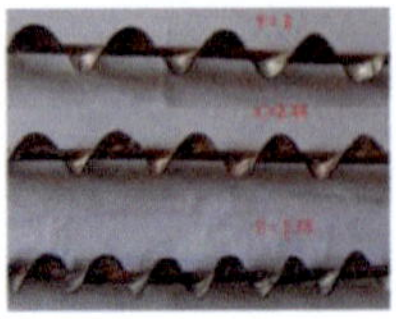 | The enhancement in Nusselt number (Nu) for water ($H_2O$) with twist ratios (TR) of 1.78, 2.44, and 3 were 156.3, 122.2, and 89.3%, respectively, compared to the smooth tube. In contrast, the improvement in Nu for $H_2O$ with pitch-to-diameter ratios (y/D) of 1.78, 2.44, and 3 reached 166.9, 128.7, and 89.3%, respectively, when utilizing $Al_2O_3/H_2O$ nanofluid |
| Mehta et al. [18] | 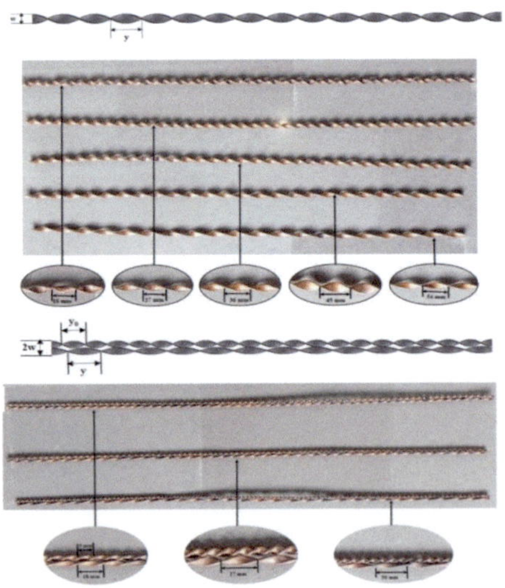 | Inserting single and dual twisted tape inserts in parabolic trough solar collectors enhances performance, with maximum efficiency at low flow rates. The dual twisted tape insert (pitch ratio 2) boosts thermal efficiency by 14%, Nusselt number by 372%, and friction factor by 1600%. Exergy efficiency increases by 20.8%, resulting in an annual $CO_2$ reduction of 1.53 tons and an enviro-economic benefit of 22.19 USD per year |

(continued)

**Table 5.1** (continued)

| Authors | Geometry used | Principle finding |
|---|---|---|
| Alnaqi et al. [19] | 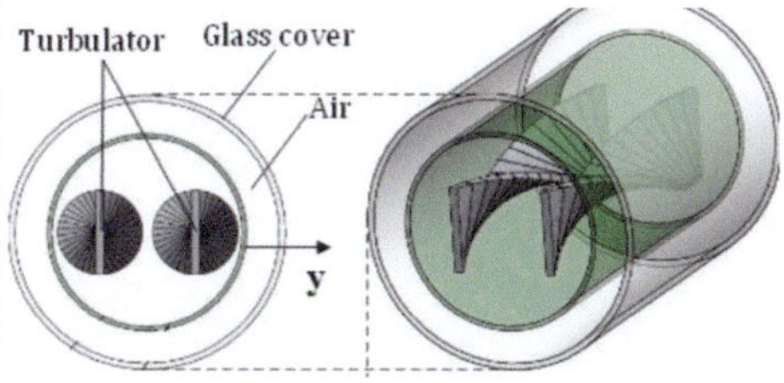 | MgO-MWCNT (80:20%) in thermal oil, with 2.0 wt% nanofluid, enhances PTSC performance. LCRW is optimal for Re < 30,873, while LWRC works best for Re > 30,873 with 15 mm turbulators |
| Jaramillo et al. [20] | 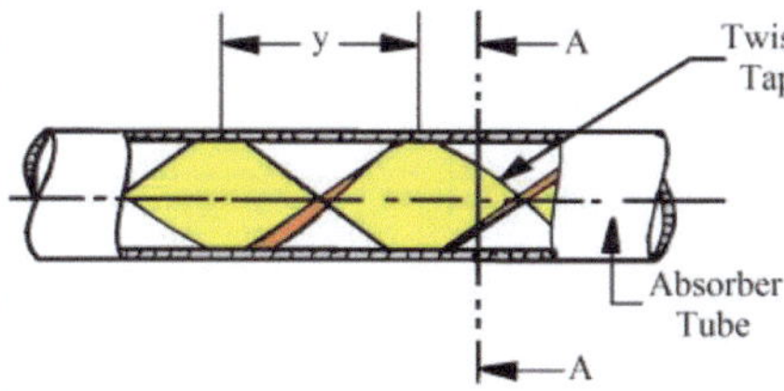 | This work develops a thermodynamic model to analyze the performance of parabolic trough collectors with twisted tape inserts. The model identifies conditions where twisted tape inserts enhance performance, particularly for devices with low twisted ratios operating at low Reynolds numbers. Experimental data support the results |
| Isravel et al. [21] | 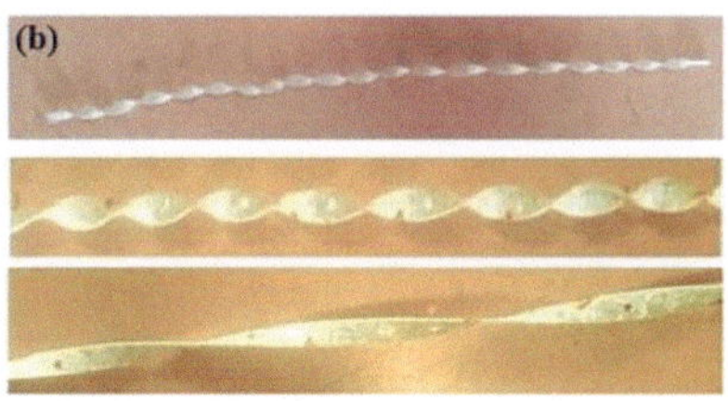 | Twisted tapes enhance heat transfer in solar water heating systems. Rings-attached twisted tapes improve heat transfer but increase pressure drop. Modifying these tapes by increasing twist ratio and adding a center cut further improves performance |

(continued)

**Table 5.1** (continued)

| Authors | Geometry used | Principle finding |
|---|---|---|
| Thapa et al. [22] | 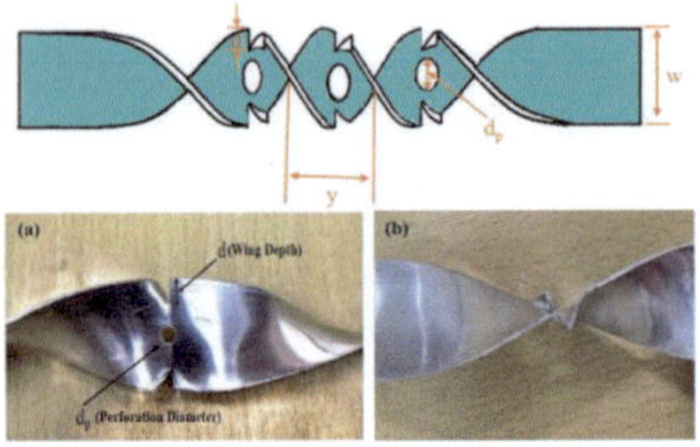  | This study used MATLAB R2018a to analyze thermal and exergetic efficiencies of parabolic trough receivers with twisted tape inserts. The optimal performance occurs with twist ratio 3, perforation ratio 0.1, and winglet ratio 0.05, yielding 84.27% energy and 4.28% exergy efficiencies |
| Abidi et al. [23] |   | This study examines parabolic trough solar collectors using twisted tape with circular holes and water–copper oxide nanofluid. The highest thermal performance occurs at a 4% nanoparticle volume fraction, enhancing energy and exergy efficiency at a Reynolds number of 10,000 |

(continued)

**Table 5.1** (continued)

| Authors | Geometry used | Principle finding |
|---|---|---|
| Razzaghi et al. [24] | 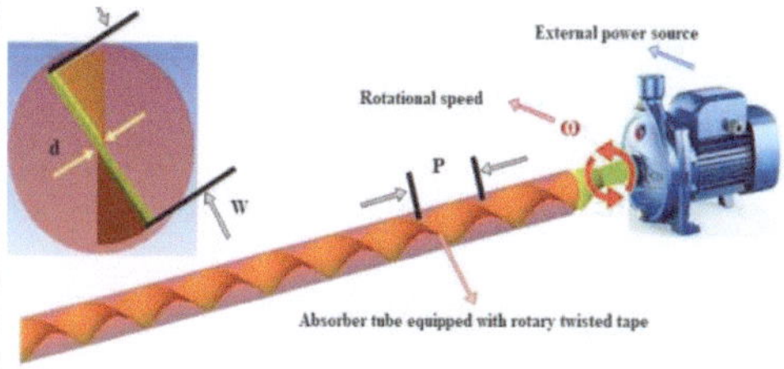  | This study numerically analyzes the LS-3 parabolic trough collector with a rotary twisted tape and line dipole magnetic field. Results show a 97% increase in heat transfer with a magnetic field strength of 100 G, and a 305% increase at a twisted ratio of 2.5 and 200 rpm. The twisted tape also reduces thermal entropy generation |
| Chakraborty et al. [25] | 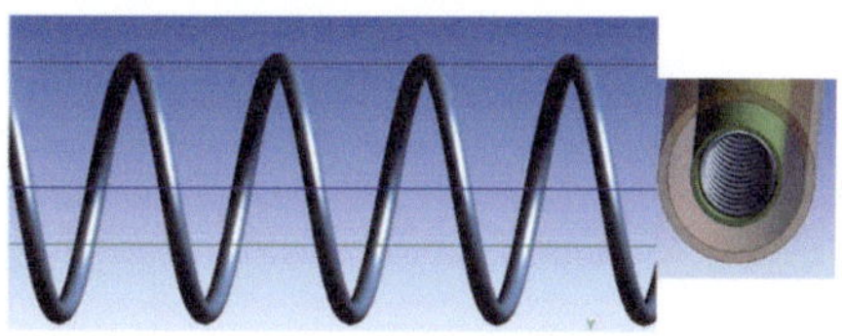 | Nanofluid (2% $Al_2O_3$) enhances thermal and exergy efficiencies by up to 12.32 and 8.72%, respectively, with 145-turn coil inserts at 0.033 kg/s flow rate. Higher coil turns improve thermal output but increase pump work |
| Jafar et al. [26] |  | The helical coiled tube absorber with 42 turns improves heat transfer in the PTC. At a 0.01 kg/s flow rate, the maximum heat gain is 800 W, with a water temperature of 64 °C and efficiency of 56% |

(continued)

**Table 5.1** (continued)

| Authors | Geometry used | Principle finding |
|---|---|---|
| Naved et al. [27] |  | The study compares "U"-tube and "helical coil" receivers in parabolic trough collectors. The helical coil receiver achieved a maximum output temperature of 90 °C and efficiency of 90.92%, outperforming the "U"-tube receiver in performance |
| Yılmaz et al. [28] | 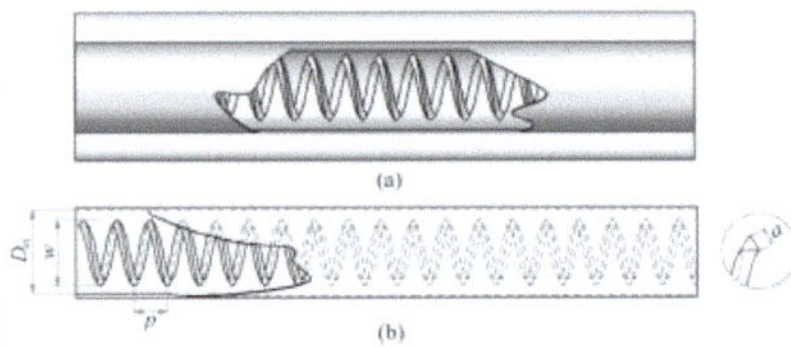 | Wire coil inserts in parabolic trough solar collectors improve thermal performance by enhancing fluid mixing and reducing absorber tube temperatures. Heat transfer performance increases by up to 183%, and thermal efficiency improves by 0.4–1.4% at low flow rates |
| Varun et al. [29] | 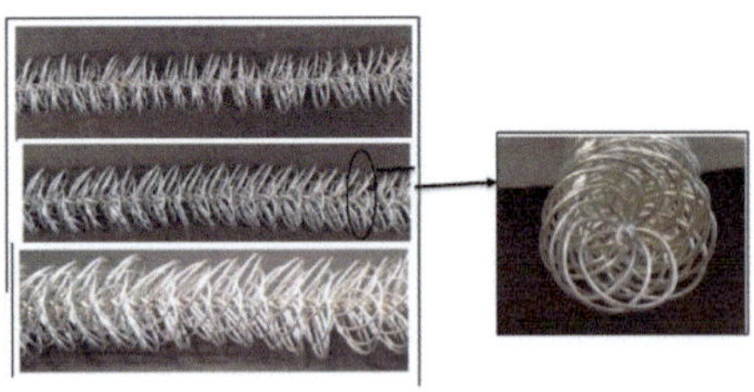 | The medium-dense wire matrix achieves maximum thermo-hydraulic efficiency, while twisted tape (twist ratio 3.37) performs better per the performance evaluation criterion. Thermo-hydraulic efficiency is preferred due to its realistic consideration of pumping power and operating conditions |

(continued)

**Table 5.1** (continued)

| Authors | Geometry used | Principle finding |
|---|---|---|
| Saini et al. [30] | 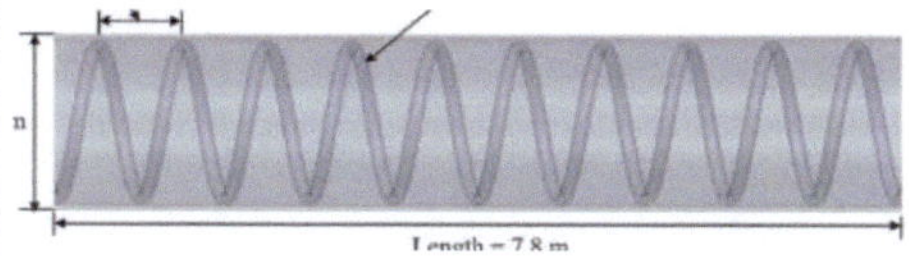 | The study shows that a helical wire coil flow insert improves heat transfer in a parabolic trough collector. At 50 L/min flow rate and 303 K inlet temperature, it enhances overall exergy and thermal efficiencies by 2.07, 2.1, and 2.2%, respectively |
| Jafaryar et al. [31] | 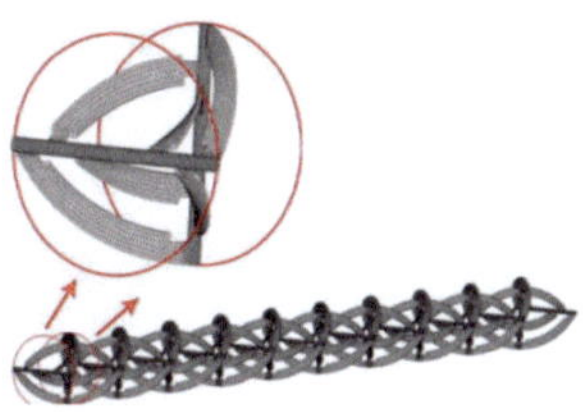 | The addition of 1% $Al_2O_3$ and MWCNT in oil, combined with multi-helical tapes, enhances Parabolic Trough Collector (PTC) performance, increasing the Nusselt number (Nu) by 135.54% and reducing the friction factor ($f$) by 13.08% |
| Ghadirijafarbeigloo et al. [32] | 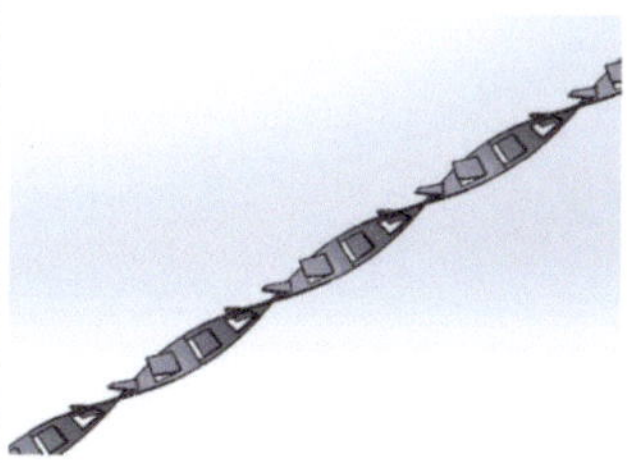 | The study investigates the use of perforated louvered twisted tape (LTT) in parabolic trough collector tubes. Results show significant improvements in heat transfer coefficient and pressure drop compared to plain twisted tape and plain tubes, enhancing convection heat transfer efficiency |

(continued)

**Table 5.1** (continued)

| Authors | Geometry used | Principle finding |
|---|---|---|
| Borunda et al. [33] | 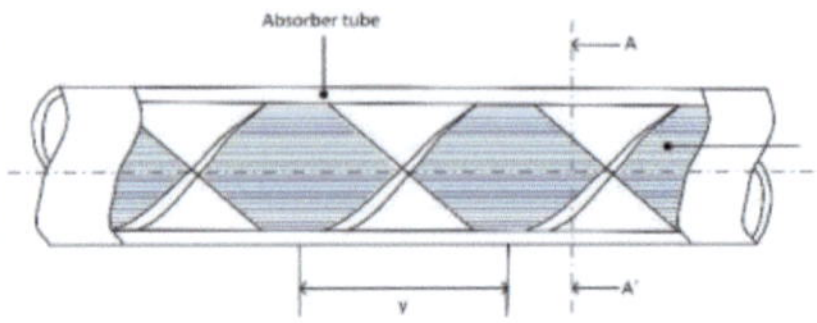 | The study analyzes the performance of a parabolic trough collector with a twisted tape insert, using thermodynamic laws to optimize operation. Results show that small twist ratios and low mass flow improve efficiency, while large twist ratios enhance performance but reduce efficiency |
| Stanek et al. [34] | 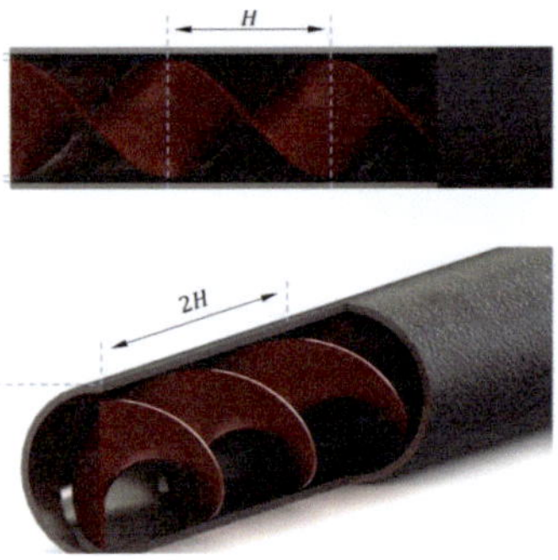 | The study analyzes heat enhancement in low-concentration parabolic trough collectors using twisted tapes with ratios 1, 2, and 4. Results show that the twisted ratio 1 is optimal for most conditions, while the ratio 2 is better at higher temperatures, improving efficiency and power gain by 0.27% |
| Diwan and Soni [35] |  | The introduction of wire-coil inserts in parabolic trough concentrators enhances turbulence, increasing the Nusselt number by 104–330% and significantly improving convective heat transfer rates |

(continued)

**Table 5.1** (continued)

| Authors | Geometry used | Principle finding |
|---|---|---|
| Afsharpanah et al. [36] | 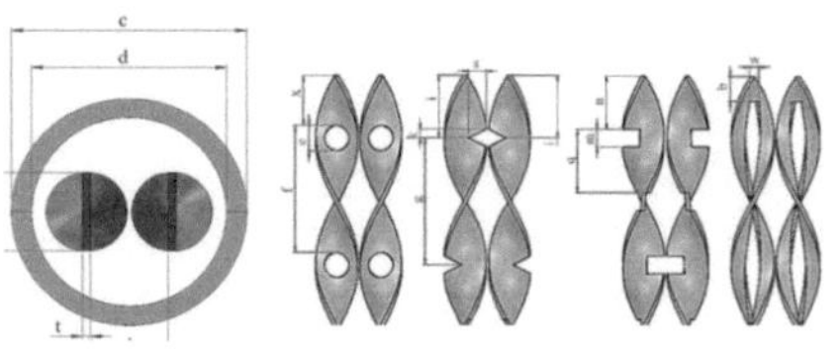 | Dual V-cut twisted tapes enhance heat transfer by 19.58 and 17.44% at Reynolds numbers 10,000 and 20,000. All twisted tapes outperform plain tubes, with dual square-cut tapes showing the highest performance improvement |
| Too and Benito [37] |  | Tubes with deeper inner wall protrusions enhance heat transfer and reduce drop in pressure in pressurized air tubular absorbers for solar applications, enabling practical lengths at gas temperatures exceeding 800 °C, with $CO_2$ and He further minimizing pressure drop |

## Design Complexity

While customization is a significant advantage of incorporating inserts, the design and integration of these geometries can complicate the system. Engineers must carefully balance the benefits of enhanced heat transfer with the potential drawbacks, such as increased pressure losses and fouling risks. Moreover, the specific characteristics of the HTF, flow rates, and operating conditions must be considered during the design phase to achieve the best performance outcomes.

## Future Directions

As the field of solar thermal technology continues to evolve, several future directions are anticipated:

1. **Hybrid Systems**: Combining PTCs with other renewable energy technologies, such as photovoltaics or energy storage systems, could enhance efficiency and reliability.

2. **Smart Materials**: Integrating smart materials that can adapt their properties based on environmental conditions has the potential to optimize heat transfer and energy capture.
3. **Advanced Computational Fluid Dynamics (CFD)**: Enhanced simulation techniques using CFD can improve estimations of fluid behavior and heat transfer, enabling the design of more efficient PTC systems.
4. **Sustainability Focus**: Research aimed at developing sustainable materials for inserts and tubes and methods to minimize environmental impact during production and disposal will be crucial as the industry moves toward greener solutions.

The integration of twisted tape and coil inserts into Parabolic Trough Collectors significantly enhances thermal performance. These innovations have demonstrated substantial efficiency improvements by promoting turbulence and increasing surface area for heat transfer. While challenges such as increased pressure drops and fouling remain, ongoing research and development facilitate the transition to enhanced efficiency and reliable solar thermal units. As the demand for renewable energy solutions continues to increase, the optimization of PTCs through innovative designs and advanced materials has become important.

**Enhancements in Thermal Performance of Parabolic Trough Collectors with Differently Structured Inserts**

Recent advancements in parabolic trough collectors (PTCs) have significantly enhanced their thermal performance through innovative insert technologies and design modifications. A notable study by Reddy et al. [39] revealed that incorporating porous disk inserts in the receiver tube improved the thermal efficiency of PTCs compared to traditional configurations, demonstrating the potential of porous materials to enhance fluid dynamics and energy capture. Similarly, Biswakarma et al. [40] explored V-grooved collector tubes using $Al_2O_3$-based water nanofluids, achieving a 13.8% increase in heat transfer coefficients, underscoring the effectiveness of nanofluids in optimizing temperature control and overall efficiency. Furthermore, Bellos et al. [41] tested the impact of converting the tube shape to a converging–diverging design, which yielded a mean efficiency improvement of 4.55% owing to inconsistent thermal flux distribution, emphasizing the importance of geometric modifications. In a parallel investigation, Kalidasan et al. [42] demonstrated that fitting hinged blade inserts could elevate thermal efficiency to 69.33%, a notable increase from the 60.82% efficiency of smooth tubes, highlighting the benefits of dynamic modifications in optimizing fluid interaction and thermal transfer. Ebrahim and Akbar [43] analyzed the incorporation of porous rings, whose numerical study indicated significant improvements in thermal performance, with the highest Nusselt number and Reynolds number recorded, reinforcing the role of porous structures in optimizing thermal dynamics. Tajik et al. [44] focused on installing porous media, finding that increasing flow rates corresponded to rising thermal efficiency, with copper foam integration reducing overall heat loss by 45%. Zheng et al. [45] reported impressive enhancements in heat transfer rates (1.8–3.6 times) through double inclined ribs, demonstrating the critical influence of design parameters on performance. Zhen et al. [46] explored surface dimples, achieving up to a 21% increase in Nusselt numbers, indicating the balance between heat transfer improvement and drop in pressure. Additionally, Bellos et al. [47] investigated several cylindrical inserts, noting that while thermal efficiency increased, pump work demands also rose, necessitating a careful design approach to optimize energy use. The introduction of star-shaped flow patterns by Saini et al. [30] showed a remarkable 60% rise in heat transfer coefficients, although it also led to increased pressure drops. Moreover, Bellos and Tzivanidis [48] utilized $TiO_2$-based nanofluids, achieving collector performance efficiency increases of 21.1%, while acknowledging the challenges of higher viscosity.

**Overview of Recent Studies on Spinning Inserts and Tubes**

The ongoing research into boosting the effectiveness of parabolic solar trough collectors (PTCs) highlights revolutionary strategies, particularly those implementing spinning inserts and tubes. These advancements address the critical issues of high surface temperatures and thermal stresses that often compromise the efficiency and longevity of traditional designs. By introducing rotational motion into the fluid dynamics of the absorber tubes, these solutions notably elevate heat transfer rates and thermal effectiveness.

One of the most notable approaches is integrating spinning star inserts and helical inserts within the absorber tubes. Implementing these rotational mechanisms induces a more turbulent flow of the heat transfer fluid, significantly augmenting the global heat transfer coefficient and facilitating a more homogeneous thermal distribution across the absorber surface. Studies have shown that, at a rotational speed of 15 rad/s, the incorporation of these spinning inserts can result in a substantial enhancement in peak thermal efficiency of 38.03%, along with a Nusselt number increase of 50.90%. This performance boost is particularly evident when using hybrid nanofluids composed of aluminum oxide ($Al_2O_3$) and copper oxide (CuO), with the optimal mixture being 75% CuO and 25% $Al_2O_3$ [49].

In addition to spinning inserts, implementing rotating tubes has proven effective in further enhancing thermal performance. For instance, a rotating helical shaft within the absorber tube has significantly improved thermal efficiency. Experimental studies indicate that at various rotational speeds, such as 21 rpm, these spinning tubes can achieve thermal efficiency enhancements of up to 46.47% compared to traditional smooth tubes. The dynamics introduced by rotating tubes help reduce the temperature gradient across the fluid-absorber boundary, leading to notable reductions in thermal stress—up to 64 K in some cases [50].

The elliptical configurations in spinning tubes also play a critical role in optimizing thermal performance. One study found that an elliptical insert with a primary diameter of 28 mm, oriented longitudinally, achieved a heat transfer coefficient increase of 44.21% and a thermal efficiency boost of 29.21%. This demonstrates that not only the rotation itself but also the geometric design of the inserts contributes significantly to enhancing the heat transfer characteristics [51].

The utilization of hybrid nanofluids further complements the effectiveness of spinning flower designs. Researchers have enhanced thermal conductivity and stability by blending nanoparticles like $Al_2O_3$ and CuO in water, allowing for improved heat absorption and distribution within the rotating systems. The flow rates employed in the simulations—extending from 0.016 kg/s up to 0.033 kg/s—show a positive correlation with thermal performance. However, it is essential to noting that while increased flow rates boost efficiency, the percentage gains diminish at higher flow rates, highlighting the intricate balance between flow dynamics and thermal efficiency.

Moreover, the economic implications of incorporating spinning inserts and rotating tubes are substantial. Not only do these innovations improve thermal performance, but they also contribute to lower annual costs of proper heat production. This makes the systems technologically advanced and economically viable for broader adoption in solar energy applications.

In summary, implementing spinning inserts and rotating tubes in parabolic solar trough collectors represents a significant advancement in solar thermal technology. By addressing the challenges of high-temperature operation and improving fluid dynamics through these mechanical innovations, researchers are laying the foundation for greater efficiency and sustainability of solar energy systems. Future research could focus on the long-term stability of hybrid nanofluids in rotating systems, scaling up studies for larger PTC configurations, and exploring alternative materials for inserts to enhance performance further while maintaining cost-effectiveness. These collective advancements contribute to a deeper understanding of PTC dynamics and promise a brighter, more sustainable energy future. Table 5.2 lists the studies based on different geometrical inserts.

**Fundamental Equations**

In the system involving fluid flow as the working medium within a rotating tube and tubes equipped with spinning inserts, the governing equations encompass mass conservation (continuity equation), momentum conservation (Navier–Stokes equations), and energy conservation (energy equation). Assuming steady, incompressible flow with constant thermo-physical properties and neglecting viscous dissipation, the governing equation is formulated as outlined in Table 5.3.

## *5.2.1 Comparative Analysis of Modified PTC Models*

The integration of flow-enhancing inserts in Parabolic Trough Collectors (PTCs) offers several significant advantages that improve the overall performance and efficiency of solar thermal systems. These inserts—such as twisted tapes, helical coils, wire coils, and ribbed inserts—work by disrupting laminar flow and promoting turbulence, which leads to better heat transfer and fluid mixing. This enhancement in heat transfer allows for more effective energy absorption and enables the system to operate at lower HTF flow rates, which reduces pumping energy and minimizes

**Table 5.2** Studies based on different geometrical inserts

| Authors | Insert type | Geometry | Method | Findings |
|---|---|---|---|---|
| Kumar and Reddy [52] | Porous disks |  | CFD | The top porous disk configuration enhanced heat transfer by A 64.3% enhancement in the Nusselt number, coupled with a pressure drop of 457 Pa, was observed relative to conventional tubular receivers |
| Chakraborty et al. [50] | Spinning star inserts |  | CFD | The top porous disk configuration enhanced heat transfer by A 64.3% enhancement in the Nusselt number, coupled with a pressure drop of 457 Pa, was observed relative to conventional tubular receivers with a hybrid nanofluid, while the increase in pump work demand remains negligible at 20.41% |

(continued)

**Table 5.2** (continued)

| Authors | Insert type | Geometry | Method | Findings |
|---|---|---|---|---|
| Kumar and Reddy [53] | Porous disks |  | CFD | The optimal porous disk receiver architecture significantly augments heat transfer rates by 221 W $m^{-1}$ (13.5%) for water and 575 W $m^{-1}$ (31.4%) for Therminol oil, with low pumping penalties compared to tubular receivers |
| Wang et al. [54] | Metal foam |  | CFD | Inserting metal foams in the receiver tube optimizes thermo-hydraulic performance, with Nu increasing 5–12 times and a 45% reduction in maximum circumferential temperature difference, significantly improving heat transfer in parabolic trough collectors |

(continued)

**Table 5.2** (continued)

| Authors | Insert type | Geometry | Method | Findings |
|---|---|---|---|---|
| Sadaghiyani and Mirzaee [55] | Closed-end eccentric cylindrical plug |  | CFD | Flow restriction devices (plugs) in the LS-2 parabolic trough collector optimize outlet temperature, showing that a non-dimensional displacement of +0.5 leads to gentler temperatures, independent of plug diameter and working fluid, while efficiency is evaluated with negligible convective losses |
| Chakraborty et al. [56] | Spherical ball inserts | 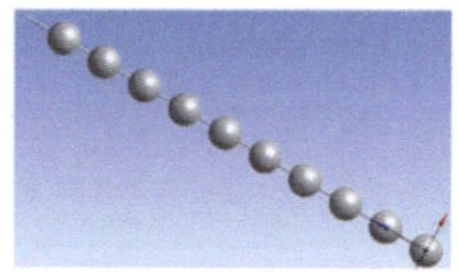 | CFD | Inserting 48 mm spherical balls into the receiver tube of a parabolic trough solar collector significantly enhances thermal performance, achieving a maximum local Nusselt number of 585.80 and improving thermal efficiency by 64.25% and exergy efficiency by 42.23% at 1000 $W/m^2$ solar energy flux and fluid flow rate of 0.36 kg/s compared to 18 mm balls |

(continued)

**Table 5.2** (continued)

| Authors | Insert type | Geometry | Method | Findings |
|---|---|---|---|---|
| Chakraborty et al. [57] | | 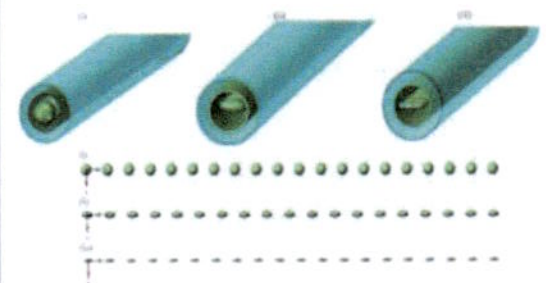 | CFD | The study found that elliptical inserts with a 12 mm minor diameter improved thermal efficiency, Nusselt number, and thermal performance index by 13.3, 10.02, and 16%, respectively, compared to spherical ball inserts in solar collectors |
| Kasperski and Nemś [58] | Multiple-fin array |  | Model | The introduced multiple-fin array technology in a solar air heater enhances thermal efficiency by as much as 14% compared to smooth pipes, while concurrently reducing the necessary air flux by a factor of 7–10, and maintains higher performance even under a 5 m/s speed limit, making it suitable for solar space heating and solar dryers |

(continued)

**Table 5.2** (continued)

| Authors | Insert type | Geometry | Method | Findings |
|---|---|---|---|---|
| Chakraborty [59] | Spinning flower inserts | 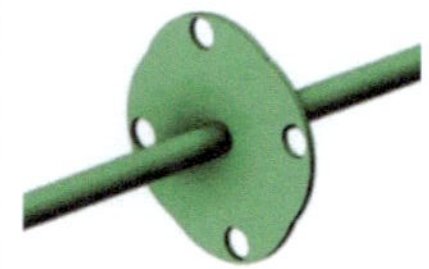 | CFD | The integration of spinning flower inserts within the LS-2 parabolic trough collector yields a 30.87% improvement in thermal efficiency and a 98.34% augmentation in the heat transfer coefficient at 15 rad/s, with only a 21.29% increase in pump work requirement, indicating significant thermal performance improvements with negligible impact on energy costs |
| Mwesigye et al. [60] | Perforated plates |  | CFD | The incorporation of perforated plate inserts within a parabolic trough solar collector results in a thermal performance enhancement ranging from 1.2 to 8% across designated Reynolds numbers, improves thermodynamic performance at flow rates below 0.01205 $m^3/s$, and reduces temperature gradients within the receiver |

(continued)

**Table 5.2** (continued)

| Authors | Insert type | Geometry | Method | Findings |
|---|---|---|---|---|
| Mwesigye et al. [61] | Perforated conics | 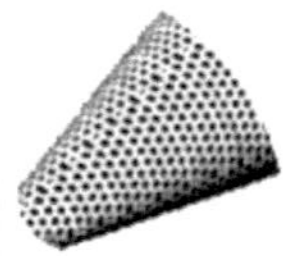 | CFD | Using perforated conical inserts in parabolic trough collectors enhances thermal efficiency by 3–8% under specific geometrical parameters, with performance gains surpassing increases in pumping power across certain Reynolds numbers |
| Panja et al. [62] | Porous strips | 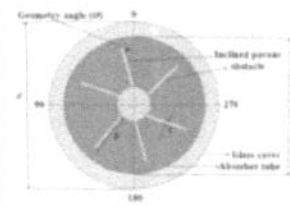 | CFD | Copper oxide (CuO) and aluminum oxide ($Al_2O_3$) nanoparticles (3–7%) in water enhance thermal efficiency in solar systems, achieving a peak of 62.9% at a 15° strip angle for optimal heat transfer |
| Esmaeili et al. [63] | Newly designed turbulator | 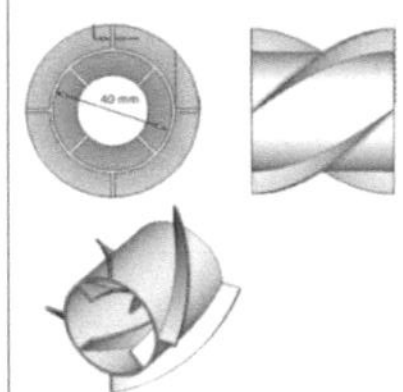 | CFD | The addition of 2% (CNT + $SiO_2$) hybrid nanoparticles in Syltherm 800 significantly enhances solar concentrator efficiency, boosting performance by 46.65% through improved heat transfer and thermal conductivity |

(continued)

**Table 5.2** (continued)

| Authors | Insert type | Geometry | Method | Findings |
|---|---|---|---|---|
| Reddy et al. [39] | Porous disks | 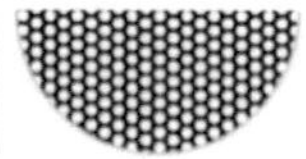 | Experimental | The porous disk-enhanced receiver in a 15 $m^2$ solar parabolic trough collector significantly improves performance, achieving efficiencies of 63.9–66.66%, stagnation temperatures up to 312 °C, and reduced heat losses, rendering it compatible with process heat applications |
| Bellos et al. [47] | Multiple cylinders | 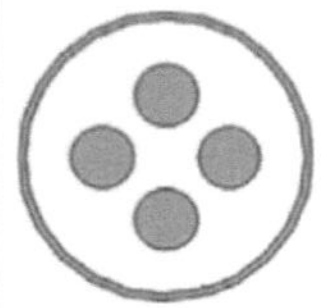 | CFD | The integration of multiple cylindrical longitudinal inserts within the LS-2 parabolic trough solar collector results in a 0.656% improvement in thermal efficiency, a 5.63% reduction in thermal losses, and a 26.88% increase in the heat transfer coefficient, with optimal performance realized at four inserts |

(continued)

**Table 5.2** (continued)

| Authors | Insert type | Geometry | Method | Findings |
|---|---|---|---|---|
| Bellos and Tzivanidis [48] | Closed-end star plug |  | CFD | The novel star-shaped flow insert for parabolic trough solar collectors enhances thermal efficiency by up to 1% with increased inlet temperatures, while larger insert dimensions improve performance despite a significant increase in pressure drop, maintaining low pumping work across all cases |
| Liu et al. [64] | Conical strip |  | CFD | The utilization of conical strip inserts in a parabolic trough receiver enhances heat transfer (Nusselt number up by 45–203%) and reduces heat loss by up to 82.1%, but increases pressure loss significantly, resulting in overall thermal–hydraulic performance of 0.70–1.33 and thermal efficiency improvements of 0.02–5.04% |

(continued)

**Table 5.2** (continued)

| Authors | Insert type | Geometry | Method | Findings |
|---|---|---|---|---|
| Ahmed and Natarajan [65] | Toroidal rings |  | CFD | Incorporating internal toroidal rings in the absorber tube of the solar parabolic trough collector optimizes thermal performance, with the optimal configuration achieving a 3.74% efficiency increase and a Nusselt number 2.33 times larger than a smooth absorber tube at 600 K, while an energy-efficient case shows a 1.88% increase and a Nusselt number 1.49 times greater |
| Kalidasan et al. [42] | | 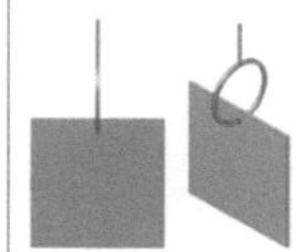 | | The modified hinged blade absorber tubes in a 1.5-m parabolic trough collector achieved an average efficiency of 69.33%, surpassing the conventional design's 60.82%, highlighting improved energy transfer and potential for domestic and industrial applications |

(continued)

**Table 5.2** (continued)

| Authors | Insert type | Geometry | Method | Findings |
|---|---|---|---|---|
| Chakraborty et al. [50] | Spinning star inserts |  | CFD | The addition of 1% (CuO 75%–$Al_2O_3$ 25%, CuO 50%–$Al_2O_3$ 50%, and CuO 25%–$Al_2O_3$ 75%) hybrid nanoparticles in water, combined with spinning star inserts, enhances thermal efficiency by 38.03% and increases the Nusselt number by 50.90% |
| Esmaeili et al. [63] | Conical turbulator |  | CFD | The use of 3% alumina and 1% copper in water ($Al_2O_3$–Cu/water) with conical helical gear rings enhances heat transfer by 35.7% and reduces entropy by 32.8%, improving the overall thermal efficiency |
| Al-Rashed et al. [66] | Finned rod | 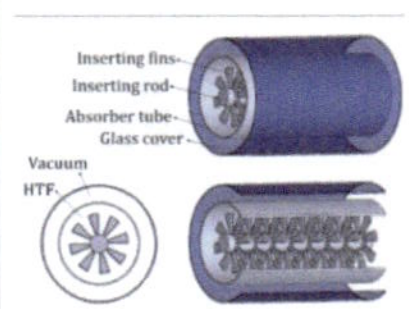  | CFD | The study improves PTSC performance using a hybrid nanofluid and finned rod turbulator. Results show benefits for Reynolds numbers below 12,000, with a 14% material cost saving, enhancing thermal efficiency and performance metrics |

(continued)

**Table 5.2** (continued)

| Authors | Insert type | Geometry | Method | Findings |
|---|---|---|---|---|
| Chakraborty and Nath [67] | Revolving fins inserts |  | CFD | The study shows that installing rotating star inserts in parabolic solar trough collectors enhances performance. The maximum improvements in thermal efficiency (38.03%) and Nusselt number (50.90%) occurred with star inserts at 15 rad/s and a 0.033 kg/s flow rate. However, pump work demand increased by 20.41% under the same conditions |
| Wang et al. [68] | Eccentric structure insert | 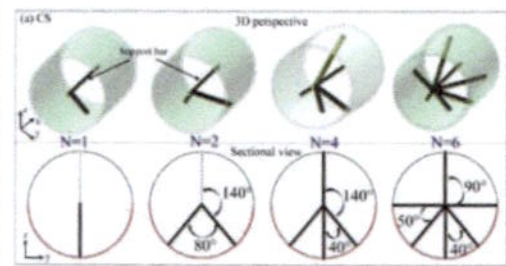 | CFD | The upper eccentric structure with vortex generators enhances heat transfer in parabolic trough solar collectors, increasing the Nusselt number by 1.16–2.34 times and friction factor by 1.47–5.38 times, with optimal performance at 6 vortex generators |

(continued)

**Table 5.2** (continued)

| Authors | Insert type | Geometry | Method | Findings |
|---|---|---|---|---|
| Saini et al. [69] | X-grid static mixer flow inserts | 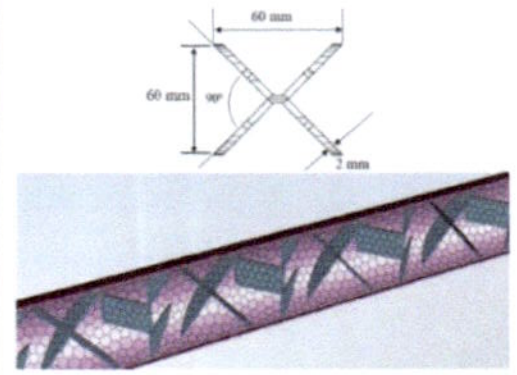 | CFD | This study investigates the use of X-grid static mixer flow inserts in Parabolic Trough Collectors (PTC) to improve thermal and exergy efficiency. Results show enhanced thermal performance, especially at higher inlet temperatures, though pressure drop increases |
| Wang et al. [70] | Inserted absorbers | 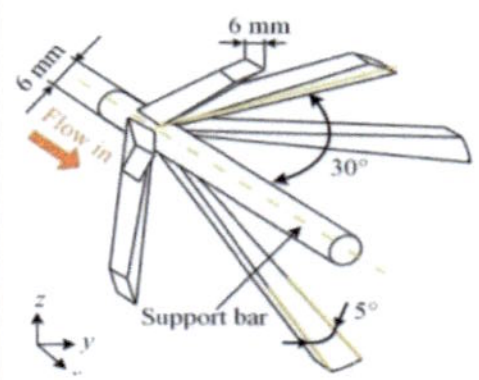  | CFD | Vortex generator inserts improve heat transfer in parabolic trough solar collectors, enhancing rates by 1.5–2.12 times. The generated high-intensity mixed vortices enhance heat transfer efficiency, particularly in regions with high heat flux density |

(continued)

**Table 5.2** (continued)

| Authors | Insert type | Geometry | Method | Findings |
|---|---|---|---|---|
| Darbari et al. [71] | Porous disk inserts | 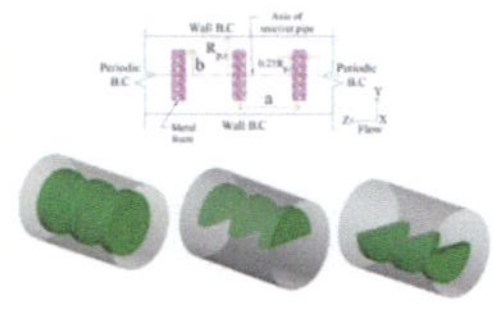 | CFD | Full-disk configurations outperform half-disk designs in thermal efficiency. Increasing disk diameter and reducing spacing improve performance, while higher inlet temperatures and lower flow rates decrease efficiency. Nusselt number and friction factor variations assess heat transfer and pumping power trade-offs |

(continued)

**Table 5.2** (continued)

| Authors | Insert type | Geometry | Method | Findings |
|---|---|---|---|---|
| Mwesigye et al. [60] | Centrally placed perforated plate inserts | 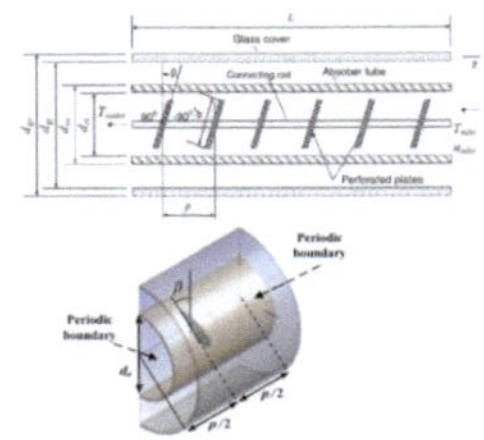 | CFD | Perforated plate inserts improve thermal efficiency by 1.2–8% within a specific Reynolds number range. Thermodynamic performance enhances flow rates below 0.01205 $m^3/s$, and temperature gradients reduce. Nusselt number and friction factor correlations were derived |
| Valizade et al. [72] | Copper metal foam inserts |  | Experimental | Copper metal foam enhances thermal performance in a direct absorption parabolic trough solar collector, improving thermal efficiency by 171.2% with full foam and 119.6% with semi-foam. Increased flow rates and lower inlet temperatures further enhance efficiency |

**Table 5.3** Fundamental relations

| Sl. No | | Equations [73] |
|---|---|---|
| 1 | Derivation of continuity | $\nabla \cdot \left(\vec{V}\right) = 0$ (1)<br>This equation encapsulates the principle of mass conservation within the fluid flow, asserting that the divergence of the velocity vector ($\vec{V}$) is zero, thereby confirming that mass remains conserved and is neither generated nor annihilated within the flow domain |
| 2 | Derivation of momentum | $\rho\left(\frac{\partial u}{\partial t} + \left(\vec{V} \cdot \nabla\right)\vec{V}\right) = -\nabla p + \rho\vec{g} - 2\rho\vec{\Omega} \times \vec{V} - \rho\nabla\Phi + \mu\nabla^2\vec{V}$ (2)<br>Here, $\rho$ = Density of liquid<br>$P$ = The pressure<br>$g$ = Gravitational acceleration vector<br>$\Omega$ = The angular velocity vector of the tube<br>$\Phi$ = Gravitational potential<br>This equation constitutes the Navier–Stokes equation, which governs the conservation of momentum within the fluid flow. It incorporates terms accounting for the pressure gradient, gravitational force, Coriolis force, centrifugal force, body forces, and viscous dissipation |
| 3 | Derivation of energy | $\vec{V} \cdot \nabla\left(h + \frac{1}{2}\vec{V}^2\right) = \nabla \cdot \left(\frac{k}{\rho c_p}\nabla T\right)$ (3)<br>$h$ = Enthalpy<br>$V$ = The magnitude of velocity<br>$k$ = Thermal conductivity<br>$T$ = Temperature<br>$c_p$ = specific heat |

thermal losses. Below are the key benefits of using flow-enhancing inserts in PTC systems:

i. **Flow Disruption and Turbulence**: Inserts create turbulence, disrupting laminar flow and improving fluid mixing.
ii. **Enhanced Heat Transfer**: Increased Nusselt number (up to 200%), improving heat transfer efficiency and maximizing solar energy absorption.
iii. **Lower Flow Rates and Energy Savings**: The system operates efficiently at lower flow rates, reducing energy consumption for pumping and lowering operational costs.
iv. **Uniform Temperature Distribution**: Improved fluid–wall interaction ensures an even temperature profile, preventing hotspots and thermal stress.
v. **Improved System Performance**: Higher fluid temperatures and better energy absorption result in increased system efficiency and performance.
vi. **Long-Term Reliability**: Stable thermal conditions extend the lifespan of the PTC system, reducing material degradation and ensuring consistent performance.

Incorporating these inserts into PTCs enhances their sustainability, efficiency, and reliability, making them a valuable innovation for solar thermal energy systems.

## 5.3 Thermal Performance Enhancements

The integration of flow-enhancing inserts such as twisted tapes, helical coils, wire coils, and ribbed inserts into Parabolic Trough Collectors (PTCs) leads to substantial improvements in thermal performance and operational efficiency by addressing the limitations posed by laminar flow. In conventional systems, the heat transfer fluid (HTF) typically flows in a laminar fashion, where the fluid moves in smooth, parallel layers, which limits the interaction between the fluid and the tube surface. This lack of interaction reduces the rate at which heat is transferred from the tube to the fluid, ultimately lowering the efficiency of the system. Flow-enhancing inserts are designed to disrupt this laminar flow and induce turbulence, which increases fluid mixing and the convective heat transfer between the fluid and the tube wall. The introduction of turbulence, as achieved through twisted tapes or helical coils, significantly increases the Nusselt number, a dimensionless measure of heat transfer, by up to 200% compared to conventional smooth tubes. This dramatic increase in heat transfer efficiency allows the system to absorb more solar energy over a given period, improving its overall thermal performance.

Another significant advantage of these inserts is their ability to enable the system to operate at lower HTF flow rates. With enhanced heat transfer capabilities, the system requires less fluid flow to achieve the desired temperature, resulting in a reduction in pumping energy consumption. This reduction not only lowers the operational costs but also minimizes the associated thermal losses, as the fluid spends more time in contact with the tube wall, leading to greater heat absorption before it exits the system. The increased fluid contact time also results in a more consistent and uniform temperature distribution across the collector, helping to prevent the formation of hotspots. Hotspots can cause localized overheating, leading to thermal stress and material degradation, which shortens the lifespan of system components. By promoting uniform heat distribution, the inserts reduce the risk of these thermal stresses and contribute to the overall durability and reliability of the system.

Moreover, the improved heat transfer performance allows the system to maintain higher fluid temperatures for extended periods. This is particularly important in maximizing the thermal energy harnessed from the sun, as higher fluid temperatures are essential for efficient energy transfer to the working liquid, increasing the overall system efficiency. The higher operational efficiency and ability to operate at lower flow rates not only reduce energy consumption but also contribute to the sustainability of the solar thermal system by maximizing energy utilization while minimizing losses. In the long term, the combination of better heat transfer, lower energy consumption, and reduced thermal stresses enhances the reliability and lifespan of the PTC system, making it a more cost-effective and sustainable solution for harnessing solar energy.

## 5.4 Disadvantages of Modified Designs

Incorporating flow-enhancing inserts like twisted tapes and helical coils into Parabolic Trough Collectors (PTCs) can significantly improve thermal performance. However, these modifications come with some challenges that must be addressed for the system to be effective and reliable. Below are the main disadvantages of these modified designs:

1. **Increased Pressure Drop**
   i. Inserts create additional turbulence, which increases flow resistance and friction, leading to a higher pressure drop across the collector system.
   ii. To maintain the desired flow rates, more powerful and energy-intensive pumps may be required, raising operational costs.
   iii. Engineers must balance enhanced heat transfer with the pressure drop, as inadequate flow rates can impact the system's efficiency.
   iv. The increased pressure drop can also lead to greater wear on components like pumps, pipes, and valves, resulting in more frequent maintenance and downtime.
2. **Higher Manufacturing Costs**
   i. The integration of specialized inserts requires advanced manufacturing techniques, adding complexity and cost to the production process.
   ii. High precision is needed to ensure the correct fit of inserts within the PTC design, increasing production time and cost.
   iii. In competitive markets, the higher initial costs may deter developers and investors, limiting the scalability of modified PTCs.
   iv. Systems with inserts often require more intensive maintenance, skilled labor, and specialized tools, increasing long-term financial costs.
3. **Mechanical Issues**
   i. The addition of inserts can create new stress points within the system, increasing the risk of mechanical failures in components like absorber tubes.
   ii. Installation and maintenance become more complex, requiring specialized training for technicians and potentially leading to longer downtimes.
   iii. Inserts can increase mechanical strain on components, leading to higher fatigue rates and more frequent inspections or replacements.
   iv. Thermal cycling can cause misalignment or stress concentrations, complicating the system's thermal management and reducing long-term reliability.

Despite the performance enhancements, these challenges must be carefully managed through optimized design and maintenance strategies to ensure the continued effectiveness and sustainability of modified PTCs.

## 5.5 Methodologies for Testing and Evaluation

Experimental testing, standardized testing procedures, computational fluid dynamics (CFD) simulations, performance modeling, field testing, comparative analysis, and statistical analysis are crucial methods for evaluating the performance of Parabolic Trough Collectors (PTCs) and optimizing their design. For experimental testing, a test rig replicating PTC operational conditions is constructed, equipped with solar irradiance sensors, flow meters, and temperature sensors to monitor key metrics like inlet/outlet temperatures, flow rates, and solar radiation. This provides empirical data on real-world performance. Standardized testing procedures, such as ASHRAE Standard 93 and ISO 9806, ensure consistency and reliability in performance measurements across different studies. CFD simulations allow for detailed analysis of liquid flow and heat transfer dynamics, enabling optimization of design parameters by visualizing flow patterns and temperature distributions. Performance modeling involves developing frameworks that predict PTC performance based on variables like solar irradiance, HTF properties, and collector geometry, with experimental data used to validate these models. Field testing in actual solar installations provides real-world insights into PTC efficiency, comparing design predictions with actual performance. Comparative analysis evaluates different collector configurations and insert designs, identifying the most effective solutions for specific environments. Finally, statistical analysis of the collected performance data, using tools like MATLAB or R, helps identify trends and correlations, guiding improvements in collector design and functionality. Together, these methods play a pivotal role in enhancing the efficiency, reliability, and sustainability of PTCs for solar energy applications.

## 5.6 Future Directions in Inserts Design for Parabolic Trough Collectors

Inserts in parabolic trough collectors (PTCs) play a crucial role in enhancing heat transfer and improving overall system efficiency. By exploring diverse geometrical modifications, engineers can develop innovative designs that leverage various physical principles to optimize thermal performance. These modifications include ribbed inserts, which balance rib height and spacing to enhance turbulence and improve heat transfer while promoting uniform heat distribution and reducing hotspot risks. Pin–fin inserts increase the surface area for heat transfer and can be customized to optimize flow characteristics with adjustable pressure drops. Porous media inserts enhance heat retention and fluid distribution through controlled porosity and high-conductivity materials. Spiral inserts create secondary flow effects, increasing turbulence and improving convective heat transfer while fitting easily into existing PTC infrastructure. Multi-layer inserts use layered materials to enhance heat transfer and structural integrity, with each layer tailored to manage temperature fluctuations effectively. Additionally, integration with other technologies such as thermal energy storage (TES) and sensor systems can further improve performance. Hybrid inserts incorporating phase change materials (PCMs) store excess heat and stabilize temperatures, while smart inserts with embedded sensors adjust performance dynamically based on real-time data. Sustainability is also a key consideration, with materials selected for recyclability, biodegradability, and minimal environmental impact. Life cycle assessments (LCA) help identify sustainable insert designs, while closed-loop systems and dry cooling capabilities reduce water consumption, particularly important in arid regions. These innovations in insert design, when combined with other technological integrations, pave the way for more efficient, sustainable, and reliable solar thermal systems.

## References

1. Kakac S, Liu H, Pramuanjaroenkij A (2020) Heat exchangers: selection, rating, and thermal design. CRC Press, New York
2. Yılmaz IH, Mwesigye A (2020) Enhancing the overall thermal performance of a large aperture parabolic trough collector using wire coil inserts. Sustain Energy Technol Assess 39:100696
3. Eiamsa-Ard S, Thianpong C, Eiamsa-Ard P (2010) Turbulent heat transfer enhancement by counter/co-swirling flow in a tube fitted with twin twisted tapes. Exper Thermal Fluid Sci 34(1):53–62
4. Gunes S, Ozceyhan V, Buyukalaca O (2010) Heat transfer enhancement in a tube with equilateral triangle cross sectioned coiled wire inserts. Exp Therm Fluid Sci 34:684–691
5. Prasad PD, Gupta AV, Sreeramulu M, Sundar LS, Singh MK, Sousa AC (2015) Experimental study of heat transfer and friction factor of $Al_2O_3$ nanofluid in U-tube heat exchanger with helical tape inserts. Exp Therm Fluid Sci 62:141–150
6. Prasad PD, Gupta AV (2016) Experimental investigation on enhancement of heat transfer using $Al_2O_3$/water nanofluid in a u-tube with twisted tape inserts. Int Commun Heat Mass Tran 75:154–161
7. Sekhar YR, Sharma KV, Karupparaj RT, Chiranjeevi C (2013) Heat transfer enhancement with $Al_2O_3$ nanofluids and twisted tapes in a pipe for solar thermal applications. Procedia Eng 64:1474–1484
8. Azmi WH, Sharma KV, Mamat R, Anuar S (2014) Turbulent forced convection heat transfer of nanofluids with twisted tape insert in a plain tube. Energy Proc 52:296–307
9. Eiamsa-Ard S, Kiatkittipong K, Jedsadaratanachai W (2015) Heat transfer enhancement of $TiO_2$/water nanofluid in a heat exchanger tube equipped with overlapped dual twisted-tapes. Eng Sci Technol Int J 18:336–350
10. Khoshvaght-Aliabadi M, Eskandari M (2015) Influence of twist length variations on thermal–hydraulic specifications of twisted-tape inserts in presence of Cu–water nanofluid. Exp Therm Fluid Sci 61:230–240
11. Maddah H, Alizadeh M, Ghasemi N, Alwi SR (2014) Experimental study of $Al_2O_3$/water nanofluid turbulent heat transfer enhancement in the horizontal double pipes fitted with modified twisted tapes. Int J Heat Mass Tran 78:1042–1054
12. Naik MT, Janardana GR, Sundar LS (2013) Experimental investigation of heat transfer and friction factor with water–propylene glycol based CuO nanofluid in a tube with twisted tape inserts. Int Commun Heat Mass Tran 46:13–21
13. Naik MT, Fahad SS, Sundar LS, Singh MK (2014) Comparative study on thermal performance of twisted tape and wire coil inserts in turbulent flow using CuO/water nanofluid. Exp Therm Fluid Sci 57:65–76

14. Wongcharee K, Eiamsa-ard S (2011) Enhancement of heat transfer using CuO/water nanofluid and twisted tape with alternate axis. Int Commun Heat Mass Transf 38:742–748
15. Wongcharee K, Eiamsa-ard S (2012) Heat transfer enhancement by using CuO/water nanofluid in corrugated tube equipped with twisted tape. Int Commun Heat Mass Transf 39:251–257
16. Zheng L, Xie Y, Zhang D (2017) Numerical investigation on heat transfer performance and flow characteristics in circular tubes with dimpled twisted tapes using $Al_2O_3$-water nanofluid. Int J Heat Mass Tran 111:962–981
17. Suresh S, Venkitaraj KP, Selvakumar P, Chandrasekar M (2012) A comparison of thermal characteristics of $Al_2O_3$/water and CuO/water nanofluids in transition flow through a straight circular duct fitted with helical screw tape inserts. Exp Therm Fluid Sci 39:37–44
18. Mehta B, Subhedar D, Panchal H, Said Z, Sharma K, Siddiqui MIH, Natrayan L, Marc A, Rosen MA (2024) Twisted tape inserts in parabolic trough solar collectors: assessment of energy, exergy, and environmental impacts. Appl Therm Eng 250:123566
19. Alnaqi AA, Alsarraf J, Al-Rashed AAAA (2021) Hydrothermal effects of using two twisted tape inserts in a parabolic trough solar collector filled with MgO-MWCNT/thermal oil hybrid nanofluid. Sustain Energy Technol Assess 47:101331
20. Jaramillo OA, Borunda M, Velazquez-Lucho KM, Robles M (2016) Parabolic trough solar collector for low enthalpy processes: an analysis of the efficiency enhancement by using twisted tape inserts. Renew Energy 93:125–141
21. Isravel RS, Raja M, Saravanan S, Vijayan V (2020) Thermal augmentation in parabolic trough collector solar water heater using rings attached twisted tapes. Mater Today Proceed 21:127–129
22. Thapa S, Kumar R, Kumar K, Samir S, Lee D, Singh T (2023) Analytical investigation of solar parabolic trough receiver with multiple twisted tapes using exergy and energy analysis. Case Stud Therm Eng 50:103493
23. Abidi A, El-Shafay AS, Degani M, Guedri K, Sajadi SM, Sharifpur M (2022) Improving the thermal-hydraulic performance of parabolic solar collectors using absorber tubes equipped with perforated twisted tape containing nanofluid. Sustain Energy Technol Assess 52:102099
24. Razzaghi MJP, Asadollahzadeh M, Tajbakhsh MR, Mohammadzadeh R, Abad MZM, Nadimi E (2023) Investigation of a temperature-sensitive ferrofluid to predict heat transfer and irreversibilities in LS-3 solar collector under line dipole magnetic field and a rotary twisted tape. Int J Therm Sci 185:108104
25. Chakraborty O, Roy S, Das B, Gupta R (2023) Computational analyses of parabolic trough solar collector in the presence of helical coil-insert. Int J Environ Sci Technol 20:683–702
26. Jafar KS, Venkatesaperumal R, Beemkumar N, Santhoshkumar M, Rajamurugan TV (2022) Performance analysis of a solar parabolic trough collector fitted with a helical coiled tube absorber. Int J Amb Energy 44:442–449
27. Naved MM, Joshi SS, Bhave NA (2020) Performance analysis of parabolic trough solar collector with 'U'-tube and helical coil receivers. Adv Energy Res 2:59–65
28. Yılmaz IH, Mwesigye A, Göksu TT (2020) Enhancing the overall thermal performance of a large aperture parabolic trough solar collector using wire coil inserts. Sustain Energy Technol Assess 39:100696
29. Varun K, Arunachala UC, Elton DN (2020) Trade-off between wire matrix and twisted tape: SOLTRACE® based indoor study of parabolic trough collector. Renew Energy 156:478–492
30. Saini P, Dhar A, Powar S (2024) Performance evaluation of a parabolic trough collector with a uniform helical wire coil flow insert. Results Eng 21:101794
31. Jafaryar M, Sheikholeslami M (2022) Efficacy of turbulator on performance of parabolic solar collector with using hybrid nanomaterial applying numerical method. Renew Energy 198:534–548
32. Ghadirijafarbeigloo S, Zamzamian AH, Yaghoubi M (2014) 3-D numerical simulation of heat transfer and turbulent flow in a receiver tube of solar parabolic trough concentrator with louvered twisted-tape inserts. Energy Proced 49:373–380
33. Borunda M, Garduno-Ramirez R, Jaramillo OA (2019) Optimal operation of a parabolic solar collector with twisted-tape insert by multi-objective genetic algorithms. Renew Energy 143:540–550

34. Stanek B, Ochmann J, Węcel D, Bartela L (2023) Study of twisted tape inserts segmental application in low-concentrated solar parabolic trough collectors. Energies 16(9):3716
35. Diwan K, Soni MS (2015) Heat transfer enhancement in absorber tube of parabolic trough concentrators using wire-coils inserts. Univ J Mech Eng 3(3):107–112
36. Afsharpanah F, Sheshpoli AZ, Pakzad K, Ajarostaghi SSM (2021) Numerical investigation of non-uniform heat transfer enhancement in parabolic trough solar collectors using dual modified twisted-tape inserts. J Therm Eng 7:133–147
37. Too YCS, Benito R (2013) Enhancing heat transfer in air tubular absorbers for concentrated solar thermal applications. Appl Therm Eng 50(1):1076–1083
38. Bhuiya MMK, Chowdhury MSU, Saha M, Islam MT (2013) Heat transfer and friction factor characteristics in turbulent flow through a tube fitted with perforated twisted tape inserts. Int Commun Heat Mass Transfer 46:49–57
39. Reddy KS, Kumar KR, Ajay CS (2015) Experimental investigation of porous disc enhanced receiver for solar parabolic trough collector. Renew Energy 77:308–319
40. Biswakarma S, Roy S, Das B, Debnath BK (2020) Performance analysis of internally helically v-grooved absorber tubes using nanofluid. Therm Sci Eng Progr 15:100538
41. Bellos E, Tzivanidis C, Antonopoulos KA, Gkinis G (2016) Thermal enhancement of solar parabolic trough collectors by using nanofluids and converging-diverging absorber tube. Renew Energy 94:213–222
42. Kalidasan B, Shankar R, Srinivas T (2016) Absorber tube with internal hinged blades for solar parabolic trough collector. Energy Proced 90:463–469
43. Ebrahim GS, Akbar RA (2017) Numerical thermal study on effect of porous rings on performance of solar parabolic trough collector. Appl Therm Eng 118:807–816
44. Tajik JM, Seyfollah S, Mohammad A (2017) Experimental investigation on a solar parabolic trough collector for absorber tube filled with porous media. Renew Energy 107:156–163
45. Zheng N, Liu W, Liu Z, Liu P, Shan F (2015) A numerical study on heat transfer enhancement and the flow structure in a heat exchanger tube with discrete double inclined ribs. Appl Therm Eng 90:232–241
46. Zhen H, Zeng-Yao L, Guang-Lei Y, Wen-Quan T (2017) Numerical investigations on fully-developed mixed turbulent convection in dimpled parabolic trough receiver tubes. Appl Therm Eng 114:1287–1299
47. Bellos E, Daniil I, Tzivanidis C (2000) Multiple cylindrical inserts for parabolic trough solar collector. Appl Therm Eng 143:80–89
48. Bellos E, Tzivanidis C (2018) Investigation of a star flow insert in a parabolic trough solar collector. Appl Energy 224:86–102
49. Hosseini SMS, Dehaj MS (2021) Assessment of $TiO_2$ water-based nanofluids with two distinct morphologies in a U type evacuated tube solar collector. Appl Therm Eng 182:116086
50. Chakraborty O, Das B, Gupta R (2023) Performance of parabolic trough solar collector with spinning star inserts in receiver tube with hybrid nanofluid. Proceed Instit Mech Eng E J Process Mech Eng 45:1252
51. Norouzi AM, Siavashi M, Ahmadi R, Milad-Tahmasbi M (2021) Experimental study of a parabolic trough solar collector with rotating absorber tube. Renew Energy 168:734–749
52. Kumar KR, Reddy K (2009) Thermal analysis of solar parabolic trough with porous disc receiver. Appl Energy 86:1804–1812
53. Kumar KR, Reddy K (2012) Effect of porous disc receiver configurations on performance of solar parabolic trough concentrator. Heat Mass Transf 48:555–571
54. Wang P, Liu D, Xu C (2013) Numerical study of heat transfer enhancement in the receiver tube of direct steam generation with parabolic trough by inserting metal foams. Appl Energy 102:449–460
55. Sadaghiyani OK, Mirzaee I (2013) Numerical simulation coupled with MCRT method to study the effect of plug diameter and its position on outlet temperature and the efficiency of LS-2 parabolic trough collector. J Sol Energy Eng 135:041001
56. Chakraborty O, Das B, Gupta R (2022) Effect of spherical balls insert in collector tube of parabolic trough collector by considering $Al_2O_3$-based water. J Brazil Soc Mech Sci Eng 44:38. https://doi.org/10.1007/s40430-021-03337-2

57. Chakraborty O, Das B, Gupta R (2024) Performance of parabolic trough solar collector with spinning star inserts in receiver tube with hybrid nanofluid. Proceed Instit Mech Eng E J Process Mech Eng 238:1687–1699
58. Kasperski J, Nemś M (2013) Investigation of thermo-hydraulic performance of concentrated solar air-heater with internal multiple-fin array. Appl Therm Eng 58:411–419
59. Chakraborty O (2023) () Influence of spinning flower structure inserts in the thermal performance of LS-2 model of parabolic trough collector with ternary hybrid nanofluid. Renewable Energy 210:215–228. https://doi.org/10.1016/j.renene.2023.03.112
60. Mwesigye A, Bello-Ochende T, Meyer JP (2014) Heat transfer and thermodynamic performance of a parabolic trough receiver with centrally placed perforated plate inserts. Appl Energy 136:989–1003
61. Mwesigye A, Bello-Ochende T, Meyer JP (2014) Thermal performance of a parabolic trough receiver with perforated conical inserts for heat transfer enhancement. In: ASME 2014 international mechanical engineering congress and exposition: American society of mechanical engineers, pp V06BT07A004–V006BT007A004
62. Panja SK, Das B, Mahesh V (2025) Performance evaluation of a novel parabolic trough solar collector with nanofluids and porous inserts. Appl Therm Eng 258:124495. https://doi.org/10.1016/j.applthermaleng.2024.124495
63. Esmaeili Z, Akbarzadeh S, Rashidi S, Valipour MS (2023) Effects of hybrid nanofluids and turbulator on efficiency improvement of parabolic trough solar collectors. Eng Anal Bound Elements 148:114–125
64. Liu P, Zheng N, Liu Z, Liu W (2019) Thermal-hydraulic performance and entropy generation analysis of a parabolic trough receiver with conical strip inserts. Energy Conv Manage 179:30–45
65. Ahmed KA, Natarajan E (2019) Thermal performance enhancement in a parabolic trough receiver tube with internal toroidal rings: a numerical investigation. Appl Therm Eng 162:114224
66. Al-Rashed AAAA, Alnaqi AA, Alsarraf J (2021) Thermo-hydraulic and economic performance of a parabolic trough solar collector equipped with finned rod turbulator and filled with oil-based hybrid nanofluid. J Taiwan Instit Chem Eng 124:192–204
67. Chakraborty O, Das B, Gupta R (2024) Thermal performance evaluation of parabolic trough collector having different inserts and working with hybrid nanofluid. Energy Environ 35(6):3010–3037
68. Wang J, Zeng L, Zhao B, Li C (2024) Entropy generation analysis and thermal performance of absorber tube in parabolic trough solar collector inserted with eccentric structure insert. Renew Energy 237:121611
69. Saini P, Dhar A, Powar S (2024) Effect of the X-grid static mixer flow inserts on the performance of parabolic trough collector. Int J Thermofluids 21:100544
70. Wang J, Zeng L, Yu S, He Y (2024) Effects of insert design and optimization on the performance of parabolic trough receivers with inserted absorbers. Sol Energy 284:113061
71. Darbari B, Derikvand M, Shabani B (2023) Thermal performance improvement of a LS-2 parabolic trough solar collector using porous disks. Appl Therm Eng 228:120546
72. Valizade M, Heyhat MM, Maerefat M (2020) Experimental study of the thermal behavior of direct absorption parabolic trough collector by applying copper metal foam as volumetric solar absorption. Renew Energy 145:261–269
73. Fox RW, McDonald AT, Pritchard PJ (2011) Introduction to fluid mechanics. Wiley. http://ftp.demec.ufpr.br/disciplinas/TM240/Marchi/Bibliografia/Pritchard-Fox-McDonalds_2011_8ed_Fluid-Mechanics.pdf

# Chapter 6
# Secondary Reflectors in Parabolic Trough Collectors

This chapter focuses on the integration of secondary reflectors in Parabolic Solar Trough Collectors (PTCs) as a strategy to overcome key optical inefficiencies. It examines the design and construction of secondary reflectors aimed at improving beam concentration and ensuring more uniform heat flux distribution along the receiver surface. Uneven heating in traditional PTC designs can lead to thermal stress, material fatigue, and reduced system lifespan—challenges this chapter addresses in detail. By analyzing the impact of secondary reflectors on thermal and optical performance, the chapter highlights their potential to enhance overall system efficiency. It also explores the technical trade-offs involved, balancing design complexity with performance gains. This discussion offers valuable insight into the optimization of PTC systems for more reliable and effective solar thermal energy conversion.

## 6.1 Thermal Stress and Material Fatigue

Non-uniform heat flux distribution significantly contributes to thermal stress in the receiver tube of conventional Parabolic Trough Collectors (PTC). As sunlight is concentrated on the cylindrical receiver, different areas of the outer surface absorb varying amounts of thermal energy, resulting in temperature gradients along the length of the tube. Areas exposed to more concentrated sunlight heat up faster than those receiving less direct sunlight, creating differential heating across the surface. This temperature difference leads to thermal expansion, as materials expand when heated. In a straight receiver tube, the hotter sections expand more than the cooler sections, creating internal stresses because the cooler parts constrain the expanding hot areas.

Additionally, the daily temperature fluctuations caused by the sun's movement and seasonal variations in solar intensity contribute to repetitive thermal cycling. These cycles exacerbate material fatigue, causing micro-cracks that can grow over time

 
O. Chakraborty and B. Das, *Parabolic Solar Trough Collector*, Mechanical Engineering Series, https://doi.org/10.1007/978-3-032-11229-3_6

into larger failures. The materials commonly used in the receiver, such as stainless steel or aluminum, have specific thermal expansion coefficients, which means if the temperature differences exceed the material's tolerance, the receiver's structural integrity can be compromised. Differential heating, thermal expansion, repetitive cycling, and material limitations result in localized stress concentrations. Over time, these stresses can cause significant damage, including warping, cracking, and even catastrophic failure of the receiver tube, reducing the efficiency and lifespan of the PTC system. Therefore, managing heat flux distribution and material properties is crucial to preventing thermal stress-related issues in PTCs.

## 6.2 Receiver Design Limitations

The design of the cylindrical receiver tube significantly impacts heat absorption and distribution, with several inherent limitations that affect overall performance. Cylindrical receivers typically incorporate selective coatings to improve thermal absorption, but the tube's curvature and the parabolic reflector's geometry can result in shadowing or less-than-optimal angles of incidence. These factors lead to uneven heating along the tube's surface, reducing thermal efficiency. Additionally, the characteristics of the heat transfer fluid (HTF) are critical for maintaining an even temperature throughout the receiver. If the HTF flow is inconsistent—due to variations in pressure or velocity—certain sections may receive insufficient heat transfer, creating cooler regions, while stagnant fluid can cause localized overheating, compounding thermal stress issues. The structural integrity of cylindrical receivers can also be compromised under thermal stress. As temperatures rise, materials lose stiffness and become more susceptible to deformation under mechanical loads. This means the structural design must balance factors such as weight, thermal performance, and resistance to environmental stresses like wind and seismic activity. Moreover, effective insulation is crucial to minimizing heat loss, but improper or uneven insulation can cause certain areas to overheat while others remain cooler, creating additional thermal stress. This uneven insulation exacerbates the challenge of maintaining consistent heat distribution and increases the risk of damage to the receiver. Overall, while cylindrical receivers are efficient for manufacturing and installation, their design limitations must be carefully managed to optimize thermal performance and prevent structural failure.

## 6.3 Additional Factors Affecting Performance

Environmental conditions, such as dust, pollution, and seasonal weather changes, can significantly impact solar intensity and the overall heat flux received by the collector, contributing to inconsistencies in heating and complicating the thermal dynamics of the receiver. These external factors, along with the inherent non-uniform heat flux distribution in parabolic solar trough collectors (PSTC), create challenges, particularly regarding thermal stress and material fatigue in cylindrical receiver tubes. The differential heating caused by varying solar incidence, coupled with material properties, can lead to significant structural issues over time. Additionally, the presence of thermal stress and the potential for material fatigue necessitate more frequent maintenance checks, leading to higher operational costs and potential downtime for repairs. Addressing these issues involves improving design practices, selecting appropriate materials, and integrating technologies like secondary reflectors to enhance uniformity in heat distribution. By considering these factors, stakeholders can optimize the efficiency, reliability, and longevity of PSTC systems, ensuring better performance and reducing the risk of failure. Understanding the impact of environmental conditions and non-uniform heat flux is essential for improving the overall design and maintenance strategies for these solar energy systems.

## 6.4 Overview of the Role of Secondary Reflectors in Parabolic Solar Trough Collectors

Parabolic solar trough collectors (PSTCs) are essential for harnessing solar energy efficiently, using a parabolic reflector to focus sunlight onto a linear receiver. This setup generates thermal energy that can be utilized for various applications, such as electricity generation and heating. However, PSTCs face challenges in achieving optimal performance due to issues like non-uniform heat flux distribution. As the solar radiation absorbed by the receiver is not uniformly distributed, areas of the receiver tube experience differing amounts of heat. This results in inefficient energy capture and uneven heating, which can lead to material fatigue and thermal stress. Secondary reflectors emerge as a valuable enhancement to address these challenges, improving the overall performance and reliability of PSTCs. These reflectors not only help concentrate light more effectively but also play a significant role in reducing thermal stress by ensuring a more even distribution of heat across the receiver surface.

**Enhanced Light Concentration**

One of the primary roles of secondary reflectors is to improve the concentration of sunlight onto the receiver, which enhances the overall thermal efficiency of the system. By redirecting sunlight that might otherwise miss the receiver due to the geometry of the system or the changing position of the sun, secondary reflectors capture more solar energy. This is particularly beneficial during times of low solar angles, such as early mornings or late afternoons, when the sun is at a lower elevation. During these periods, the parabolic reflectors alone may not capture all the available sunlight, but secondary reflectors can help redirect this missed energy back onto the receiver. This redirection increases the concentration of sunlight, thus improving the system's optical efficiency and ensuring that more solar energy is converted into thermal energy. With improved light concentration, PSTCs can harness more energy, even during periods when the solar angle is less favorable, maximizing their potential throughout the day.

**Mitigating Non-uniform Heat Flux Distribution**

Non-uniform heat flux distribution is a critical challenge for PSTCs, as varying solar incidence and design limitations can cause uneven heating along the receiver tube. This uneven heating results in hot spots and temperature gradients that contribute to thermal stress and can ultimately cause material failure. Secondary reflectors help to mitigate these issues by distributing light more evenly across the receiver surface. By reflecting sunlight onto cooler sections of the receiver, they ensure a more uniform temperature profile. This uniform heating prevents the formation of hot spots, which are areas of concentrated heat that can cause localized thermal expansion. Hot spots lead to structural damage over time, as the receiver tube material undergoes stress from the differing expansion rates between hot and cool areas. By reducing these hot spots and promoting even heating, secondary reflectors help to maintain the integrity of the receiver tube, prolonging its lifespan and enhancing the overall durability of the system.

**Increased System Efficiency**

The integration of secondary reflectors leads to significant improvements in the overall efficiency of PSTCs. Research has shown that systems equipped with secondary reflectors can achieve thermal efficiencies that are 15–20% higher compared to those without. This increase is especially noticeable during peak sunlight hours, when the sun's intensity is at its highest. With improved light concentration, more solar energy is captured and converted into thermal energy, increasing the system's overall performance. In addition to improving energy capture during peak times, secondary reflectors also enable the system to maintain higher efficiencies under varying environmental conditions. Whether in regions with high solar variability, such as areas with intermittent cloud cover, or in places with less consistent solar intensity, secondary reflectors adapt the system's performance, ensuring that it continues to operate efficiently regardless of external factors. This adaptability makes PSTCs with secondary reflectors more reliable and capable of generating consistent thermal energy over time.

**Design Flexibility**

Secondary reflectors offer significant design flexibility, allowing engineers to customize their configuration to fit the specific needs of a project. Reflectors can be designed in various shapes, sizes, and materials, depending on the collector's configuration and the environmental conditions in which it operates. This flexibility allows for optimal integration into the system, ensuring that each installation maximizes performance based on its unique requirements. For example, in regions with frequent cloud cover or varying solar angles, secondary reflectors can be designed to adjust the angle and positioning to capture the most sunlight possible. Additionally, secondary reflectors can be integrated with advanced tracking systems, which follow the sun's movement throughout the day. By working in tandem with these tracking systems, secondary reflectors can further optimize the system's performance. This integration allows for real-time adjustments, ensuring that sunlight is always focused on the receiver for maximum energy capture, even as the sun moves across the sky.

**Economic Benefits**

The economic advantages of integrating secondary reflectors into PSTCs are substantial. While there may be initial costs associated with their installation, the increase in energy capture and overall system efficiency leads to reduced operational costs over time. The enhanced efficiency of systems with secondary reflectors means that more energy can be generated with the same amount of solar radiation, making the system more cost-effective. Additionally, the reduction in thermal stress and material fatigue provided by secondary reflectors translates to less frequent maintenance and fewer repairs. This decrease in maintenance requirements not only reduces operational costs but also extends the lifespan of the system. Over time, these benefits lead to faster returns on investment, making secondary reflectors a financially attractive option for solar energy projects. Their ability to reduce long-term costs while enhancing system performance makes them a key technology for improving the economic viability of solar thermal systems.

**Environmental Impact**

The use of secondary reflectors not only boosts the performance of PSTCs but also supports broader environmental objectives. By enhancing the system's efficiency, secondary reflectors help maximize the use of renewable energy, reducing reliance on fossil fuels and contributing to lower carbon emissions. This aligns with global efforts to promote sustainable energy practices and combat climate change. Moreover, the improved thermal performance enabled by secondary reflectors can make it easier to integrate energy storage solutions into the system. Effective energy storage is crucial for ensuring a reliable energy supply even when sunlight is not available, such as during cloudy days or at night. With better thermal performance, PSTCs with secondary reflectors can store more energy for later use, reducing the need for backup power from non-renewable sources and improving grid stability. Thus, secondary reflectors not only enhance the performance and efficiency of solar thermal systems but also play a vital role in promoting the use of clean, renewable energy and supporting global sustainability goals.

## 6.5 Design of Parabolic Solar Trough Collectors with Secondary Reflector

PTCs consist of a parabolic reflector, linear receiver, and secondary reflector. The parabolic reflector concentrates sunlight onto the receiver using reflective materials like polished aluminum or silvered glass. Its size and curvature optimize sunlight concentration. The linear receiver, a heat-resistant cylindrical tube, absorbs solar energy and is coated to minimize heat loss. Made from durable materials like stainless steel, it withstands high temperatures. The secondary reflector captures stray sunlight, redirecting it to the receiver, improving efficiency by increasing energy concentration on the heat transfer fluid (HTF), ultimately enhancing the overall thermal performance of the system.

The operational components of PTCs include the heat transfer fluid (HTF), pumping systems, heat exchangers, and thermal energy storage. The HTF circulates through the linear receiver, absorbing thermal energy from concentrated sunlight. Various HTFs, such as water, thermal oils, and molten salts, are chosen based on their thermal properties. The heated HTF is pumped at an optimal flow rate for efficient heat absorption. It is then directed to a heat exchanger, where it transfers energy to another medium, such as water or steam, for applications like heating or electricity generation. Thermal energy storage, using insulated tanks with molten salts or water, retains excess heat, enabling energy availability even without sunlight, enhancing system reliability.

## 6.6 Operation of Parabolic Solar Trough Collectors with Secondary Reflector

**Working of PTC with Secondary Reflector**

The working principle of PSTCs with secondary reflectors enhances solar energy conversion efficiency. The parabolic reflector concentrates sunlight onto a linear receiver. The secondary reflector captures stray sunlight and redirects it to the receiver, increasing the total solar energy focused on the heat transfer fluid (HTF). The HTF, circulating through the receiver, absorbs this concentrated heat, with the receiver's selective surface aiding in efficient temperature rise. The heated HTF then flows to a heat exchanger, transferring energy to water or steam for heating or electricity generation. Additionally, thermal energy storage systems retain excess heat for use when sunlight is unavailable, improving system reliability and energy output.

**Challenges in Heat Flux Distribution in Parabolic Solar Trough Collectors**

Parabolic solar trough collectors (PSTCs) are essential for converting solar energy into thermal energy, yet they face significant challenges related to heat flux distribution. Efficient heat distribution is crucial for maximizing performance and ensuring the longevity of these systems. This overview discusses the primary challenges: geometric imperfections, variability in solar incidence angles, and atmospheric conditions.

1. **Geometric Imperfections**

The parabolic reflector is the core component of PSTCs, designed to focus sunlight onto a linear receiver. Any deviations from the ideal parabolic shape can significantly affect how effectively sunlight is concentrated. These geometric imperfections can arise from

i. **Manufacturing Tolerances**: Slight errors during production can lead to misalignment in the reflector's shape.
ii. **Installation Errors**: Improper installation can further misalign the collector, disrupting the focal point.
iii. **Material Deformation**: Over time, environmental factors such as temperature changes and wind can deform the reflector, altering its effectiveness.

The consequences of these imperfections include the formation of hot spots, reduced energy capture efficiency, and increased maintenance requirements due to thermal stress on the receiver.

2. **Solar Incidence Angle**

The solar incidence angle, or the angle at which sunlight strikes the collector, varies throughout the day and across seasons. This variability poses challenges such as

i. **Suboptimal Focusing**: If the tracking system does not align accurately with the sun, some parts of the receiver may receive less concentrated sunlight.
ii. **Shadowing Effects**: Nearby objects can cast shadows on the collectors, further diminishing solar input during certain times of the day.

Seasonal changes also impact performance; for example, lower sun angles in winter can reduce solar energy capture, while higher angles in summer can enhance it if the system is properly adjusted.

The variability in solar incidence leads to inefficient energy capture, complicating integration into energy grids and increasing thermal stress on receiver materials.

3. **Atmospheric Conditions**

Atmospheric factors significantly influence the performance of PSTCs. Key considerations include

i. **Cloud Cover**: Overcast skies can drastically reduce solar irradiance, limiting energy available for conversion.
ii. **Dust and Pollution**: Accumulation on reflective surfaces decreases efficiency by diminishing reflectivity.
iii. **Humidity**: High humidity can scatter sunlight, impacting energy concentration on the receiver.

Spinning inserts enhance heat transfer by converting laminar flow to turbulence, with adjustable speed to optimize performance across varying conditions.

## 6.7 Types of Secondary Reflectors

Secondary reflectors are vital components in solar concentrating systems, used to enhance the efficiency of capturing and focusing solar radiation onto the receiver. These reflectors are positioned to intercept stray sunlight that might miss the primary receiver, redirecting it toward the focal point, thus increasing the system's energy capture efficiency. There are two primary types of secondary reflectors: fixed and adjustable. Each offers unique advantages and challenges depending on the application, geographic location, and energy capture requirements.

**Fixed Secondary Reflectors**

Fixed secondary reflectors are stationary components designed to operate without the need for tracking the sun's movement. These reflectors maintain a consistent orientation throughout the day, redirecting sunlight that would otherwise bypass the receiver. Fixed reflectors are often used in simpler, less expensive systems where mechanical complexity is undesirable, or in locations where ease of maintenance and durability are prioritized.

Fixed secondary reflectors are placed in such a way as to intercept and redirect sunlight that is not captured by the primary concentrating system, such as a parabolic trough or dish. By focusing this stray sunlight back toward the receiver, fixed reflectors can help increase the total energy captured by the system.

Fixed reflectors come in various shapes, each designed to optimize the redirection of sunlight in specific ways:

i. **Flat Reflectors**: These are simple, planar surfaces that redirect sunlight at a fixed angle. They are ideal for installations with limited space or where light needs to be redirected from particular angles. Flat reflectors are easy to manufacture and install, and they require minimal adjustment after installation. Their straightforward design also makes them easier to maintain.
ii. **Curved Reflectors**: Curved reflectors, such as parabolic or elliptical designs, are more effective at focusing sunlight across a wider range of incident angles. They can capture sunlight from more diffuse sources or when the sun is at a lower angle in the sky. This increased capture range makes curved reflectors particularly valuable in installations where maximizing solar gain throughout the day is important.

**Advantages of Fixed Secondary Reflectors**

i. **Simplicity**: One of the key benefits of fixed secondary reflectors is their simplicity. Because they have no moving parts and do not require complex tracking mechanisms, they are easier and cheaper to manufacture and install. This lack of mechanical components also makes them less prone to failure or degradation over time, which is an important consideration for long-term solar installations.
ii. **Low Maintenance**: The absence of moving parts means that fixed reflectors generally require very little maintenance. Once installed, they can continue

functioning with minimal intervention, making them an ideal choice for remote installations where access for regular maintenance may be limited or costly.

iii. **Cost-Effective**: Fixed reflectors are often less expensive both in terms of initial investment and operational costs. Their simplicity allows for faster and cheaper installation, and their low maintenance needs translate to reduced lifetime costs. This makes them an attractive option for projects with tight budgets or in locations where reliability and durability are prioritized over absolute efficiency.

**Limitations of Fixed Secondary Reflectors**

i. **Limited Adaptability**: The major limitation of fixed secondary reflectors is their inability to adjust to the sun's changing position throughout the day and year. Because they are stationary, they are optimized for a specific angle or range of angles, which means they perform best when the sun is within a certain range of positions in the sky. Outside of these optimal conditions, their performance declines.

ii. **Suboptimal Performance**: While fixed reflectors do increase the overall energy capture of a solar concentrating system, they are not as efficient as adjustable reflectors, which can track the sun's movement. Fixed reflectors may not be able to capture all available sunlight, especially in the early morning or late afternoon when the sun is low on the horizon. As a result, fixed systems may generate less energy overall, particularly in locations with significant seasonal variation in sunlight.

### Adjustable Secondary Reflectors

Adjustable secondary reflectors are more advanced than their fixed counterparts. They are designed to dynamically change their orientation throughout the day, following the sun's path to maximize solar energy capture. These reflectors can tilt, rotate, or otherwise adjust their position to ensure sunlight is always optimally redirected toward the receiver. While adjustable reflectors offer enhanced efficiency, they also have increased mechanical complexity and higher costs.

The primary function of adjustable secondary reflectors is to enhance the efficiency of solar energy capture by tracking the sun's movement. As the sun moves across the sky, the reflectors adjust their orientation to maintain the optimal angle for light redirection. This allows the system to capture sunlight more effectively during times when fixed reflectors struggle to maintain peak efficiency, such as early in the morning, late in the afternoon, and during the winter months when the sun is lower in the sky.

Adjustable reflectors are often integrated with automated tracking systems, which use sensors or algorithms to adjust the reflectors' positions continuously. These systems can be designed to follow the sun on both azimuth and elevation axes, allowing for precise control over the direction of reflected sunlight.

**Advantages of Adjustable Secondary Reflectors**

i. **Enhanced Efficiency**: By tracking the sun's movement and maintaining the optimal angle for reflecting sunlight toward the receiver, adjustable secondary reflectors can significantly increase a solar concentrating system's overall thermal energy output. This makes them particularly valuable in high-performance solar applications where maximizing energy capture throughout the day is critical.

ii. **Improved Heat Distribution**: Adjustable reflectors can help create a more uniform temperature distribution across the receiver. By ensuring that sunlight is evenly spread over the receiver's surface, these reflectors can reduce the risk of localized overheating, which can cause thermal stress and reduce the system's lifespan.

iii. **Versatility**: Adjustable secondary reflectors can be tailored to a solar installation's specific geographic location and seasonal variations. This flexibility allows for optimized performance in a wide range of environmental conditions, making them suitable for both high-latitude locations with significant seasonal variation and low-latitude locations with more consistent sunlight.

**Limitations of Adjustable Secondary Reflectors**

i. **Increased Complexity**: The major downside of adjustable secondary reflectors is their mechanical complexity. These systems require motors, sensors, and control systems to adjust the reflectors' position throughout the day continuously. This added complexity increases the cost of the system's initial installation and long-term operation.

ii. **Maintenance Requirements**: Because adjustable reflectors rely on moving parts, they require more regular maintenance than fixed reflectors. Motors and tracking systems can wear out over time, and sensors or control systems may need recalibration or replacement. This additional maintenance can increase the system's lifetime costs and may be a significant concern for installations in remote areas where access to maintenance personnel is limited.

iii. **Higher Upfront Costs**: Adjustable systems' added mechanical components and complexity make them generally more expensive to design, build, and install. While their increased efficiency may offset these costs over time, the higher upfront investment can be a barrier for projects with limited capital.

The choice between fixed and adjustable secondary reflectors depends on the specific needs and constraints of the solar concentrating system. Fixed reflectors offer simplicity, low maintenance, and cost-effectiveness, making them suitable for installations where reliability and durability are prioritized. However, their limited adaptability means they may not capture all available sunlight, leading to lower overall efficiency.

On the other hand, adjustable secondary reflectors provide enhanced efficiency, versatility, and improved heat distribution by tracking the sun's movement. These systems are ideal for maximizing energy capture in high-performance applications but come with increased mechanical complexity, higher costs, and greater maintenance.

## 6.8 Design Considerations for Secondary Reflectors

The shape and size of fixed secondary reflectors are crucial for their effectiveness. Common designs include the following.

**Flat Secondary Reflectors**: Flat secondary reflectors are commonly employed in solar concentrating systems to enhance the light concentration onto a receiver by redirecting light from a primary concentrator. Their simple, flat geometry makes them easy to manufacture and install, offering a cost-effective solution for increasing system efficiency. By reflecting light at specific angles, they help capture more sunlight that would otherwise miss the receiver, thus improving the overall energy conversion. Flat secondary reflectors are particularly beneficial in compact designs where space is limited, as they do not require the curvature or bulk associated with other types of reflectors, such as parabolic or hyperbolic shapes.

However, one of the key limitations of flat secondary reflectors is their sensitivity to the sun's position. Since they are designed to reflect light at fixed angles, their performance can decline as the sun moves throughout the day, especially if the system lacks sun tracking capabilities. The amount of redirected sunlight reaching the receiver may be reduced at lower sun angles (e.g., in the morning or late afternoon). Furthermore, flat reflectors are less effective under diffuse light conditions, as they rely primarily on direct solar rays for optimal reflection and redirection. Therefore, while flat secondary reflectors offer a practical and space-saving solution, their efficiency is highly dependent on the sun's position and the availability of direct sunlight. Flat secondary reflectors are illustrated in Fig. 6.1, and the study of flat secondary reflectors by Rodriguez-Sanchez and Rosengarten [1] is presented in Fig. 6.2.

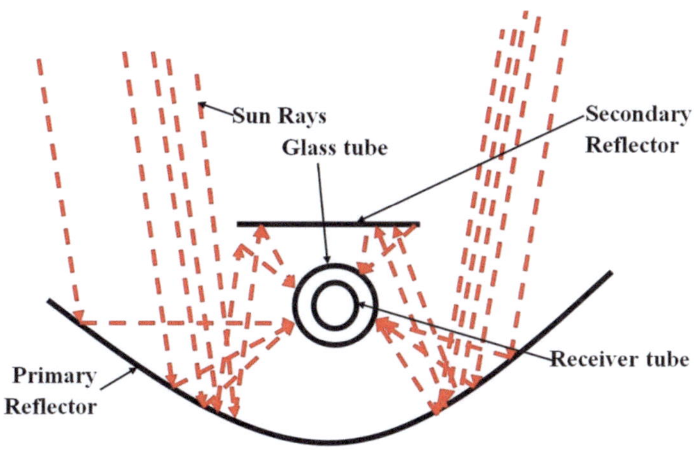


**Fig. 6.1** Flat secondary reflectors

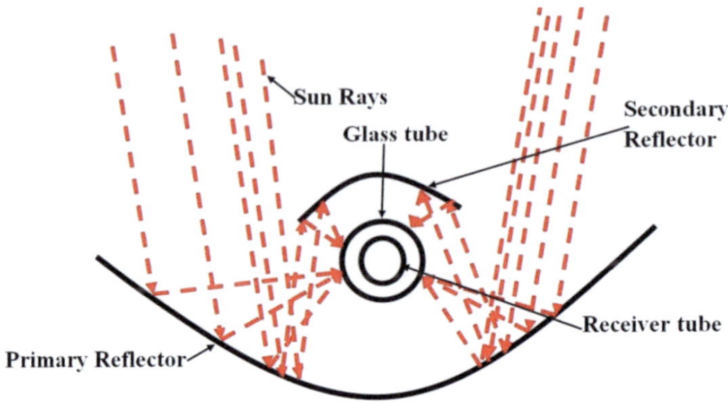


**Fig. 6.2** Curved secondary reflectors

### Curved Secondary Reflectors

Curved reflectors, including parabolic and elliptical designs, are crucial in solar concentrating systems due to their superior ability to focus sunlight. Parabolic reflectors are highly efficient, capturing parallel sunlight and reflecting it to a single focal point. This concentrated energy is ideal for high-temperature applications, such as solar thermal power or concentrated photovoltaic (CPV) systems. The parabolic shape ensures consistent sunlight direction, even as the sun moves, especially when paired with a sun-tracking system, optimizing efficiency throughout the day. Elliptical reflectors, which focus light between two focal points, are helpful for specific configurations, such as transferring light to secondary concentrators or different targets. Both curved reflectors are more efficient than flat ones, maintaining light concentration as the sun's position changes. However, they are more complex, expensive to manufacture, and require more space. Their precision makes them ideal for dynamic solar systems requiring consistent performance. Curved secondary reflectors are illustrated in Fig. 6.3.

### Triangular Secondary Reflectors

Triangular secondary reflectors are an innovative design used in solar thermal systems, particularly in conjunction with parabolic trough collectors. Their triangular shape, often equilateral or right-angled, allows them to capture sunlight from multiple angles, minimizing shadows and boosting energy collection. This makes them especially useful in space-limited installations where maximizing efficiency is key. These reflectors are versatile, fitting fixed and adjustable systems, and complement primary reflectors for enhanced performance. The reflectivity of triangular reflectors is critical, often utilizing high-reflectivity materials like polished aluminum or silver coatings, with durable protective layers to withstand environmental conditions. Additionally, robust mounting and support systems are essential for maintaining structural integrity and proper alignment with the primary reflector. Easy

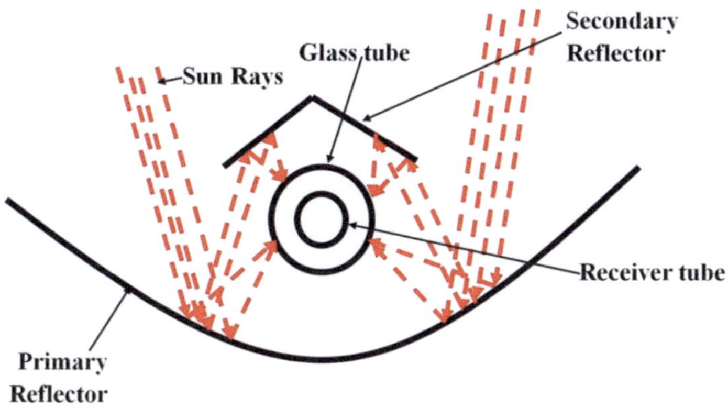


**Fig. 6.3** Triangular secondary reflectors

access for cleaning is also essential for maintaining long-term performance. Triangular secondary reflectors contribute significantly to solar thermal power plants, hybrid systems, and various residential or commercial applications, advancing solar technology in growing renewable energy market. Triangular secondary reflectors are illustrated in Fig. 6.3.

## 6.9 Optimizing the Design and Performance of Secondary Reflectors in Solar Thermal Systems

The size of the reflector is crucial in maximizing energy collection; larger reflectors can intercept more stray sunlight, enhancing overall performance. However, the design must balance the size with structural constraints and installation costs to ensure that larger reflectors remain feasible. Proper positioning is also essential for the effectiveness of fixed secondary reflectors. Key considerations include optimal orientation to intercept sunlight that the primary reflector misses, requiring precise calculation of angles for different times of the day and year. Additionally, reflectors must be designed to account for seasonal variations in the sun's path to ensure efficient operation throughout the year. The choice of reflective material is critical for both performance and durability, with high reflectivity materials such as polished aluminum or silvered glass being preferred for their excellent reflective properties. In contrast, durability is ensured by materials capable of withstanding environmental factors like UV exposure, temperature fluctuations, and wear. Structural integrity is vital for the longevity of fixed secondary reflectors, requiring robust support structures that can endure environmental stresses, such as wind loads and thermal expansion, while also considering weight to maintain stability without compromising strength. Additionally, the design should prioritize manufacturing simplicity and installation efficiency, reducing production costs and labor time. Overall, the design of secondary reflectors must balance size, positioning, material selection, structural strength, and manufacturing considerations to optimize energy capture and system performance.

## 6.10 Design Considerations for Secondary Reflectors in Solar Thermal Systems

Secondary reflectors play a crucial role in enhancing the performance of solar thermal systems, particularly parabolic trough collectors (PTCs). By optimizing light capture and directing it toward the receiver, secondary reflectors can significantly increase the efficiency of solar energy systems. This document discusses key design considerations, focusing on geometric design, material selection, and integration with trough collectors, while incorporating mathematical expressions to elucidate the concepts.

**Geometric Design**

The geometric configuration of secondary reflectors directly influences their ability to capture and redirect sunlight effectively. The shape and orientation of the reflector determine how well it can intercept stray solar radiation that would otherwise bypass the receiver. The design must consider the sun's position throughout the day and across different seasons to maximize efficiency.

**Mathematical Model for Light Capture**

We can utilize geometric optics principles to quantify the performance of secondary reflectors. The angle of incidence ($\theta_i$) and the angle of reflection ($\theta_r$) can be expressed as

$$\theta_i = \theta_r \tag{6.1}$$

This fundamental law of reflection ensures that the angle at which sunlight strikes the reflector is equal to the angle at which it reflects off the surface.

The effectiveness of light capture can be evaluated using the solid angle ($\Omega$) subtended by the reflector, given by

$$\Omega = 2\pi(1 - \cos(\emptyset)) \tag{6.2}$$

where $\phi$ is the half-angle subtended by the reflector at the focal point. The larger the solid angle, the more sunlight the reflector can capture.

**Ray-Tracing Simulations**

To optimize geometric designs, ray-tracing simulations are invaluable. These simulations model the path of individual rays of sunlight as they interact with the reflector's surface. The primary parameters to be analyzed include the following:

i. Parabolic, elliptical, or flat geometries can be tested to determine which configuration maximizes light capture for specific locations and seasons.
ii. Adjusting the tilt and azimuth angles of the reflector can enhance performance based on seasonal solar trajectories.

Ray-tracing algorithms typically employ the following steps:

i. Define the geometry of the reflector.
ii. Simulate the sun's path throughout the year.
iii. Calculate the number of rays that strike the reflector.
iv. Determine the proportion of rays redirected toward the receiver.

Mathematically, the efficiency ($\eta$) of the reflector can be defined as

$$\eta = \frac{N_{\text{captured}}}{N_{\text{incident}}} \tag{6.3}$$

$N_{\text{captured}}$ is the number of rays captured by the receiver, and $N_{\text{incident}}$ is the total number of rays on the reflector.

**Material Selection**

The materials used for secondary reflectors must be highly reflective and durable to withstand environmental conditions. The choice of materials impacts the overall efficiency and longevity of the solar thermal system.

1. **Polished Aluminum**

Polished aluminum is a widely used material for reflective surfaces in solar applications due to its high reflectivity, typically ranging 85–90%. This makes it highly efficient at capturing and redirecting sunlight, which is essential for maximizing energy collection. One of the key advantages of polished aluminum is its cost-effectiveness, especially in large-scale applications, where budget considerations are crucial. Despite its affordability, polished aluminum is durable, as it is resistant to corrosion, which helps maintain its structural integrity over time. However, to preserve its reflective properties and protect it from wear and environmental factors, polished aluminum may require a protective coating. This additional layer helps to maintain its high reflectivity and extend the material's lifespan, ensuring reliable performance in solar energy systems. Overall, polished aluminum offers an ideal balance of reflectivity, cost-effectiveness, and durability, making it a popular choice in solar thermal and photovoltaic systems.

2. **Silver-Coated Surfaces**

Silver-coated surfaces are highly efficient for solar applications, offering reflectivity values exceeding 95%, which allows them to capture and redirect more sunlight compared to other materials. This makes them an excellent choice for maximizing energy collection. However, silver coatings come with higher upfront costs due to the material expense and potential maintenance needs. Over time, silver coatings may tarnish, which can reduce their reflectivity. As a result, regular maintenance is required to preserve their high performance and ensure continued efficiency. Despite the increased cost and maintenance requirements, the superior reflectivity and performance of silver-coated surfaces make them a valuable option for applications where maximum efficiency is crucial.

**Mathematical Expressions for Reflectivity**

Reflectivity ($R$) can be quantitatively expressed in terms of the incident light intensity ($I_i$) and the reflected light intensity ($I_r$):

$$R = \frac{I_r}{I_i} \tag{6.4}$$

where

$I_r$ = intensity of reflected light

$I_i$ = intensity of incident light.

This equation helps in determining the effectiveness of various materials under real-world conditions.

## 6.11 Integration with Trough Collectors

The integration of secondary reflectors with parabolic trough collectors requires a strong focus on structural integrity and stability to maintain long-term performance and efficiency. Proper mounting and alignment are essential in ensuring the system functions optimally under varying environmental conditions. A key structural consideration is the mounting and support of the reflectors. The reflectors must be securely fixed to withstand environmental forces such as wind, thermal expansion, and other mechanical loads. The design must incorporate robust frames that minimize deflection, ensuring the reflectors stay in place and maintain consistent alignment with the collector. These structural elements are critical in preventing performance degradation due to misalignment or deformation.

Alignment between the secondary reflectors and the primary trough is another crucial factor for achieving optimal performance. The reflectors must be carefully positioned to direct sunlight onto the receiver for maximum light redirection accurately. This alignment is typically determined using geometric principles to ensure the correct angle relative to the sun's rays and the trough collector. Optimizing this angle is essential for maximizing solar energy capture throughout the day, as even small misalignments can lead to significant efficiency losses. Proper alignment and mounting are integral to ensuring that the entire system operates smoothly, enhancing energy collection and the long-term reliability of the solar thermal system. The angle of the reflector relative to the sun's rays and the collector should be optimized, generally represented by

$$\theta = \arctan\left(\frac{h}{d}\right) \tag{6.5}$$

where

$h$ = height of the collector

$d$ = distance from the reflector to the collector.

This alignment must be adjustable to accommodate seasonal changes in solar angle.

**Mathematical Model for Structural Stability**

The structural integrity of the reflector can be evaluated using the bending moment ($M$) and shear force ($V$) equations:

$$M = \frac{F \times L}{4} \tag{6.6}$$

$$V = \frac{F}{2} \tag{6.7}$$

where, the applied force ($F$) on a parabolic trough collector includes the weight of the reflectors, any additional equipment such as motors for tracking, and environmental loads, primarily from wind. The length ($L$) of the collector plays a critical role in determining bending moments and deflections along the trough. These factors must be carefully calculated to understand how the load is distributed across the collector. In a parabolic design, the load distribution is typically not uniform, meaning adjustments to the length ($L$) may be necessary depending on where the forces are applied to ensure the system's structural integrity and optimal performance.

The maximum deflection ($\delta$) of the reflector can be calculated using the formula for a supported beam:

$$\delta = \frac{5FL^3}{48EI} \tag{6.8}$$

where

$E$ = modulus of elasticity of the material

$I$ = moment of inertia of the reflector's cross-section.

## 6.12 Literature Review

In recent years, considerable research has been conducted on improving parabolic trough collectors' optical and thermal performance (PTCs) by incorporating secondary reflectors. A secondary reflector at the end or around the trough helps capture and redirect otherwise lost solar radiation back to the absorber tube, thus reducing optical losses and increasing system efficiency. These studies aim to improve PTC systems' overall performance, particularly by addressing challenges such as non-uniform heat flux distribution, optical losses, and concentration ratio limitations.

Bellos and Tzivanidis [2] investigated using a vertical secondary booster reflector at the end of the parabolic trough. This reflector aimed to reduce the end optical losses by reflecting otherwise lost radiation back to the receiver. Their study found that the booster reflector enhanced the system's optical performance, especially during periods with larger solar incident angles. Over a year-long period, the use of the secondary reflector improved optical efficiency by 21.7%. They reported monthly increases of 8.7% in June, 24.2% in September, and 77.1% in December, demonstrating that the enhancement was more significant during months with larger solar angles. Furthermore, the thermal analysis indicated even greater improvements in thermal performance, particularly when the heat transfer fluid operated at higher temperatures.

Bharti et al. [3] conducted experimental tests on a small-scale parabolic trough collector with different secondary reflector configurations: parabolic (PSR), triangular (TSR), and without secondary reflectors (WSR). The results indicated that the maximum outlet temperature of the heat transfer fluid reached 49.2 °C for the PSR configuration, 47.3 °C for the TSR, and 44.2 °C for the WSR. Correspondingly, the maximum thermal efficiency observed was 24.3, 22.5, and 17.8% for PSR, TSR, and WSR, respectively. Moreover, secondary reflectors helped achieve a uniform circumferential temperature distribution on the absorber tube surface, indicating reduced thermal stress and longer component lifetimes.

Reda and Abdelylah [4] addressed the problem of non-uniform heat flux distribution, which often leads to temperature gradients and mechanical stress in PTCs. They moved the absorber tube slightly below the focal line of the parabolic trough and introduced a secondary reflector to improve the uniformity of the heat flux. Through numerical simulations using SolTrace software, they demonstrated that the heat flux gradient could be reduced by over 70%, and the temperature gradient on the absorber tube could be decreased from 159.39 to 24.16 K, significantly improving the durability and performance of the PTC system.

Sundaram and Senthil [5] focused on reducing heat losses and improving thermal efficiency using a secondary reflector above the receiver tube. They observed that the secondary reflector increased thermal efficiency by 10% while reducing heat loss from the PTC system by 0.5 kJ/s. Their findings emphasized that even slight improvements in optical efficiency can result in meaningful thermal gains, which are critical in solar energy applications.

Rodriguez-Sanchez and Rosengarten [1] explored the potential for increasing the concentration ratio of PTCs by using a secondary flat reflector. They noted that conventional PTCs faced challenges in capturing all direct solar radiation, particularly at higher concentration ratios, which led to structural issues related to thermal expansion and wind resistance. By incorporating a secondary flat reflector, they increased the concentration ratio by up to 80% while minimizing the shadow area on the primary reflector. They also found that the improved system efficiency outweighed the minor increase in pumping power, making the system more cost-effective and efficient.

Gong et al. [6] developed an innovative design method for secondary reflectors based on an adaptive approach. The reflector geometry was optimized to reflect back solar radiation that was not initially captured by the absorber tube. Their ray-tracing simulations indicated that the optical efficiency of the PTC system could be improved by 5.2%, while the thermal efficiency increased by 4.9%. Furthermore, they demonstrated that the solar flux distribution on the absorber tube became significantly more uniform, reducing temperature gradients and improving system reliability.

Tang et al. [7] introduced a segmented, broken-line type secondary reflector composed of multiple plane mirrors, aimed at balancing the heat flux between the upper and lower surfaces of the absorber tube. Their study showed that the uniformity of the heat flux distribution improved by over 90%, leading to higher optical and thermal efficiency than other secondary reflector designs. Their work highlighted the importance of optimizing secondary reflector designs for specific system configurations. Table 6.1 elaborates the details of different studies for PTC with a secondary reflector.

Integrating secondary reflectors in parabolic trough collectors have consistently enhanced optical and thermal performance. The studies reviewed indicate that secondary reflectors significantly capture additional solar radiation, minimize optical losses, and improve the uniformity of heat distribution on the absorber tube. These improvements lead to higher thermal efficiencies, reduced temperature gradients, and extended system lifespans, making secondary reflectors a vital innovation in solar thermal technology. However, despite these benefits, there has been limited research on the application of secondary reflectors in parabolic trough collectors.

## 6.13 Considerations for Solar Collectors

### Dynamic Loads

Maximum wind speeds should be assessed based on local weather data. Standards, such as ASCE 7, are utilized to calculate wind pressure on the collector surfaces. The wind load distribution must be understood, as the parabolic shape affects how wind interacts with the collector. The effects of wind during different operational states, such as tracking versus stationary positions, should also be considered.

### Thermal Expansion

Different materials' thermal expansion coefficient (CTE) must be accounted for, as varying expansion rates are experienced. The expected expansion should be calculated based on the temperature ranges encountered by the collector. Movement should

**Table 6.1** Studies on PTC with secondary reflector

| Author | Type of secondary reflector | Findings |
|---|---|---|
| Shajan and Baiju [8] | 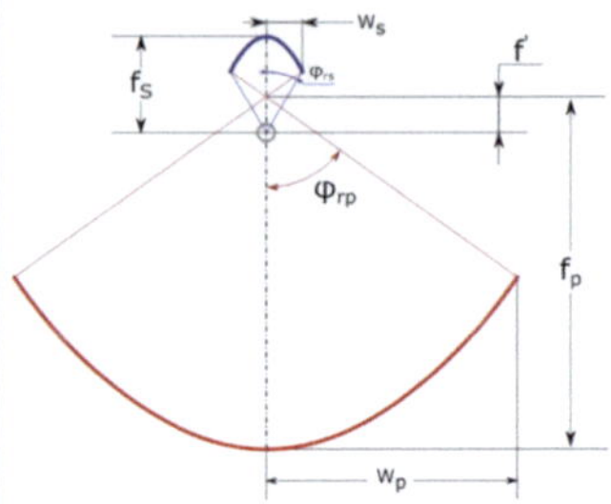  | The study found that a secondary reflector improves heat flux uniformity from 1.0836 to 0.58. The optimal configuration achieved a desirability value of 0.974 for a 5.5 kW output. CFD analysis showed reduced thermal gradients and improved heat flux distribution |
| Tang et al. [7] | 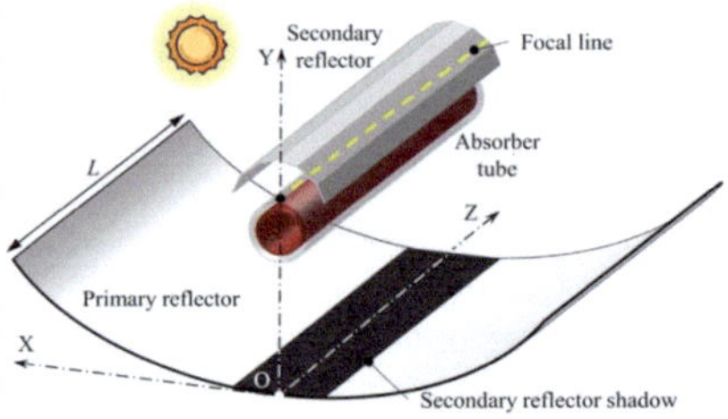  | The proposed method for designing a secondary reflector (SR) in a parabolic trough concentrator (PTC) improves heat flux uniformity on the absorber tube surface, achieving over 90% uniformity. The SR design also enhances optical efficiency, outperforming existing designs |
| Tian et al. [9] | 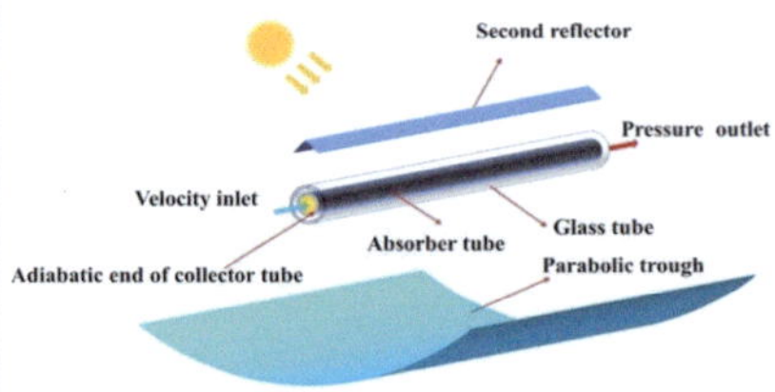  | Adding a secondary reflector (SR) and microencapsulated phase change material (MEPCM) slurries improves heat flux uniformity, enhances local heat transfer, and reduces thermal stress by 81.8%. Higher MEPCM slurry concentrations further enhance heat transfer and reduce thermal stress compared to water |

(continued)

**Table 6.1** (continued)

| Author | Type of secondary reflector | Findings |
|---|---|---|
| Taher et al. [10] | 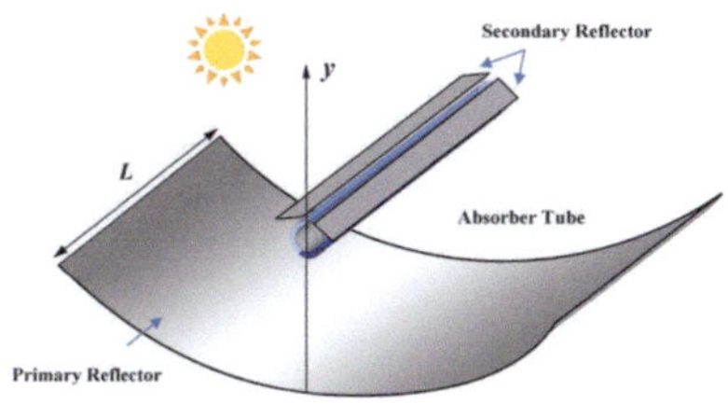  | The study presents a novel parabolic trough collector (PTC) design using the Monte Carlo Ray-Trace (MCRT) method to predict solar flux distribution. The design achieves 83.01% optical efficiency and 92.24% heat flux uniformity, optimizing both factors for improved performance in solar energy applications |
| Rodriguez-Sanchez and Rosengarten [1] | 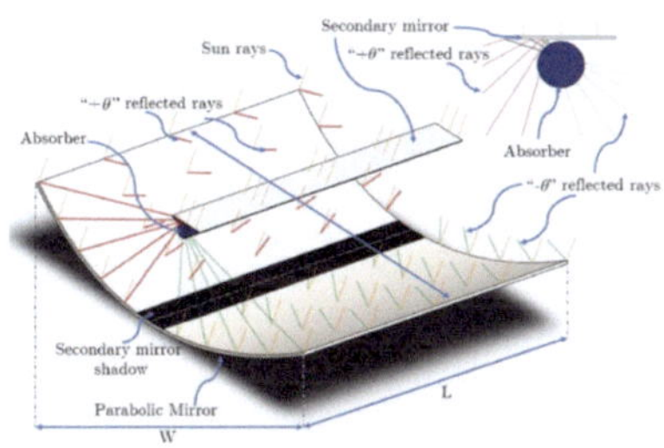  | The study found that secondary flat reflectors increase concentration in parabolic trough collectors, with minimal optical efficiency reduction (<3.5%) and improved misalignment tolerance. A truncated reflector version reduces shading, enhancing performance, especially for high-temperature applications |
| Reda and Abdelylah [4] | 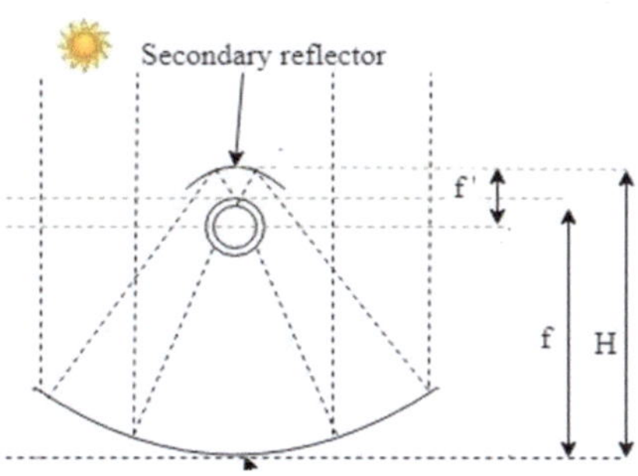  | The study found that moving the absorber tube downward and adding a secondary reflector reduced the heat flux gradient by 70.37% and lowered the temperature gradient from 159.39 to 24.16 K, enhancing the performance and longevity of the Parabolic Trough Collector |

(continued)

**Table 6.1** (continued)

| Author | Type of secondary reflector | Findings |
|---|---|---|
| Baiju and Shajan [11] | 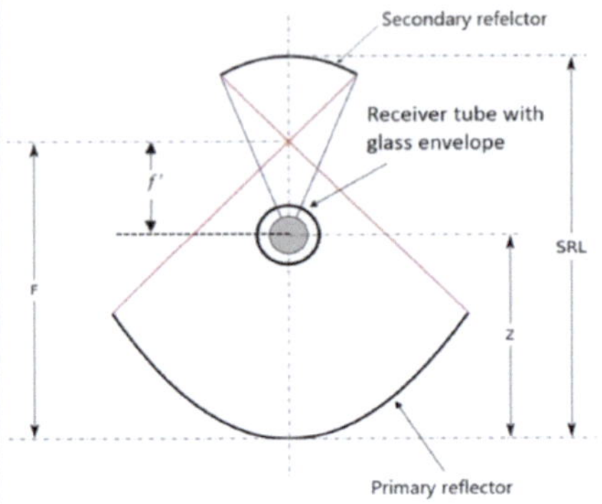 | The study analyzed the effect of receiver tube position and secondary reflector configuration on heat flux distribution, temperature gradient, and power output. Using the Tonatiuh simulation tool, results showed a 4.83% power output reduction, but significant improvements in flux uniformity and reduced heat gradients |
| Kun et al. [12] | 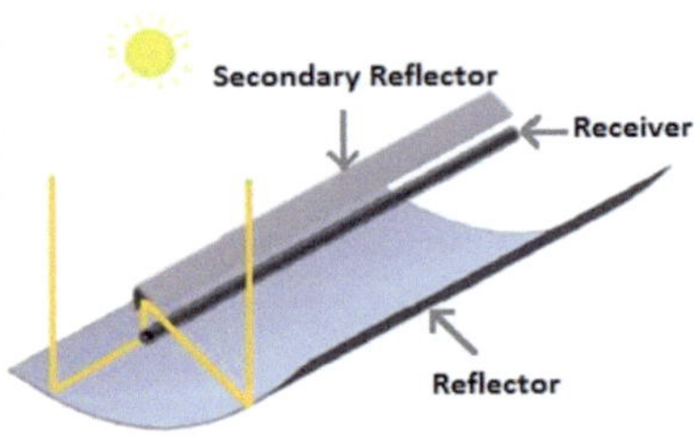 | Researchers found that adding a secondary reflector to traditional parabolic trough concentrators improves the intercept factor by capturing missed solar radiation. This results in a uniform heat flux distribution on the receiver, mitigating high local temperatures and temperature gradients |

(continued)

**Table 6.1** (continued)

| Author | Type of secondary reflector | Findings |
|---|---|---|
| Shyam et al. [13] | 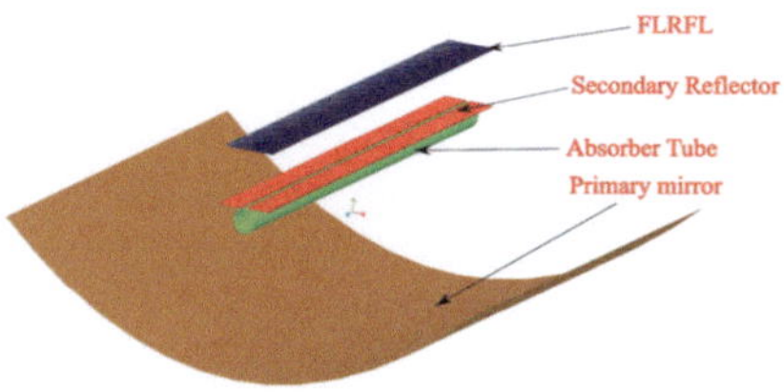 | The study presents a numerical analysis of a solar parabolic trough collector with optimized secondary optics to improve thermal performance. Simulation results show that the proposed model outperforms the standard LS-3 collector, achieving a maximum thermal efficiency of 92.6%, a heat transfer coefficient of 1481 W/m$^2$K, and an exergy efficiency of 52.8% |
| Bharti et al. [3] | 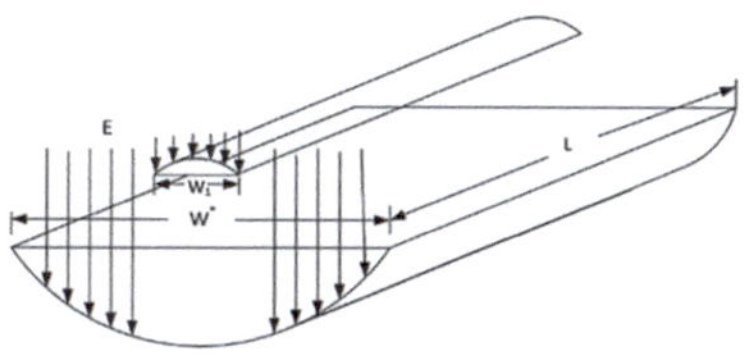 | The study finds that using secondary reflectors (parabolic and triangular) improves the thermal performance of solar PTCs. Secondary reflectors raise the maximum outlet temperatures and thermal efficiencies, and they also lead to a more uniform temperature distribution on the receiver tube |

be allowed in joints or connections to accommodate thermal expansion and contraction without inducing stress. Temperature gradients across the collector surface must be recognized, as they can lead to differential expansion and potential warping.

**Tracking Mechanism**

Additional forces that occur due to the movement of the tracking system should be evaluated. Dynamic forces must be considered, including inertial forces during acceleration or deceleration. Mechanical connections need to be ensured to handle the stresses applied during operation. Regular checks should be conducted to confirm that the tracking system remains properly aligned, minimizing unnecessary stress on the structure.

**Safety Factors**

Variability in material strength and properties must be accounted for through rigorous testing and conservative estimates. Load assumptions should be based on conservative values, factoring in higher-than-expected wind speeds or weights. Local and international building codes and standards should be followed, specifying minimum safety factors for solar structures. Design margins should consider factors such as fatigue, creep, and unexpected impacts over the structure's lifetime, with typical safety factors of 1.5–2.0 being employed depending on the application and materials.

Engineers can confirm the structural viability of the reflector design by ensuring that the deflection remains within acceptable limits.

The design considerations for secondary reflectors in solar thermal systems encompass geometric design, material selection, and integration with trough collectors. By understanding the mathematical principles behind these elements, engineers can optimize the performance and efficiency of solar energy systems.

Through careful geometric modeling, material evaluation, and structural analysis, secondary reflectors can be effectively designed to enhance the energy capture of solar thermal systems. As the demand for renewable energy solutions continues to grow, optimizing these components will be paramount in advancing solar technology.

Future research should focus on developing advanced materials and innovative designs that further improve the efficiency and reliability of secondary reflectors, ensuring their critical role in the transition to sustainable energy systems.

## 6.14 Optical Performance Analysis of Solar Collectors

The efficiency of solar collectors is heavily dependent on their optical performance, which encompasses how effectively sunlight is captured, concentrated, and converted into usable thermal energy. This analysis focuses on three critical aspects: ray-tracing simulations to model light interaction, metrics for evaluating performance, and case studies that illustrate real-world improvements.

**Ray-Tracing Simulations**

Ray-tracing simulations are employed to analyze the interaction of light with the reflectors in solar collectors. These simulations provide insights into the optimal configurations for reflectors, allowing designers to maximize the amount of sunlight focused onto the receiver. Tracing the paths of individual rays of sunlight makes it possible to visualize and quantify how effectively the solar energy is concentrated.

**Methodology**

i. **Ray Generation**: The simulation begins with generating rays that represent sunlight. These rays are modeled as straight lines emanating from a virtual sun source, which is typically defined by parameters such as its position in the sky and intensity. For a stationary solar collector, the rays are generated based on the sun's angle relative to the Earth's surface.

ii. **Interaction with Reflectors**: As the rays encounter the reflector surface, their behavior is analyzed based on reflection principles. The law of reflection states:

$$\theta_i = \theta_r$$

where $\theta_i$ is the angle of incidence and $\theta_r$ is the angle of reflection. For parabolic reflectors, this principle is crucial because the curvature of the reflector influences how effectively rays are directed toward the receiver. The equations governing the reflective properties are used to calculate the new directions of the rays after they strike the surface.

iii. **Energy Collection**: The energy focused at the receiver is calculated once the rays have been traced through the simulation. The total energy captured by the receiver can be expressed as

$$E_{\text{focused}} = \sum E_i \tag{6.9}$$

$E_i$ represents each individual ray's energy that successfully reaches the receiver. This summation allows for quantifying how much solar energy has been effectively concentrated.

The outcome of ray-tracing simulations is a comprehensive analysis of light distribution on the receiver. By adjusting the design parameters of the reflectors—such as shape, size, and orientation—optimal configurations that maximize solar energy capture can be identified. The data generated from these simulations provides invaluable insights into design adjustments that can enhance overall system performance.

## 6.15 Metrics of Performance

Several key metrics are considered to effectively evaluate the optical performance of solar collectors. These metrics help quantify how well a solar collector captures and utilizes solar energy.

**Concentration Ratio (CR)**

The concentration ratio (CR) is a crucial performance indicator that measures how effectively sunlight is focused on the receiver. It is defined mathematically as

$$\text{CR} = \frac{P_{\text{receiver}}}{P_{\text{incident}}} \tag{6.10}$$

where

$P_{\text{receiver}}$ is the power received at the receiver (W).

$P_{\text{incident}}$ is the power incident on the collector's area (W).

A higher concentration ratio indicates better performance in focusing solar energy, directly correlating to increased thermal energy output. The concentration ratio can vary depending on the geometry of the reflector and its alignment with the sun.

### Uniformity of Heat Distribution

Uniformity of heat distribution across the receiver is vital for ensuring efficient thermal energy capture. This can be measured using the uniformity ratio:

$$\text{UR} = \frac{P_{\text{max}}}{P_{\text{avg}}} \tag{6.11}$$

where

$P_{\text{max}}$ is the maximum power density on the receiver ($\text{W/m}^2$).

$P_{\text{avg}}$ is the average power density on the receiver ($\text{W/m}^2$).

A uniformity ratio close to 1 indicates an even distribution of heat, which minimizes the risk of thermal stress on the receiver and improves the system's overall efficiency. Uneven heating can lead to localized overheating, which may damage the receiver and reduce its lifespan.

### Overall Optical Efficiency ($\eta_{\text{optical}}$)

Overall optical efficiency is a critical metric that measures the effectiveness of converting solar energy into thermal energy. It is defined as

$$\eta_{\text{optical}} = \frac{P_{\text{thermal}}}{P_{\text{solar}}} \tag{6.12}$$

where

$P_{\text{thermal}}$ is the useful thermal power output (W).

$P_{\text{solar}}$ is the total incident solar power (W).

This efficiency metric accounts for various losses due to reflection, absorption, and other factors, providing a comprehensive view of collector performance. Higher optical efficiency indicates a better capability to convert sunlight into useful heat, making it a vital target for design optimization.

### Case Studies

Empirical case studies provide practical insights into the effectiveness of design optimizations in enhancing the performance of solar collectors. Here, we present several notable studies that illustrate significant improvements achieved through tailored reflector configurations.

**Case Study 1: Parabolic Trough Collector Optimization**

In a study optimizing a parabolic trough collector, researchers adjusted the reflector's curvature. Initial simulations indicated a concentration ratio of 20:1. After modifying the design to improve the reflector's parabolic profile, the concentration ratio increased to 25:1. This improvement can be quantified as

$$\Delta CR = CR_{new} - CR_{baseline} = 25 - 20 = 5$$

Furthermore, the overall optical efficiency of the system improved from 75 to 87% due to better light focus and heat distribution. This efficiency improvement can be expressed as

$$\Delta \eta_{optical} = \eta_{new} - \eta_{baseline} = 0.87 - 0.75 = 0.12$$

$$\text{Percentage Increase} = \left(\frac{\Delta \eta_{optical}}{\eta_{baseline}}\right) \times 100$$

Calculating the percentage increase:

$$\text{Percentage Increase} = \left(\frac{0.12}{0.75}\right) \times 100 \approx 16\%$$

These results underscore the importance of precise geometric configurations in enhancing collector performance.

**Case Study 2: Thermal Distribution Uniformity**

Another study focused on the uniformity of heat distribution across the receiver. Initial tests revealed a maximum power density of 800 W/m$^2$ and an average of 600 W/m$^2$, leading to a uniformity ratio of

$$UR = \frac{800}{600} \approx 1.33$$

After implementing a secondary reflector designed to direct additional sunlight toward the receiver, the maximum power density improved to 700 W/m$^2$, while the average remained at 600 W/m$^2$. The new uniformity ratio was calculated as

$$UR = \frac{700}{600} \approx 1.17$$

This adjustment resulted in a more uniform heat distribution, reducing the risk of thermal stress and increasing the system's longevity.

**Case Study 3: Impact of Tracking Systems**

A third case study evaluated the impact of a dual-axis tracking system on the optical performance of solar collectors. The tracking system significantly improved energy capture by allowing the collector to follow the sun's path throughout the day. The study found that implementing the tracking system resulted in a 15% increase in overall energy capture compared to a fixed system.

Before the installation of the tracking system, the average incident solar power was measured at 900 W/m$^2$. After the tracking system was implemented, the average increased to 1035 W/m$^2$:

$$\Delta P_{\text{solar}} = P_{\text{solar,tracking}} - P_{\text{solar,fixed}} = 1035 - 900 = 135\,\text{W/m}^2$$

This significant improvement highlights the effectiveness of tracking systems in maximizing solar energy capture throughout the day.

**Additional Considerations**

In addition to the aforementioned metrics and case studies, several factors must be considered when assessing the optical performance of solar collectors:

i. **Material Properties**: The choice of reflective material impacts the reflection coefficient and durability. High-quality reflective coatings can enhance optical efficiency by minimizing absorption losses.
ii. **Environmental Factors**: The impact of weather conditions, such as clouds and atmospheric dust, on solar performance should also be evaluated. Regular maintenance is required to keep the collectors clean and operational.
iii. **System Integration**: The design of the entire solar thermal system, including storage and heat transfer components, should be optimized alongside the collectors to ensure overall efficiency.

## 6.16 Thermal Performance Evaluation of Solar Collectors

The thermal performance of solar collectors is essential for maximizing the conversion of solar energy into usable thermal energy. This evaluation will delve deeply into the heat transfer analysis of solar collectors, the experimental results showcasing their efficacy, and the impact of secondary reflectors in improving thermal performance.

**Heat Transfer Analysis**

Heat transfer is a critical aspect of solar collectors, as it directly influences their ability to convert solar energy into thermal energy. Solar collectors are designed to absorb sunlight and transfer the energy to a working fluid, typically water or a glycol mixture. This fluid then carries the heat to where it is needed for various applications such as space heating, industrial processes, and electricity generation in concentrated solar power systems. To maximize the efficiency of solar collectors, it is essential to understanding and optimizing the three primary mechanisms of heat transfer: conduction, convection, and radiation. Each mechanism uniquely ensures that the maximum amount of solar energy is transferred to the working fluid.

Conduction is the process by which heat is transferred through a solid material. In solar collectors, conduction occurs between the absorber plate and the heat transfer fluid. The absorber plate is designed to capture solar radiation, converting it into heat. For efficient heat transfer, the material of the absorber plate must have high thermal conductivity, allowing heat to flow easily from the plate into the fluid. Because of their excellent thermal conductivity, copper and aluminum are commonly used for absorber plates. The collector design must also minimize thermal resistance at the interface between the absorber and the fluid, ensuring efficient heat transfer and reducing energy losses. This is particularly important for enhancing the overall efficiency of the solar collector.

Convection is the process by which heat is transferred through a fluid, whether it is a liquid or a gas. In solar collectors, convection occurs in two stages: natural convection and forced convection. In natural convection, when the absorber plate heats the heat transfer fluid, it becomes less dense and rises, while cooler, denser fluid moves downward to replace it. This natural circulation enhances heat transfer, but natural convection alone may not be sufficient in larger systems or those with higher energy demands. Forced convection, which is achieved by using pumps to circulate the heat transfer fluid, improves heat transfer rates significantly. By constantly moving the fluid over the absorber plate, forced convection helps to prevent the buildup of thermal gradients and ensures uniform heating of the fluid. This is particularly useful in solar thermal systems requiring larger surface areas or higher thermal outputs.

Radiation is the process by which energy is transferred through electromagnetic waves. Solar energy reaches Earth through solar radiation, which is absorbed by the absorber plate in the solar collector. The efficiency of this radiation absorption depends on the surface properties of the absorber, particularly its absorptance (the ability to absorb solar radiation) and **emissivity** (the ability to re-radiate heat). To maximize absorption while minimizing energy loss, many absorber plates are coated with selective coatings that improve their absorptance and reduce emissivity. These coatings allow the absorber to capture more solar energy and prevent excessive heat loss, especially during cooler periods when retaining heat is essential for maintaining collector efficiency. By enhancing the radiation absorption and minimizing re-radiation, the collector becomes more efficient in converting solar energy into usable thermal energy.

## 6.17 Heat Loss Considerations

Minimizing heat losses in solar collectors is crucial for maximizing their efficiency and overall performance. Heat losses occur primarily through three mechanisms: conduction, convection, and radiation. Each of these heat transfer processes can significantly reduce the amount of thermal energy that is retained within the solar collector, which in turn lowers the efficiency of the system. Addressing these losses requires careful selection of materials, design optimization, and the application of advanced technologies.

Conduction losses are related to the transfer of heat through solid materials, such as the absorber plate, pipes, and surrounding insulation. In a solar collector, heat is absorbed by the absorber plate, and if the materials surrounding it are not sufficiently insulated, this heat will transfer away from the collector and into the environment. This can significantly reduce the amount of heat available for use. To minimize conduction losses, selecting materials with low thermal conductivity for insulation is essential. Insulating materials like fiberglass, rock wool, or polyurethane foam are commonly used because they prevent heat from escaping through the sides and back of the collector. In addition, ensuring that the absorber plate itself is made from a material with high thermal conductivity, such as copper or aluminum, allows the absorbed heat to be efficiently transferred to the heat transfer fluid. However, the back of the absorber plate must also be properly insulated to prevent heat from being lost through conduction. The overall design of the collector is also crucial; systems using double-walled or vacuum-insulated panels further reduce conduction losses by providing an additional layer of insulation between the collector's internal components and the external environment.

Convection losses occur when heat is transferred from the surface of the collector to the surrounding air through the movement of warm air. As the absorber plate heats up, the air adjacent to the collector surface warms and rises, carrying heat away. This can be a significant issue, especially in high-temperature collectors or those exposed to wind. Convection losses can be reduced using a cover or glazing that minimizes air movement around the collector. Materials such as tempered glass or polycarbonate are commonly used for this purpose. These materials trap heat close to the surface of the collector, reducing convective heat loss. Additionally, these covers can have selective coatings that reduce the emissivity of the glazing, helping to retain more heat. Another design consideration is the use of a deeper or enclosed absorber plate or back plate, which can reduce the surface area exposed to air movement, further minimizing convection losses.

Radiative losses are a result of the collector's surface emitting infrared radiation to the cooler surrounding environment. As the temperature of the collector increases, its surface radiates heat in the form of infrared radiation, which can be lost if not effectively retained. This type of loss is particularly significant at higher operating temperatures, and it can be minimized by using selective coatings on the absorber plates. These coatings are designed to have high absorptance for solar radiation (ensuring that the plate absorbs as much sunlight as possible) but low emissivity (minimizing the amount of infrared radiation that the plate radiates back into the environment). Materials such as black chrome or anodized aluminum are commonly used for these coatings, as they provide high efficiency in both absorbing solar radiation and reducing heat loss through radiation. Additionally, design features like evacuated tubes or double-glazing can further reduce radiative losses by limiting the surface area exposed to cooler air and by providing additional layers of insulation.

By addressing all three types of heat loss—conduction, convection, and radiation—solar collectors can operate at higher efficiencies, capturing and retaining more of the thermal energy from the sun. Proper material selection, advanced coatings, and optimized design all play critical roles in minimizing heat losses and ensuring that solar thermal systems can deliver the maximum amount of usable heat. This not only improves the performance of solar collectors but also enhances the overall feasibility and effectiveness of solar thermal technology as a renewable energy solution.

## 6.18 Overview of Experimental Studies

Numerous experimental studies have been conducted to evaluate the thermal performance of solar collectors, particularly those equipped with secondary reflectors. These studies focus on key metrics such as thermal efficiency, temperature rise, and system stability under varying environmental conditions, including clear skies, partial clouds, and full cloud cover. Overall, the results indicate that solar collectors with secondary reflectors perform better than traditional designs, largely due to their enhanced ability to concentrate solar energy onto the absorber, resulting in improved energy capture and system performance.

**Experimental Setup by Bharti et al.** [3]

In a study conducted by Bharti et al., a parabolic trough collector (PTC) system was tested using different configurations of secondary reflectors: without secondary reflectors (WSR), parabolic secondary reflectors (PSR), and trapezoidal secondary reflectors (TSR). The primary reflector in this setup was made of stainless steel with 88% reflectivity, while the receiver tube was coated with matte black paint for 92% absorptivity. To minimize heat losses, an acrylic tube was used to surround the receiver, providing high transmissivity and cost-effectiveness. The inclusion of secondary reflectors allowed for more concentrated solar energy and improved heat absorption.

**Direct Normal Irradiance (DNI) and Temperature Rise**

The experiments showed that the PSR configuration achieved the highest direct normal irradiance (DNI) at 862.59 W/m$^2$, outperforming the TSR and WSR setups. The PSR setup's enhanced ability to focus solar energy during peak daylight hours resulted in greater heat transfer to the heat transfer fluid (HTF). Consequently, the PSR system recorded the highest outlet temperature for the HTF, reaching 49.2 °C, compared to 47.3 °C for TSR and 44.2 °C for WSR. The greater temperature rise observed in the PSR configuration demonstrates its superior ability to enhance thermal performance.

**Thermal Efficiency Improvements**

Solar collectors with secondary reflectors consistently demonstrated better thermal efficiency than those without. Bharti et al.'s experiments revealed that the PSR configuration achieved the highest thermal efficiency at 24.3%, followed by 22.5% for TSR and 17.8% for WSR. This improvement can be attributed to the PSR system's ability to better concentrate solar energy onto the receiver tube, resulting in increased heat gain. Similarly, in broader studies, thermal efficiency improvements of 10–20% have been observed with secondary reflectors, particularly under peak solar irradiance.

**Uniform Heat Distribution and System Stability**

The PSR configuration showed the most uniform temperature distribution along the circumference of the receiver tube, with minimal temperature differences. Uniform heat distribution is critical for preventing thermal stress, which can negatively impact the system's reliability and lifespan. In contrast, the WSR setup exhibited larger temperature variations along the receiver, indicating uneven heat distribution and potentially higher thermal stress. With their better heat concentration, the PSR and TSR setups contributed to greater system durability and more consistent thermal performance, even under fluctuating weather conditions.

**Effect of Wind Speed on System Performance**

Wind speed was measured during the experiments, but it had a negligible impact on the overall performance of the PTC systems. Despite variations in wind speed, the thermal efficiency and heat transfer rates remained largely unaffected. This suggests that the performance differences between the various configurations were primarily due to the design and inclusion of secondary reflectors, rather than external environmental factors such as wind.

Both Bharti et al.'s study and the broader body of research demonstrate that secondary reflectors significantly enhance the thermal performance of PTC systems. The PSR configuration, in particular, consistently outperformed the other setups in key areas such as thermal efficiency, temperature rise, and heat distribution. These findings underscore the importance of integrating secondary reflectors—especially parabolic ones—into solar thermal systems to optimize energy capture, improve system reliability, and increase overall efficiency.

## 6.19 Thermal Performance Under Different Conditions

**Seasonal Variability**

The performance of solar collectors can vary significantly with seasonal changes. Experimental studies have shown the following.

1. **Summer Versus Winter Performance**

In the summer, when solar irradiance is naturally higher, all solar collectors benefit from increased sunlight. However, the **advantage of secondary reflectors becomes particularly significant during winter months**. At lower solar angles, the sunlight tends to be more diffuse, and collectors without secondary reflectors may struggle to capture and concentrate enough solar energy efficiently.

Secondary reflectors are specifically designed to **redirect and concentrate this diffuse radiation** onto the absorber, maximizing the amount of sunlight captured. This results in more efficient heat absorption during winter, when irradiance is lower and sunlight is less direct. Consequently, systems with secondary reflectors can maintain higher thermal efficiency and better overall performance even under less favorable conditions. This makes them especially useful in regions with seasonal variability in sunlight angles and intensity.

2. **Temperature Effects**

High ambient temperatures reduce the thermal gradient between the heat transfer fluid (HTF) and the environment, diminishing the efficiency of solar collectors. Secondary reflectors mitigate this by enhancing sunlight concentration onto the absorber, ensuring more consistent and intense solar radiation exposure for the HTF. This helps maintain higher fluid temperatures despite the reduced gradient, improving performance. Additionally, secondary reflectors minimize thermal losses by redirecting scattered or diffuse sunlight back onto the collector, reducing energy lost to radiation and convection. This feature is especially beneficial in hot climates or summer months, optimizing solar collector efficiency and maintaining effective energy capture.

**Impact of Weather Conditions**

Weather conditions greatly impact the performance of solar thermal systems, and secondary reflectors play a crucial role in mitigating some of these effects, thereby enhancing overall efficiency.

i. **Cloudy Days**: On overcast days, solar irradiance is reduced as clouds scatter sunlight, leading to a significant drop in the performance of conventional solar collectors that rely on direct sunlight. However, solar systems equipped with secondary reflectors are more effective in capturing diffused solar radiation. These reflectors can redirect scattered sunlight onto the receiver, helping maintain a higher thermal efficiency even under cloudy conditions. Studies show that solar systems with secondary reflectors can achieve thermal efficiencies of around 60% on overcast days, compared to ~40% for conventional systems. This ability to optimize the collection of diffused light ensures a more reliable and consistent energy generation, particularly in regions with frequent cloud cover.
ii. **Wind Effects**: Wind can increase convective heat losses from the absorber surface, reducing the overall thermal performance of solar collectors, especially in exposed installations. Air currents strip away heat from the absorber, leading to inefficient energy transfer. Secondary reflectors, designed with appropriate

shielding, can mitigate these losses. By creating a barrier between the absorber and wind, secondary reflectors limit the heat carried away by air currents, improving thermal stability. Additionally, they can concentrate sunlight, compensating for heat loss due to wind exposure, making the system more efficient in windy environments where traditional systems struggle to maintain performance.

Longitudinal studies highlight the importance of ongoing monitoring and maintenance for solar systems with secondary reflectors. Regular cleaning and inspection of reflectors and maintenance of the heat transfer fluid system are crucial for ensuring long-term reliability. Experimental findings confirm that secondary reflectors enhance energy capture, improve thermal absorption, and stabilize system performance across various environmental conditions. Their use contributes to the growing viability of solar energy as a reliable and sustainable power source.

## 6.20 Advantages and Disadvantages of Secondary Reflectors

The use of secondary reflectors in solar collector systems offers both significant advantages and some potential disadvantages. Understanding these factors is crucial for optimizing solar thermal systems and making informed design decisions.

**Advantages**

**Improved Heat Flux Uniformity**

One of the primary benefits of incorporating secondary reflectors is the enhancement of heat flux uniformity across the receiver surface.

i. Secondary reflectors help distribute sunlight more evenly, minimizing hotspots where certain areas of the receiver absorb significantly more heat than others. This uniformity reduces the risk of thermal stress and damage to the receiver material.
ii. A more uniform temperature distribution improves the overall thermal efficiency of the system. This consistency is essential for applications requiring stable heat output, such as industrial processes or district heating systems.

**Increased Energy Capture**

Secondary reflectors play a vital role in maximizing the energy collection capabilities of solar thermal systems.

i. By redirecting additional sunlight toward the receiver, secondary reflectors can increase the effective solar irradiance received by the collector. This leads to higher thermal energy absorption and ultimately enhances the overall energy capture of the system.

ii. Secondary reflectors are particularly beneficial during mornings, evenings, or cloudy days, when the solar angle is low or sunlight diffuses. They can effectively capture and focus this available solar energy, increasing overall efficiency during less-than-ideal conditions.

**Flexibility in Design**

The incorporation of secondary reflectors allows for greater design flexibility in solar collector systems.

i. Secondary reflectors can be designed to meet specific application requirements, such as size, shape, and orientation. This adaptability enables the creation of more efficient systems tailored to particular geographic locations, climate conditions, or energy demands.
ii. Secondary reflectors can be added to existing solar collector systems without the need for complete redesign, providing a straightforward way to enhance performance and energy capture.

**Disadvantages**

**Increased Cost**

While the advantages of secondary reflectors are substantial, there are also financial considerations.

i. The integration of secondary reflectors into solar collector systems typically involves higher upfront costs. This includes the cost of materials, fabrication, and installation. While the long-term energy savings may justify the investment, the initial expenditure can be a barrier for some users.
ii. Potential users must weigh the increased costs against the expected efficiency and energy output gains to determine if secondary reflectors are a worthwhile investment for their specific situation.

**Maintenance Requirements**

Secondary reflectors introduce additional maintenance challenges.

i. The presence of secondary reflectors means regular inspections and cleaning are necessary to ensure optimal performance. Dust, dirt, or debris can accumulate on the reflective surfaces, reducing their effectiveness and potentially leading to lower energy capture.
ii. Ongoing maintenance can result in increased labor costs, as the system may require more attention than conventional systems without secondary reflectors.

**Potential Structural Issues**

The addition of secondary reflectors can impact the structural integrity of solar collector systems.

i. Secondary reflectors add weight and can affect the overall load distribution within the collector system. This change may necessitate modifications to the support structures to ensure they can handle the additional stresses.
ii. Improper installation or inadequate support for secondary reflectors can lead to structural issues, including bending or deformation. These problems may compromise the performance and lifespan of the system, leading to higher maintenance costs and potential failures.

## 6.21 Future Directions in Solar Collector Technology for Secondary Reflectors

The incorporation of secondary reflectors in solar collector systems offers significant potential for improving energy efficiency and overall performance, particularly as the demand for renewable energy continues to rise. Technological advancements in this area could lead to more effective solar thermal solutions. One key development is the use of advanced reflective materials, such as high-reflectivity coatings like multi-layer dielectric films or advanced metallic coatings, which can enhance energy capture, especially in low-angle sunlight conditions, while maintaining durability. Additionally, lightweight materials, such as composites or polymers, can reduce the structural load, making it easier to integrate secondary reflectors without compromising the integrity of the system. Another promising direction is the integration of innovative reflector systems, including automated tracking systems that adjust the position of the reflectors to follow the sun's path, maximizing energy capture throughout the day. Sensors and control systems can also be used to monitor solar irradiance and temperature, enabling real-time adjustments to optimize performance. Further, the integration of thermal energy storage solutions such as phase change materials (PCMs) can store excess energy captured by secondary reflectors and release it when needed, smoothing out energy supply and improving overall efficiency. Hybrid systems combining solar collectors with secondary reflectors and other renewable energy sources like photovoltaics offer additional potential to maximize energy output and ensure a more reliable energy supply.

## 6.22 Energy Management Systems (EMS) for Parabolic Solar Trough Collectors with Secondary Reflectors

Energy Management Systems (EMS), when integrated with Parabolic Solar Trough Collectors **(PSTCs)** featuring secondary reflectors, significantly optimize energy capture, storage, and utilization. The secondary reflectors enhance the solar system's ability to collect and redistribute sunlight toward the receiver, leading to improved thermal efficiency and energy output. Through intelligent control and dynamic management, EMS plays a critical role in maximizing the efficiency and sustainability of these solar thermal systems.

**Load Management**

**Dynamic Load Balancing**

i. **Matching Supply with Demand**: In a parabolic solar trough system, the amount of energy generated depends on solar irradiance, which fluctuates throughout the day. An Energy Management System (EMS) can manage the energy produced by the solar system to match consumption dynamically. For example, during peak sunlight hours, EMS can direct more energy to industrial processes or charge thermal storage systems, while during off-peak hours, it can pull energy from storage systems.
ii. **Managing Fluctuating Energy Demand**: In commercial or industrial applications, energy demand is often unpredictable, with spikes in usage at certain times of the day. EMS monitors real-time energy consumption and ensures that the energy produced by the solar trough is used directly, stored for later use, or fed back into the grid. This is especially critical in systems with secondary reflectors, as they increase energy capture, resulting in more energy availability.
iii. **Demand Response Mechanisms**: EMS can also incorporate demand response strategies, where it shifts energy-intensive processes to times when solar energy production is high. For example, heating systems or air compressors can be scheduled to operate when the parabolic trough collector is at its peak solar concentration.

**Environmental Considerations**

**Sustainable Manufacturing Practices**

The future development of secondary reflectors should emphasize sustainability in materials and manufacturing processes.

i. **Recyclable Materials**: Developing secondary reflectors using recyclable or biodegradable materials can reduce their environmental impact. Emphasizing life cycle assessments during the design phase can help identify the most sustainable options.
ii. **Eco-friendly Coatings**: Researching environmentally friendly reflective coatings that do not compromise performance can contribute to more sustainable solar technologies.

**Life Cycle Assessments**

Conducting comprehensive life cycle assessments (LCAs) for solar collector systems with secondary reflectors will be crucial.

i. **Assessment of Environmental Impact**: LCAs can provide insights into the environmental impacts of secondary reflectors throughout their life cycle, from production to disposal. This data can guide manufacturers in making more sustainable choices.

ii. **Improvement of Efficiency**: By analyzing energy inputs and emissions associated with different materials and designs, manufacturers can optimize their processes to reduce waste and energy consumption.

The future of solar collector technology, especially with secondary reflectors and energy storage, promises enhanced efficiency, stability, and sustainability.

## References

1. Rodriguez-Sanchez D, Rosengarten G (2015) Improving the concentration ratio of parabolic troughs using a second-stage flat mirror. Appl Energy 159:620–632
2. Bellos E, Tzivanidis C (2019) Investigation of a booster secondary reflector for a parabolic trough solar collector. Sol Energy 179:174–185
3. Bharti A, Mishra A, Paul B (2019) Thermal performance analysis of small-sized solar parabolic trough collector using secondary reflectors. Int J Sustain Energy 38:1002–1022
4. Reda HM, Abdelylah B (2019) Numerical investigation and solar flux distribution analysis of parabolic trough solar collector by adding secondary reflector. Instrum Mesur Metrol 18:275–280
5. Sundaram P, Senthil R (2017) Thermal performance enhancement of solar parabolic trough collector using secondary reflector. Int J Eng Technol 15:985
6. Gong J, Wang J, Lund PD, Hu E, Xu Z, Liu G, Li G (2020) Improving the performance of a 2-stage large aperture parabolic trough solar concentrator using a secondary reflector designed by adaptive method. Renew Energy 152:23–33
7. Tang XY, Yang WW, Yang Y, Jiao YH, Zhang T (2021) A design method for optimizing the secondary reflector of a parabolic trough solar concentrator to achieve uniform heat flux distribution. Energy 229:120749
8. Shajan S, Baiju V (2022) Designing a novel small-scale parabolic trough solar thermal collector with secondary reflector for uniform heat flux distribution. Appl Therm Eng 213:118660
9. Tian L, Liang M, Guo J, Zhang Y, Wu Z, Wang K (2024) Heat transfer and thermal stress of a parabolic trough solar collector with secondary reflector using microencapsulated phase change material slurries. Therm Sci Eng Progr 48:102364
10. Taher MAB, Pelay U, Russeil S, Bougeard D (2023) A novel design to optimize the optical performances of parabolic trough collector using Taguchi, ANOVA and grey relational analysis methods. Renew Energy 216:119105
11. Baiju V, Shajan S (2022) Secondary reflector and receiver positions for uniform heat flux distribution in parabolic trough solar thermal collector. J Solar Energy Eng 15:144
12. Kun W, YaLing H, ZeDong C (2014) A design method and numerical study for a new type parabolic trough solar collector with uniform solar flux distribution. Sci China Technol Sci 57:531–540. https://doi.org/10.1007/s11431-013-5452-6
13. Shyam A, Ahmed KRA, Christopher SS, Iniyan S (2024) Thermal performance enhancement in a solar parabolic trough collector with optimized secondary optics. Clean Technol Environ Policy 26:771–784

# Chapter 7
# Nanofluid Applications in Parabolic Solar Trough Collectors

This chapter explores the emerging role of nanofluids as advanced heat transfer media in Parabolic Solar Trough Collectors (PTCs), offering a promising path to enhance the efficiency of solar thermal systems. Traditional heat transfer fluids, such as water, oils, and molten salts, often suffer from low thermal conductivity, which limits system performance. To address this, researchers have developed nanofluids—engineered by dispersing nanoparticles into conventional fluids—to significantly improve thermal conductivity, heat transfer, and overall system efficiency. The chapter discusses various types of nanoparticles, base fluids, and the thermo-physical properties that influence their performance. It also examines hybrid nanofluids, recent advancements, and the practical challenges involved, such as stability, cost, and compatibility. Concluding the book, this chapter presents a forward-looking view on how nanofluid technology could shape the next generation of high-efficiency solar thermal systems in the global transition to clean energy.

## 7.1 Role of Nanofluids in Addressing the Limitations of Traditional Heat Transfer Liquids (HTFs) in Parabolic Trough Collectors (PTCs)

Standardized heat exchange mediums like water, oil, and molten salts, implemented in Parabolic Trough Collector (PTC) systems, are affected by low thermal conductivity, limiting their efficiency in transferring heat, especially at high temperatures. Nanofluids, formulated by incorporating nanoparticles (e.g., metals, metal oxides, carbon materials) into a base fluid, offer a solution by significantly enhancing thermal properties. These nanoparticles improve thermal transport by intensifying thermal conductivity and surface area and stimulating micro-convection. As a result, nanofluids boost PTC system efficiency, enabling better solar energy capture and transfer across varied operating conditions.

 
O. Chakraborty and B. Das, *Parabolic Solar Trough Collector*, Mechanical Engineering Series, https://doi.org/10.1007/978-3-032-11229-3_7

Hybrid nanofluids offer enhanced heat flow capabilities for Parabolic Trough Collectors (PTCs) by combining various nanoparticles in a single base fluid. This approach leverages the strengths of each nanoparticle type, such as the elevated thermal conductivity exhibited by metallic nanoparticles (e.g., copper, silver) and the stability of metal oxides (e.g., alumina, titanium dioxide), to achieve superior thermal performance. Hybrid nanofluids also address challenges like sedimentation and agglomeration, improving long-term stability. Their customizable nature allows them to be optimized for specific applications, including high-temperature operations, making them ideal for advanced solar thermal systems.

The enhancement of heat flow in nanofluids is driven by several mechanisms, primarily due to the unique properties of nanoparticles. When metallic elements or their oxides are dispersed in traditional heat transfer fluids (HTFs), they significantly improve the liquid's thermal conductivity. These nanoparticles provide additional pathways for heat to transfer more efficiently through the fluid. Brownian motion, the random movement of nanoparticles, creates micro-convection currents that improve fluid mixing and distribute thermal energy more evenly. This enhanced mixing leads to better heat absorption and transfer from the Parabolic Trough Collectors (PTCs) receiver tube. Additionally, the extensive surface area of nanoparticles increases the interaction sites between the fluid and the heat source, further improving thermal performance. The scale and configuration of nanoparticles are also crucial, with smaller particles and specific forms, such as rods or platelets, generating more turbulence, which promotes heat transfer. By optimizing these factors, nanofluids can substantially maximize the performance of solar thermal systems.

## 7.2 Experimental Studies and Numerical Simulations on Nanofluids in PTC Systems

### Overview of Parabolic Trough Collectors (PTCs) with Nanofluids

Experimental and computational research has shown that nanofluids significantly boost the thermal effectiveness of Parabolic Trough Collectors (PTCs). Numerous studies have indicated increased heat flow efficiency with nanofluids compared to traditional heat transfer fluids (HTFs). Nanofluids, incorporating metallic nanoparticles (e.g., copper, silver) or metal oxides (e.g., alumina, titanium dioxide), demonstrate up to a 40% improvement in thermal conductivity. These enhancements lead to better heat transfer, increasing energy output from PTC systems and improving the efficiency of solar thermal power plants.

Numerical simulations further support these findings by modeling heat transfer processes and revealing how multiple factors, like nanoparticle type and concentration, and flow conditions, affect system performance. These computational studies offer an understanding of the mechanisms of heat transfer and help identify optimal design and operational parameters, accelerating the implementation of nanofluids in solar thermal technologies.

Parabolic solar trough collectors have proven effectiveness in harnessing solar energy; they are not without challenges. Conventional HTFs often exhibit limitations that affect the efficiency of PTC systems, particularly at high operational temperatures. Recent advancements in nanotechnology, specifically the development of nanofluids—fluids engineered by suspending nanoparticles in a base liquid—hold promise for improving the thermal performance of PTCs. This overview aims to discuss the role of HTFs in PTC systems, highlight the limitations of conventional HTFs, and explore how nanofluids can improve the performance and effectiveness of PTCs.

**Role of HTFs in Absorbing and Transporting Heat in PTC Systems**

The role of heat transfer fluids (HTFs) in parabolic trough collector (PTC) systems is essential for capturing solar energy and efficiently transporting the thermal energy to the locations where it is used, such as in power generation or industrial heating applications. In a PTC system, parabolic mirrors are designed to focus and concentrate solar radiation onto a receiver tube filled with the HTF. This receiver tube is typically coated with a selective surface to enhance the absorption of solar radiation. The concentrated sunlight heats the HTF within the tube, converting the absorbed radiation into thermal energy. The effectiveness of this absorption process is closely tied to the thermal properties of the HTF, such as its thermal conductivity, viscosity, and specific heat capacity. A fluid with high thermal conductivity allows for rapid heat absorption. At the same time, low viscosity helps reduce pumping power requirements, and a high specific heat capacity ensures that the HTF can store and transfer a large amount of heat. Therefore, selecting an HTF with these desirable properties is critical to the efficiency of energy capture in a PTC system.

Once the HTF absorbs thermal energy from the concentrated sunlight, it must transport this heat to a secondary system, such as a heat exchanger, steam generator, and thermal storage system. The HTF must maintain its temperature during transport, minimizing heat losses to the environment to ensure the heat remains effective for power generation or other usage. The ability of the HTF to efficiently transfer heat from the receiver to the secondary system is crucial for optimizing the overall efficiency of the PTC system. Effective heat transfer reduces thermal losses, maintains consistent temperatures in the power generation cycle, and ensures the system operates at peak efficiency. In addition, the HTF must have good thermal stability at the elevated temperatures typically encountered in PTC systems, which usually operate within a range of 300–400 °C. If the HTF degrades at these temperatures, it can lead to reduced efficiency, increased maintenance costs, or even system failure.

The choice of HTF plays a significant role in the system's operational success. Due to its high heat capacity, water is one of the simplest HTFs used in lower temperature applications. Still, it has limitations, especially at high operating temperatures, as it boils at 100 °C and requires pressurization to avoid vaporization. Synthetic oils, such as mineral oils or synthetic esters, are commonly used in PTC systems because they can operate at higher temperatures (up to 400 °C) without degrading and are relatively easy to handle. However, they can have lower heat capacities than other fluids, which may affect overall energy storage. Molten salts, a mixture of sodium nitrate and potassium nitrate, are increasingly used in high-temperature solar thermal systems due to their excellent thermal stability at high temperatures and high heat capacity, allowing them to store and transfer large amounts of thermal energy efficiently. Molten salts can operate above 500 °C, making them ideal for large-scale power generation and thermal energy storage. Still, their use requires specialized materials and handling procedures due to their high operating temperatures and corrosive nature.

Overall, the HTF's ability to efficiently absorb, store, and transfer heat is a key factor in the overall performance of a PTC system. It impacts not only the immediate energy efficiency of the system but also its long-term operational stability and effectiveness in producing thermal power or providing industrial heat. Thus, selecting the right HTF—based on thermal properties, stability at high temperatures, and system design requirements—is essential for maximizing the performance and lifespan of PTC solar thermal systems.

**Limitations of PTCs Using Conventional HTFs**

Despite their widespread use, conventional HTFs face several limitations that can adversely affect the efficiency and performance of PTC systems:

1. **Low Thermal Conductivity**: Many traditional HTFs have relatively low thermal conductivities, restricting their ability to absorb and transfer heat efficiently. For instance, liquids like water and synthetic oils have lower thermal conductivities compared to the nanoparticles used in nanofluids. This limitation results in decreased heat transfer rates and a reduced capacity for energy capture from the concentrated solar radiation.
2. **Thermal Degradation at High Temperatures**: Conventional HTFs often suffer from thermal degradation when exposed to high operational temperatures. For example, synthetic oils can begin to decompose at temperatures above **400 °C**, resulting in diminished efficiency and potential failures in the PTC system. The degradation of the HTF can lead to increased maintenance costs and reduced reliability of solar power generation systems.
3. **Heat Losses**: As the operating temperatures of PTC systems increase, so do the effects of convective and radiative heat losses. Heat loss occurs when the receiver tube radiates heat to the surrounding environment, reducing the amount of usable thermal energy. These losses can significantly diminish the performance of PTC systems, especially during cooler periods or at night when solar energy is unavailable.
4. **High Viscosity and Pumping Energy**: Many conventional HTFs exhibit higher viscosities, leading to increased energy usage for pumping the liquid through the system. This additional energy expenditure can diminish the overall efficiency of the PTC, as more energy is used for fluid circulation rather than for useful energy generation.
5. **Environmental and Safety Issues**: Some conventional HTFs present environmental hazards if spilled or leaked. For example, synthetic oils can be toxic and flammable, raising safety concerns during operation. This risk necessitates

careful handling and monitoring to prevent accidents, adding complexity to the management of PTC systems.

## 7.3 Performance Improvements in PTCs with Enhanced HTFs, Including Nanofluid Applications

To overcome the limitations associated with conventional heat transfer fluids (HTFs), researchers have turned to nanofluids as an innovative solution to enhance the thermal efficiency of parabolic trough collector (PTC) systems. Nanofluids are engineered by suspending nanoparticles, such as metals, metal oxides, and carbon-based materials, in a base liquid, offering several key advantages. First, incorporating nanofluids significantly boosts heat transfer performance due to increased thermal conductivity. Nanoparticles made from materials like copper, silver, or aluminum oxide have higher thermal conductivities than traditional HTFs, and when dispersed throughout the fluid, they create additional pathways for heat transfer, resulting in thermal conductivity improvements of up to 40%. Second, the random movement of nanoparticles, known as Brownian motion, generates micro-convection currents within the fluid, enhancing mixing and thermal distribution. This improved fluid circulation further enhances heat transfer rates and ensures more even energy distribution between the fluid and the receiver tube's surface. Additionally, the elevated surface area-to-volume ratio of nanoparticles improves the interfacial interaction between the fluid and the receiver tube, resulting in more efficient heat absorption. The improved heat absorption efficiency allows the system to capture more thermal energy from concentrated solar radiation, leading to higher overall thermal efficiencies in PTC systems. As numerous studies have shown, the use of nanofluids in PTC systems can significantly enhance their thermal performance, leading to greater energy output and improved solar power generation capabilities.

**Hybrid Nanofluids**

In addition to single-component nanofluids, hybrid nanofluids—composed of two or more different types of nanoparticles—have gained attention to further enhance thermal performance. Hybrid nanofluids combine the distinct advantages of various nanoparticles to achieve superior heat transfer capabilities and stability. The use of multiple nanoparticles can result in synergistic effects, where the unique properties of each type complement one another. For example, combining metallic nanoparticles, such as copper or silver, with metal oxide nanoparticles, like alumina or titanium dioxide, creates fluids that exhibit high thermal conductivity and offer excellent dispersion stability. This combination takes advantage of metals' superior heat transfer properties and the improved robustness and stability of metal oxides. Hybrid nanofluids are also highly customizable, allowing researchers to select nanoparticles based on specific thermal properties, flow characteristics, and temperature ranges to optimize performance in parabolic trough collector (PTC) applications. This customizability enables fine-tuning fluid performance to suit a wide range of operational conditions. Moreover, hybrid nanofluids help overcome challenges faced by single-component nanofluids, such as sedimentation and agglomeration, by enhancing the dispersion and stability of the nanoparticles within the fluid, ultimately leading to improved performance and efficiency in PTC systems.

**Experimental and Numerical Studies on Nanofluids in PTC Systems**

A growing body of experimental and numerical research has confirmed the potential benefits of nanofluids in Parabolic Trough Collector (PTC) systems. Experimental studies have demonstrated that adding nanoparticles to conventional heat transfer fluids (HTFs) significantly enhances thermal conductivity and heat transfer efficiency. For instance, research comparing synthetic oils with nanofluids containing metallic nanoparticles has shown improvements in heat absorption and overall thermal performance, with some studies reporting efficiency increases of 10–30% in PTC systems that utilize nanofluids, resulting in higher energy output and better system performance. Additionally, advanced numerical simulations have been employed to model the behavior of nanofluids in PTC systems, allowing researchers to analyze the effects of parameters such as particle concentration, flow rates, and temperature variations on system performance. These computational models help identify the optimal conditions for maximizing the thermal performance of PTC systems and provide valuable insights into the heat transfer mechanisms in nanofluids, which can inform the design and operation of more efficient solar thermal systems. The combination of experimental validation and numerical modeling underscores the promising potential of nanofluids to improve the efficiency of solar thermal units. As research progresses, understanding how nanofluids can optimize PTC performance will continue to evolve, leading to more efficient and cost-effective solar energy solutions.

## 7.4 Hybrid Nanofluids: Definition and Composition

Hybrid nanofluids represent a breakthrough in heat transfer liquids, offering significant advantages in applications like solar thermal units, where enhancing thermal efficiency is essential for improving energy output. These fluids are created by suspending two or more types of nanoparticles, such as metals, metal oxides, and carbon-based materials, in a base liquid like water, oil, or other commonly used heat transfer fluids (HTFs). Combining nanoparticles with different properties allows hybrid nanofluids to offer superior thermal conductivity, improved heat flow rates, and better stability than traditional HTFs. One of the key advantages of hybrid nanofluids is their enhanced thermal conductivity. Hybrid nanofluids exhibit superior heat transfer properties by combining nanoparticles with varying thermal conductivities, such as metals like copper or aluminum, with metal oxides like alumina or titanium dioxide. This leads to more efficient heat exchange in systems like Parabolic Trough Collectors (PTCs), critical for solar thermal energy production. Another significant benefit of hybrid nanofluids is mitigating stability issues common in single-component nanofluids. Nanoparticles in single-component fluids often experience sedimentation and agglomeration, deleting the fluid's long-term effectiveness and requiring frequent maintenance. Hybrid nanofluids address these issues using nanoparticles with complementary properties that prevent aggregation, ensuring sustained thermal performance over extended periods. Additionally, hybrid nanofluids can be customized to meet specific thermal performance requirements. By adjusting the types, concentrations, and properties of nanoparticles, researchers can optimize the fluid's heat transfer capabilities for varying applications, such as solar thermal systems, where temperature and flow rate conditions may fluctuate. This customization makes hybrid nanofluids a versatile and efficient solution for improving solar thermal energy systems.

**Types of Nanoparticles Used in Hybrid Nanofluids**

Metal nanoparticles play a vital role in improving the thermal performance of hybrid nanofluids. Copper (Cu) is widely recognized for its excellent thermal conductivity, making it highly effective in enhancing heat transfer efficiency in systems that require efficient heat dissipation. Aluminum (Al) nanoparticles also offer superior thermal conductivity while being lightweight. This is particularly beneficial for applications where minimizing fluid weight is critical, such as in high-efficiency heat transport systems. Silver (Ag) nanoparticles improve thermal conductivity and possess antimicrobial properties, making them ideal for systems where hygiene is essential, such as in cooling applications for sensitive equipment. Metal oxide nanoparticles also provide significant advantages, with Titanium Dioxide ($TiO_2$) being valued for its thermal stability and ability to improve heat transfer in high-temperature applications. $TiO_2$ exhibits photocatalytic properties, which can be beneficial in applications requiring pollutant degradation. Aluminum Oxide ($Al_2O_3$) nanoparticles are well-known for their high thermal conductivity, excellent stability at elevated temperatures, and compatibility with various base fluids, making them a popular choice for high-temperature solar thermal systems. Carbon-based nanoparticles, such as Graphene, are gaining attention for their exceptional thermal conductivity, mechanical strength, and large surface area, significantly enhancing hybrid nanofluids' overall thermal performance. Carbon Nanotubes (CNTs) also provide high thermal conductivity and improve both heat transfer and the structural integrity of hybrid nanofluids. Finally, magnetic nanoparticles, like Iron Oxide ($Fe_2O_3$), offer the advantage of being manipulated by external magnetic fields, enabling more precise temperature management and fluid flow control, which is useful in applications requiring fine-tuned thermal regulation.

**Mechanisms of Heat Transfer Enhancement in Hybrid Nanofluids**

The enhanced thermal effectiveness of hybrid nanofluids is attributed to several key mechanisms that significantly improve heat transfer efficiency. The most prominent factor is the increased thermal conductivity resulting from the presence of nanoparticles. Nanoparticles, especially metals like copper and aluminum, and metal oxides such as aluminum oxide and titanium dioxide, possess much higher thermal conductivities than conventional heat transfer fluids (HTFs). By dispersing these nanoparticles in the base fluid, additional pathways for heat transfer are created, which results in more efficient heat conduction, boosting the overall performance of the fluid. Another critical mechanism is Brownian motion, which refers to the random motion of nanoparticles caused by thermal agitation. This motion generates micro-convection currents within the fluid that promote better mixing and heat distribution. As these currents help to spread thermal energy throughout the fluid evenly, the heat transfer process becomes more uniform, further improving the fluid's functionality and efficiency. In addition, nanoparticles' high surface area-to-volume ratio plays a pivotal role in enhancing heat transfer. Due to their small size, nanoparticles provide more contact points with the heat source, facilitating better heat exchange between the fluid and the receiver tube or heat exchange surfaces. This increased interaction between the fluid and the heat source improves thermal efficiency. The size and shape of nanoparticles also contribute to better heat transfer. Smaller nanoparticles have larger surface areas, which increase the fluid's ability to absorb and conduct heat. Moreover, nanoparticles with unique shapes, such as platelets or rods, can create fluid turbulence, further enhancing the heat transfer by disrupting the flow and promoting more effective mixing. Last, including various types of nanoparticles improves the thermal stability of the fluid by reducing issues like sedimentation and agglomeration, which helps hybrid nanofluids maintain consistent performance over time, ensuring prolonged and reliable operation in high-temperature applications.

**Experimental Studies and Numerical Simulations on Hybrid Nanofluids**

A growing body of research has concentrated on the effectiveness of hybrid nanofluids in improving thermal performance, particularly in PTC systems. Experimental studies and numerical simulations have offered essential perspectives on the heat transfer features of hybrid nanofluids under various operating conditions.

1. **Experimental Studies**

Numerous experimental investigations have been conducted to evaluate the thermal performance of hybrid nanofluids in PTC systems. These investigations have shown that the addition of nanoparticles to conventional heat transfer fluids can substantially increase thermal conductivity and heat transfer coefficients.

For example, experiments involving hybrid nanofluids containing metal-based nanoparticles, including copper and silver, have reported increases in thermal conductivity of up to 50% compared to base fluids without nanoparticles. Similarly, hybrid nanofluids incorporating nanoparticles of metal oxides, including alumina and titanium dioxide, have demonstrated enhancements in heat transfer performance by up to 40%.

These improvements in heat transfer efficiency directly translate into higher energy output from PTC systems, making hybrid nanofluids a promising strategy for increasing the efficiency of solar thermal power plants.

2. **Numerical Simulations**

Numerical simulations have played a crucial role in advancing the understanding of hybrid nanofluids' thermal performance. Computational fluid dynamics (CFD) techniques are frequently employed to model the flow behavior and thermal transfer of hybrid nanofluids in various thermal systems.

Simulations allow researchers to explore the influence of various nanoparticle concentrations, sizes, and shapes on the thermal performance of hybrid nanofluids. Results from these simulations have confirmed the experimental findings, demonstrating that the incorporation of hybrid nanoparticles substantially boosts heat transfer rates and reduces temperature gradients in thermal systems.

Moreover, numerical models can be utilized to upgrade the design and operating conditions of PTC, causing to enhance energy efficiency and overall performance.

## 7.5 Preparing Nanofluids: A Detailed Guide

Nanofluids are engineered colloidal suspensions of nanoparticles in a base fluid, designed to improve the thermal properties of the fluid. These fluids are particularly valuable in applications like heat transfer, cooling systems, and solar thermal collectors due to their enhanced heat conductivity and efficiency. Key considerations in nanofluid preparation include the selection of appropriate nanoparticles, base fluids, stabilizers, and characterization to ensure thermal performance and stability.

The one-step process involves directly synthesizing nanoparticles within the base fluid. This is done by dissolving precursor materials, such as silver nitrate, in the fluid, heating the solution, and adding reducing agents to form nanoparticles. Stabilizers like polyvinyl alcohol (PVA) or polyethylene glycol (PEG) are then added to prevent nanoparticle agglomeration. Ultrasonication ensures uniform dispersion of nanoparticles in the fluid, which is then cooled and stored.

The two-step process starts with synthesizing nanoparticles independently using chemical reduction or the sol–gel method. After synthesis, the nanoparticles are dried, ground, and dispersed in the base fluid using a mechanical or high-shear mixer. Stabilizers are added to improve fluid stability, and the mixture is sonicated to achieve a uniform suspension.

Both methods result in stable, high-performance nanofluids that enhance heat transfer efficiency. The two-step method allows for greater control over nanoparticle characteristics.

## 7.6 Classification and Key Applications of Organic, Polymeric, and Inorganic Nanoparticles

### Organic Nanoparticles

Organic nanoparticles are made from organic materials such as lipids, polymers, and other biodegradable and biocompatible substances, which make them ideal for applications in various fields, including biomedical and energy systems. These nanoparticles are non-toxic, lightweight, and versatile, making them suitable for various uses, including drug delivery, stabilizing other nanoparticles, and improving the thermal efficiency of fluids in solar systems. Some examples of organic nanoparticles include liposomes, dendrimers, and polymeric nanoparticles. These materials offer unique properties, such as ease of functionalization and adaptability, which enhance their performance in specific applications, like stabilizing nanofluids or serving as carriers for controlled drug release.

### Polymeric Nanoparticles

Polymeric nanoparticles are a subtype of organic nanoparticles, derived from synthetic or natural polymers. They are often classified into three types: polymeric nanospheres, polymeric nanocapsules, and polymeric micelles. Polymeric nanospheres are solid particles where polymers are uniformly dispersed throughout, and they can be used for controlled release in drug delivery systems. Polymeric nanocapsules, which feature a core–shell structure, can encapsulate active substances for more controlled and targeted release, making them ideal for various therapeutic applications, including phase-change materials (PCMs) for thermal energy storage. Polymeric micelles, on the other hand, are self-assembled structures that can encapsulate hydrophobic molecules inside their hydrophobic core, which are widely utilized in drug delivery for hydrophobic drugs. These structures can also serve as stabilizers in nanofluids, enhancing heat transfer performance in solar thermal systems.

### Liposomes and Dendrimers

Liposomes are spherical vesicles composed of phospholipid bilayers, which can encapsulate both hydrophilic and hydrophobic substances, making them highly versatile in drug delivery and other industrial applications. Dendrimers, which are highly branched, tree-like polymers, offer multiple functional sites for encapsulating small molecules and have a high surface area, making them ideal for drug delivery, imaging, and as stabilizers in nanofluids. These organic nanoparticles provide great flexibility in their applications due to their structure and functional properties.

### Inorganic Nanoparticles

Inorganic nanoparticles comprise non-organic materials like metals, metal oxides, and carbon. These particles, including gold, iron oxide, mesoporous silica, and carbon nanotubes, are particularly known for their high thermal conductivity and stability, making them ideal for applications in nanofluids, which are used to enhance heat transfer in solar thermal systems like Parabolic Trough Collectors (PTCs). Gold nanoparticles, for instance, are highly conductive and possess unique optical properties, making them valuable in electronics, medical diagnostics, and nanofluids. Iron oxide nanoparticles are magnetic, allowing targeted drug delivery and applications in magnetic nanofluids. Mesoporous silica nanoparticles, with their large surface area, are ideal for catalysis, drug delivery, and enhancing heat transfer efficiency in nanofluids. Carbon nanotubes, with their exceptional thermal, electrical, and mechanical properties, are widely used in nanofluids to enhance solar collectors' heat transfer performance. Quantum dots, small semiconductor nanoparticles, exhibit tunable optical properties and have applications in solar cells and fluorescent biological imaging.

These diverse nanoparticles, dispersed in heat transfer fluids, play an integral role in optimizing the thermal efficiency of PTCs by enhancing heat conduction, improving system performance, and increasing the overall efficiency of solar thermal systems. As shown in Fig. 7.1, their role in nanofluid systems is crucial for optimizing the energy output from concentrated solar power systems.

## 7.7 Characterization and Quality Control

Once the nanofluid is prepared, it is essential to carry out characterization to assess its stability and performance, ensuring that the desired thermal enhancements are achieved. Standard techniques used for characterization include Dynamic Light Scattering (DLS), which measures the size distribution of nanoparticles within the liquid, helping to detect any agglomeration that may have occurred. Zeta potential measurement is another key technique, as it evaluates the stability of the nanofluid; a higher zeta potential (typically >30 mV) indicates better stability and less agglomeration. Viscosity measurement is also crucial, as it assesses the fluid's flow characteristics and the energy required to pump the nanofluid through the system. It uses a viscometer to determine any changes in its flow properties. Thermal conductivity measurement is performed to evaluate how well the nanofluid conducts heat, using tools such as a thermal conductivity meter or a guarded hot plate method. This is critical to ensuring that the nanofluid achieves the intended enhancement in heat transfer. Finally, stability testing is conducted by storing samples of the nanofluid under various conditions, such as exposure to different temperatures and light, and monitoring them over time to evaluate any changes or degradation in performance, ensuring the long-term stability of the fluid. These characterization methods are vital for confirming that the nanofluid will perform as expected and maintain its efficiency in practical applications like solar thermal systems.

**Safety and Environmental Considerations**

When preparing nanofluids, it is essential to prioritizing safety and considering their environmental impacts. Personal protective equipment (PPE) is a key safety measure, including gloves, goggles, and lab coats, to protect individuals from exposure to potentially hazardous chemicals and nanoparticles. Additionally, proper ventilation

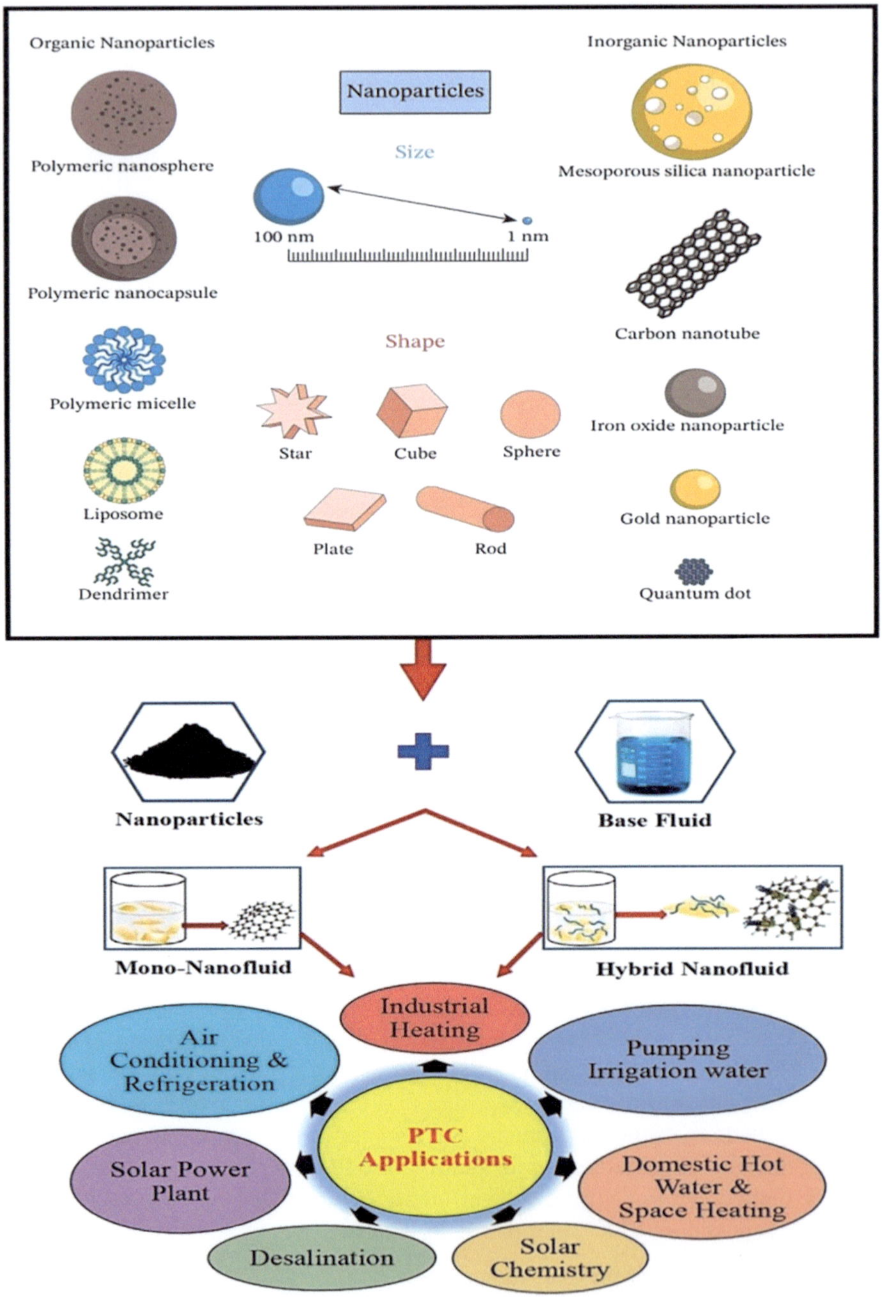


**Fig. 7.1** Schematic representation of nanoparticles in heat transfer liquid and their role in improving thermal performance in Parabolic Trough Collectors (PTCs) [1, 2]

is crucial for mitigating inhalation risks, so experiments should be conducted in a well-ventilated environment or within a fume hood. This ensures that harmful particles and vapors are not inhaled, minimizing health risks. The disposal of chemicals and nanoparticles must also be handled with care. Adhering to regional waste disposal protocols ensures that chemical waste, including nanoparticles, is managed in an environmentally responsible manner, preventing contamination of the surrounding ecosystem. The preparation of nanofluids, whether through the one-step or two-step method, involves precise planning and execution to create stable suspensions that enhance thermal properties. The process demands careful attention to the selection of base fluids, nanoparticles, and the methods used for dispersion and stabilization. Ensuring that the nanofluid remains stable over time is essential for its effectiveness in various applications. Characterization techniques, such as dynamic light scattering, zeta potential measurements, viscosity tests, and thermal conductivity evaluations, help confirm that the nanofluid performs as intended. These techniques provide insights into particle size, stability, and thermal efficiency, crucial for applications like renewable energy, particularly in solar thermal systems. With ongoing advancements in nanofluid research, the potential for more efficient energy solutions and enhanced thermal management continues to grow. As a result, nanofluids are expected to play an increasingly important role in improving the performance and sustainability of various energy systems.

## 7.8 Thermo-Physical Properties of Nanofluids (Expanded)

Nanofluids, consisting of nanoscale particles dispersed in a base liquid, offer significant advantages in thermal systems because of their enhanced thermo-physical properties. These enhancements make them particularly useful in solar energy sectors, including Parabolic Trough Collectors (PTCs), where improved heat transfer efficiency is essential for performance optimization.

**Thermal Conductivity**

**Thermal conductivity** represents a material's thermal conductivity. Nanofluids exhibit higher thermal conductivity than base fluids because of the high thermal conductivity of the nanoparticles themselves (e.g., metals, oxides, carbides). The mechanisms that explain this enhancement in thermal conductivity include the following.

i. **Nanoparticle Material**

**Metals** like copper (Cu) and silver (Ag) have larger thermal conductivity values (Cu: ~400 W/m K; Ag: ~430 W/m K), significantly boosting the overall thermal conductivity of the nanofluid when integrated into base liquids such as water or ethylene glycol.

**Metal oxides**, like alumina ($Al_2O_3$) and titanium dioxide ($TiO_2$), also contribute to enhanced thermal conductivity but to a lesser extent compared to metals.

ii. **Nanoparticle Size**

Reducing particle size increases the surface area-to-volume ratio, resulting in greater effective heat transfer between the particles and the surrounding fluid. For example, smaller nanoparticles (typically <100 nm) have higher contact surface areas, leading to more efficient energy transfer.

iii. **Particle Shape**

The shape of the nanoparticles influences the thermal conductivity. Spherical particles are commonly used due to their ease of suspension, but rod-like or tubular shapes (e.g., carbon nanotubes) offer higher enhancements due to their elongated shape, which provides more pathways for heat transfer.

iv. **Volume Fraction**

The density of nanoparticles in the base liquid (volume fraction, $\phi$) is crucial in improving thermal conductivity. As $\phi$ increases, more nanoparticles contribute to heat conduction, affecting other properties such as viscosity and flow behavior. Typically, $\phi$ ranges from 0.01 to 5% in most applications.

v. **Interfacial Layering**

Nanoparticles create an interfacial layer with the surrounding base fluid, where the fluid's properties are modified. This layer contributes to better heat transfer and increased conductivity.

vi. **Brownian Motion**

At the nanoscale, particles are in constant random motion due to thermal energy, known as **Brownian motion**. This dynamic behavior enhances energy transfer within the fluid. Brownian motion becomes especially significant at higher temperatures, further boosting the apparent thermal conductivity of the nanofluid.

**Specific Heat Capacity**

Specific heat capacity is the amount of heat required to raise the temperature of a fluid by one degree Celsius, and in nanofluids, it is influenced by the nanoparticles added to the base fluid. Nanoparticles, especially metals like copper (Cu) and silver (Ag), generally have lower specific heat capacities than fluids such as water. When nanoparticles are incorporated into the fluid, the overall specific heat capacity of the nanofluid tends to decrease. This means the fluid will store less heat per unit mass for the same temperature increase. While this reduction in specific heat may seem disadvantageous, it is compensated by the enhanced thermal conductivity of the nanofluid. The improved heat transfer properties allow for faster heat dissipation, which is crucial in applications like Parabolic Trough Collectors (PTC), where efficient and rapid heat transfer is more important than the fluid's ability to store heat. The impact of nanoparticles on specific heat also depends on their concentration in the fluid. Higher concentrations lead to a more pronounced decrease in particular heat, making it necessary to optimize the nanoparticle concentration based on the particular application. In systems like PTCs, where the goal is efficient heat transfer for solar thermal energy, balancing the reduced specific heat with enhanced thermal conductivity is essential for maximizing system performance. Therefore, understanding and adjusting nanoparticle concentration and the resulting changes in particular heat are crucial to optimizing the efficiency of nanofluid-based systems.

**Viscosity**

Viscosity is a key property that determines how easily a fluid flows, and the addition of nanoparticles to a base fluid can significantly alter its viscosity. The size and shape of the nanoparticles play an important role in how much the viscosity is affected. Smaller nanoparticles typically cause less disruption to fluid flow, leading to smaller increases in viscosity. However, non-spherical nanoparticles, such as rods or platelets, increase viscosity more due to their larger surface area and enhanced interaction with the fluid. Additionally, the concentration of nanoparticles in the base fluid is another critical factor. Higher nanoparticle concentrations lead to a more pronounced increase in viscosity. For instance, a 1% volume fraction of nanoparticles might cause a 10–20% increase in viscosity, while a 5% concentration could result in a 50–60% increase, depending on the type of base fluid and nanoparticles used. Temperature also influences the viscosity of nanofluids, with nanofluids generally exhibiting lower viscosity at higher temperatures, similar to conventional fluids. However, the impact of temperature on viscosity can be more significant in nanofluids, especially when nanoparticle interactions become more pronounced at elevated temperatures. As temperature rises, nanoparticle behavior becomes more dynamic, which can lead to greater changes in viscosity compared to traditional fluids. Understanding these factors is essential for optimizing the performance of nanofluids, especially in applications where efficient heat transfer and minimal pump power are critical, such as in solar thermal systems.

**Density and Dispersion Stability**

The density of nanofluids is influenced by both the base fluid and the density of the nanoparticles, which are typically much denser than the fluid—such as copper with a density of 8920 $kg/m^3$ compared to water's 1000 $kg/m^3$. One of the key challenges associated with nanofluids is ensuring long-term dispersion stability, as nanoparticles can agglomerate and settle over time due to gravity, leading to clogging and inconsistent thermal properties. To address this, surfactants and other stabilizing agents are often added to prevent nanoparticles from clumping together. These agents alter the surface chemistry of the particles, ensuring uniform dispersion and preventing sedimentation. Additionally, surface functionalization is another technique used to enhance the stability of nanofluids. This process involves chemically modifying the surface morphology of nanoparticles to improve their compatibility with the base fluid, enhancing both dispersal stability and thermal performance. By optimizing these strategies, nanofluids can maintain better stability and provide reliable thermal enhancement in applications such as heat transfer and solar thermal systems.

## 7.9 Models for Predicting Effective Thermal Conductivity and Viscosity

Effective Thermal Conductivity Models.

**Thermal Conductivity Enhancement**

Nanofluids exhibit increased thermal conductivity because of the superior conductivity of the nanoparticles compared to the base liquid. The improvement in thermal conductivity can be illustrated with various modeling approaches depending on the characteristics of the nanoparticles and base liquid.

Maxwell's Model (Homogeneous Spheres in a Base Fluid).

In the case of a dilute dispersion of spherical particles within a base fluid, the resultant effective thermal conductivity.

$k_{\text{eff}}$ is based on the Maxwell's equation:

$$k_{\text{eff}} = k_f \left( \frac{k_p + 2k_f + 2\phi(k_p - k_f)}{k_p + 2k_f - \phi(k_p - k_f)} \right) \tag{7.1}$$

where

$k_{\text{eff}} =$ nanofluid's effective thermal conductivity

$k_f =$ base liquid's thermal conductivity

$k_p =$ nanoparticle's thermal conductivity

$\phi =$ nanoparticle's volume fraction.

For small $\phi$ (low nanoparticle concentration), this equation simplifies and gives a reasonable approximation of thermal conductivity enhancement.

Hamilton–Crosser Model (Non-spherical Particles).

For non-spherical particles, the Hamilton–Crosser model accounts for the shape of particles using a shape factor $n$:

$$k_{\text{eff}} = k_f \left( \frac{k_p + (n-1)k_f + (n-1)\phi(k_p - k_f)}{k_p + (n-1)k_f - \phi(k_p - k_f)} \right) \tag{7.2}$$

where

$n =$ shape factor (for spheres, $n = 3$; for rods or platelets, $n > 3$).

This model is more applicable to systems where particles are not spherical, as different shapes provide more surface area for heat transfer.

Nan et al. Model (Incorporating Brownian Motion).

The **Nan et al.** model accounts for the dynamic contribution of Brownian motion to heat transfer in nanofluids. This is particularly relevant at higher temperatures:

$$k_{\text{eff}} = k_f \left( 1 + C_1\phi + C_2\frac{\phi}{d}T \right) \tag{7.3}$$

where

$d$ = diameter of the nanoparticles

$T$ = temperature of the nanofluid

$C_1, C_2$ = empirical constants.

This equation reflects that as temperature increases, the Brownian motion of the nanoparticles improves the overall rate of heat transfer.

**Specific Heat Capacity**

Specific heat capacity $C_{p,\text{eff}}$ of nanofluids is influenced by the specific heat capacities of both the base liquid as well as the nanoparticles:

$$C_{p,\text{eff}} = (1 - \phi)C_{p,f} + \phi C_{p,p} \tag{7.4}$$

where

$C_{p,\text{eff}}$ = nanofluid's effective specific heat capacity

$C_{p,f}$ = base liquid's specific heat capacity

$C_{p,p}$ = nanoparticle's specific heat capacity.

Typically, nanoparticles have a diminished specific heat capacity in comparison to the base fluid, which can reduce the overall $C_{p,\text{eff}.}$ However, this decrease can be counterbalanced by the increased thermal conductivity, which improves heat transfer efficiency.

**Viscosity of Nanofluids**

Viscosity is a pivotal factor influencing the flow dynamics and heat transfer properties of nanofluids. Various models are employed to estimate the effective viscosity of these liquids.

Einstein's Model (Dilute Suspensions).

For very low concentrations of spherical nanoparticles ($\phi \ll 1$):

$$\mu_{\text{eff}} = \mu_f(1 + 2.5\phi) \tag{7.5}$$

where

$\mu_{\text{eff}}$ = nanofluid's effective viscosity

$\mu_f$ = base liquid's viscosity

$\phi$ = nanoparticle's volume fraction.

This model works well for dilute systems but does not account for higher nanoparticle concentrations or non-spherical particles.

Brinkman's Model (Higher Concentrations).

For moderate concentrations of spherical particles:

This model extends Einstein's equation to higher particle concentrations, where the interaction between particles becomes more significant:

$$\mu_{\text{eff}} = \mu_f\left(1 + 2.5\phi + 6.2\phi^2\right) \tag{7.6}$$

Batchelor's Model (Particle–Particle Interaction).

For even higher concentrations of nanoparticles, Batchelor's model incorporates both primary and secondary interactions between nanoparticles:

$$k_{eff} = k_f\left(1 + 2.5\phi + 6.2\phi^2\right) \tag{7.7}$$

Here, the $\phi^2$ term accounts for the increased viscosity due to particle–particle interactions in concentrated nanofluids.

**Density of Nanofluids**

The density of nanofluids is based on the densities of the base liquid and nanoparticles, as well as their respective volume fractions:

$$\rho = (1 - \phi)\rho_f + \phi\rho_p \tag{7.8}$$

where

$\rho_{\text{eff}}$ = effective density of the nanofluid

$\rho_f$= density of the base fluid

$\rho_p$= density of the nanoparticles

$\phi$ = nanoparticle's volume fraction.

Nanoparticles are typically much denser than base fluids (e.g., metals vs. water), so even small additions can significantly increase the density of the fluid.

**Dispersion Stability**

Nanoparticles tend to agglomerate and settle over time due to gravitational forces, affecting dispersion stability. Stabilizing nanofluids can be enhanced by integrating surfactants or stabilizers, which prevent the nanoparticles from clumping together. The stability can also be modeled using the Derjaguin–Landau–Verwey–Overbeek (DLVO) theory, which predicts the forces between particles in a colloidal suspension based on electrostatic and van der Waals interactions:

$$F_{\text{total}} = F_{\text{electrostatic}} + F_{\text{van der Waals}} \tag{7.9}$$

where

$F_{\text{total}}$ = total force between two nanoparticles

$F_{\text{electrostatic}}$ = repulsive electrostatic force (due to surface charges)

$F_{\text{van der Waals}}$ = attractive van der Waals force (due to molecular interactions).

Stabilizers increase the repulsive $F_{\text{electrostatic}}$, reducing agglomeration and sedimentation.

In nanofluids, enhanced thermal conductivity, modified specific heat, increased viscosity, and density changes depend heavily on the size, shape, and content of nanoparticles. The theoretical models discussed above provide important tools for predicting these enhanced properties, which are essential for designing effective nanofluid systems for applications like solar thermal collectors. These models allow engineers to balance heat transfer enhancement with flow behavior to optimize performance in practical applications.

## 7.10 Influence of Nanoparticles on Thermal Performance of PTCs

**Comparison of Nanoparticle Types**

Nanoparticles in nanofluids can be categorized into metallic, ceramic, and carbon-based types, each offering distinct advantages and challenges.

i. **Metallic Nanoparticles**: Materials like copper, silver, and aluminum are noted for their excellent thermal conductivity. These nanoparticles enhance heat transfer significantly but are more prone to oxidation in high-temperature environments like PTCs, which can reduce their long-term effectiveness.
ii. **Ceramic Nanoparticles**: Ceramic substances like aluminum oxide ($Al_2O_3$), titanium dioxide ($TiO_2$), and silicon dioxide ($SiO_2$) offer exceptional thermal stability under high-temperature conditions. Although their thermal conductivity is lower than that of metallic nanoparticles, they are more resistant to corrosion and oxidation, making them ideal for long-term use in PTC systems.
iii. **Carbon-Based Nanoparticles**: Carbon nanotubes (CNTs) and graphene exhibit exceptional thermal conductivity and high surface area, making them highly effective in boosting heat transfer. However, challenges like dispersion stability and agglomeration need to be addressed to optimize their performance.

Experimental studies show that carbon-based nanoparticles generally offer the highest thermal enhancement, followed by metallic and ceramic nanoparticles. However, the optimal choice depends on the operating conditions, such as temperature range, stability requirements, and cost considerations.

**Influence of Nanoparticle Shape and Size**

The shape and size of nanoparticles significantly impact the heat transfer performance of nanofluids.

1. **Nanoparticle Shape**

**Spherical Nanoparticles**: These are commonly used due to their simple production and stable suspension in fluids. However, they possess a reduced surface area-to-volume ratio, limiting their heat transfer efficiency.

**Cylindrical Nanoparticles (e.g., CNTs)**: These particles offer a higher surface area and better contact with the fluid, leading to improved heat transfer.

**Platelet-Shaped Nanoparticles (e.g., Graphene)**: Platelets provide the highest surface area, resulting in significantly enhanced heat transfer, especially in laminar and turbulent flows.

2. **Nanoparticle Size**

Smaller nanoparticles typically offer better heat transfer performance because of their larger surface area relative to their volume. However, as size decreases, issues such as poor dispersion stability and increased viscosity can arise, especially if the nanoparticles agglomerate.

Experimental studies confirm that non-spherical nanoparticles, such as tubes or platelets, consistently outperform spherical particles in heat transfer enhancement.

**Hybrid Nanoparticles in Heat Transfer**

Hybrid nanoparticles, which combine two or more materials, offer superior heat transfer properties by overcoming the limitations of single-component nanoparticles. These hybrids combine the strengths of each material, resulting in improved thermal conductivity and stability. For example, metal–ceramic hybrids merge the high thermal conductivity of metals with the thermal stability of ceramics, providing enhanced performance in high-temperature environments. Similarly, carbon–metal hybrids integrate the exceptional conductivity of carbon-based materials, such as carbon nanotubes (CNTs) or graphene, with the stability and dispersibility of metals, leading to improved thermal conductivity and better dispersion stability. Studies have shown that hybrid nanoparticles can improve heat transfer by an additional 10–20% compared to single-material nanofluids. This enhancement is attributed to the synergistic effect of combining complementary materials, which optimizes both thermal conductivity and stability, making hybrid nanofluids a promising solution for applications requiring efficient heat transfer and long-term performance.

## 7.11 Base Liquids Used in PTC Nanofluids

In Parabolic Trough Collectors (PTCs), the selection of base liquid for nanofluids significantly impacts thermal efficiency. Base liquids not only determine the operating temperature range but also influence the liquid's heat transfer capacity, viscosity, and compatibility with nanoparticles. Below is a detailed exploration of the most commonly used base fluids—water, oils, molten salts, and ethylene glycol—followed by an analysis of trade-offs and the future scope in this field.

**Water**

Advantages: High thermal capacity and conductivity, good nanoparticle compatibility.

Limitations: Restricted to low temperatures (<100 °C), corrosion risks.

Trade-offs: Excellent heat transfer but unsuitable for high-temperature PTCs.

Future Scope: Research into corrosion-resistant nanoparticles could allow water-based fluids in higher temperature systems.

**Oils**

Advantages: Suitable for high temperatures (300 °C+), stable with many nanoparticles.

Limitations: Lower thermal conductivity, higher viscosity increase pumping costs.

Trade-offs: Good thermal stability, but lower efficiency and higher energy use.

Future Scope: Hybrid nanofluids could improve both viscosity and conductivity at high temperatures.

**Molten Salts**

Advantages: Operates at very high temperatures (>500 °C), excellent for thermal storage.

Limitations: Low thermal conductivity, complex handling due to solidification risks.

Trade-offs: High thermal stability but challenging system requirements.

Future Scope: Nanoparticle additives could boost conductivity, and hybrid salts may optimize fluidity.

**Ethylene Glycol**

Advantages: Low freezing point, good nanoparticle compatibility.

Limitations: Lower thermal conductivity, toxicity concerns.

Trade-offs: Ideal for cold climates, but lower efficiency and environmental risks.

Future Scope: Safer alternatives or improved nanofluid formulations could enhance performance.

## 7.12 Trade-Offs Between Thermal Conductivity, Viscosity, and Heat Transfer Efficiency

Each base liquid used in nanofluids comes with inherent trade-offs. High thermal conductivity fluids, like water, are efficient at heat transfer but are limited by a narrow temperature range. Conversely, high-temperature fluids like oils and molten salts offer excellent temperature stability but compromise thermal conductivity and present handling challenges due to higher viscosities. The key is balancing a fluid's ability to conduct heat with its flow characteristics, such as viscosity, and the operational requirements of Parabolic Trough Collectors (PTC) systems. For example, water offers excellent heat transfer and low viscosity but struggles at higher temperatures, while oils provide stability at high temperatures but have lower thermal conductivity and higher viscosity. Molten salts can handle high temperatures and store energy effectively, but come with handling complexity and lower thermal conductivity. Ethylene glycol works well at low temperatures with manageable viscosity but suffers from lower thermal efficiency and environmental concerns. Future research will likely focus on advanced nanoparticle additives to improve properties like thermal conductivity in oils and molten salts, hybrid nanofluids that combine different materials for better performance, and green fluids that replace harmful substances like ethylene glycol while maintaining efficiency. Furthermore, optimizing viscosity will be key to improving heat transfer in high-temperature fluids. Addressing these challenges will help develop more efficient and sustainable nanofluids, enhancing the performance of PTC systems for solar thermal applications.

## 7.13 Heat Transfer Enhancement Mechanisms in Nanofluids

Nanofluids, consisting of nanoparticle dispersions within a fluidic medium, are engineered to enhance thermal properties, and offer elevated heat transfer performance compared to conventional fluids. These enhancements are primarily driven by methods such as Brownian motion, thermophoresis, and aggregation effects.

Brownian motion refers to the random movement of nanoparticles within a fluid, caused by collisions with fluid molecules. This motion is significant in enhancing the heat transfer capabilities of nanofluids. The continuous energy exchange between the nanoparticles and the surrounding fluid molecules increases the overall thermal conductivity of the fluid. Additionally, Brownian motion induces microscale convection currents that help distribute heat more evenly throughout the fluid, which further contributes to better heat transfer efficiency. The effect of Brownian motion is more pronounced in smaller nanoparticles due to their higher surface area-to-volume ratio, which allows them to interact more effectively with the fluid. However, at higher temperatures, molecular interactions in the fluid become more dominant, which can reduce the impact of Brownian motion.

Thermophoresis, another phenomenon affecting heat transfer in nanofluids, refers to the directed migration of nanoparticles from higher temperature regions to cooler regions due to temperature gradients. This movement helps to enhance heat dissipation and stabilize the thermal flow, which is particularly useful in systems with temperature variations, such as Parabolic Trough Collectors (PTCs). Smaller nanoparticles are more responsive to these temperature differences, making thermophoresis especially effective in these systems. This process improves the overall heat transfer efficiency by ensuring a more uniform temperature distribution across the fluid.

However, nanoparticle aggregation can significantly impact the thermal performance of nanofluids. Controlled nanoparticle clustering can increase thermal conductivity by expanding the effective heat transfer surface area. On the other hand, uncontrolled aggregation can lead to larger particle clusters that block fluid flow, clog heat exchangers, and decrease heat transfer efficiency. To ensure optimal performance and stability, it is essential to preventing agglomeration, typically achieved by using surfactants or surface coatings to maintain a stable dispersion of nanoparticles in the fluid.

## 7.14 Performance Analysis of PTCs Using Nanofluids

The integration of nanofluids in Parabolic Trough Collectors (PTCs) presents a promising frontier for optimizing thermal efficiency. This section discusses the methodologies used to analyze PTC performance with nanofluids, including Computational Fluid Dynamics (CFD) models, sensitivity analysis, and real-world performance comparisons.

**Computational Fluid Dynamics (CFD) Models**

Brownian motion refers to the random movement of nanoparticles in a fluid, driven by collisions with fluid molecules, and plays a significant role in enhancing heat transfer in nanofluids by promoting energy exchange between nanoparticles and the fluid, improving thermal conductivity. This motion also induces microscale convection currents, aiding heat distribution within the fluid. Smaller nanoparticles with higher surface area-to-volume ratios experience more pronounced Brownian motion, leading to better heat transfer. However, at higher temperatures, molecular interactions become more dominant, diminishing Brownian motion's impact. Thermophoresis, the directed movement of nanoparticles from hot to cold regions due to temperature gradients, also contributes to heat transfer by enhancing heat dissipation and stabilizing thermal flow, particularly in systems with temperature variations, like Parabolic Trough Collectors (PTCs). Smaller nanoparticles are more sensitive to these temperature differences, making thermophoresis particularly effective in such systems. However, nanoparticle aggregation can affect nanofluid performance. Controlled clustering can enhance thermal conductivity by expanding the heat transfer surface area, while uncontrolled aggregation may reduce fluid flow, block heat exchangers, and decrease heat transfer efficiency. Preventing agglomeration is crucial for maintaining optimal performance, typically achieved using surfactants or surface coatings to stabilize the nanoparticle dispersion.

**Sensitivity Analysis of Nanofluid Properties**

Sensitivity analysis is a crucial process for evaluating how the thermal efficiency of Parabolic Trough Collectors (PTCs) is influenced by variations in nanofluid properties. The analysis focuses on key parameters such as nanoparticle concentration, morphology, dimensions, and the thermal properties of the base fluid, including thermal conductivity, specific heat, and viscosity. By understanding how these properties impact the heat transfer coefficient and overall system performance, sensitivity analysis plays an essential role in optimizing the use of nanofluids in PTCs. Additionally, sensitivity analysis can be performed under various operational conditions, including different flow rates, inlet temperatures, and ambient conditions, which helps identify the optimal parameters for maximum thermal efficiency. The insights gained from this analysis also reveal how robust the PTC system is to changes in both operational factors and nanofluid characteristics, providing valuable guidance for improving system performance and efficiency.

**Real-World Performance Comparisons**

Comparative studies between Parabolic Trough Collectors (PTCs) using nanofluids and those using conventional fluids provide empirical evidence of the advantages nanofluids offer in enhancing thermal performance. These experimental studies are often conducted under similar weather conditions to ensure a fair comparison, with thermal efficiency, heat transfer rates, and overall energy output being monitored over extended periods. Data from both types of systems are analyzed to evaluate the influence of nanofluids on thermal transfer dynamics, efficiency, and system stability. Key findings from these studies indicate that PTCs using nanofluids can achieve significantly higher thermal efficiencies compared to those using conventional fluids, particularly under high solar irradiance conditions. Moreover, nanofluids' ability to maintain stability and enhance heat transfer under varying environmental conditions, such as cloudy days, demonstrates their potential to improve the reliability and overall performance of solar thermal systems.

## 7.15 Advantages of Using Nanofluids in PTCs

The incorporation of nanofluids in PTCs offers several advantages that contribute to improved thermal performance and system efficiency.

i. **Increased Thermal Efficiency**

Nanofluids exhibit larger thermal conductivity than conventional fluids, leading to enhanced heat transfer rates. This increased thermal efficiency allows for more effective energy conversion from solar radiation to thermal energy, improving the overall performance of PTC systems.

ii. **Reduced Collector Size**

With enhanced heat transfer capabilities, PTCs using nanofluids can achieve the same power output with smaller collector sizes. This reduction in collector area not only lowers material costs but also minimizes space requirements, making PTC installations more feasible in urban or constrained environments.

iii. **Improved Heat Storage Capacity**

Hybrid nanofluids, which incorporate multiple types of nanoparticles, can be engineered to have tailored specific heat capacities. This property enhances the heat storage capacity of the fluid, allowing for more efficient thermal energy storage systems (TES) that can store excess heat for later use.

iv. **Better Performance Under High-Temperature Operation**

Nanofluids maintain their thermal efficiency at elevated temperatures, enabling more effective operation in high-temperature environments. This characteristic is particularly advantageous for thermal energy storage systems, which require stable performance over a range of temperatures.

v. **Enhanced Stability During Fluctuating Solar Conditions**

Nanofluids demonstrate improved thermal performance during fluctuating solar conditions, such as overcast days or transient shading. Their ability to maintain stable heat transfer rates under variable conditions enhances the reliability of solar thermal systems.

## 7.16 Disadvantages and Challenges

Despite the numerous advantages, the use of nanofluids in PTCs presents certain challenges and disadvantages that must be addressed.

i. **Higher Production Costs**

The production of nanoparticles can be expensive, particularly for high-quality, uniformly sized particles. This cost factor can limit the scalability of nanofluids for large-scale applications and solar thermal systems.

ii. **Long-Term Stability Concerns**

Maintaining the stability of nanofluids over time is a critical challenge. Issues such as sedimentation and agglomeration can lead to reduced thermal performance and system reliability. Research is ongoing to develop better stabilizers and methods for preventing these issues.

iii. **Increased Viscosity**

Nanofluids often exhibit higher viscosity than conventional liquids, which can lead to increased pumping energy requirements. This can introduce operational inefficiencies and increase energy consumption for fluid circulation.

iv. **Fouling and Clogging Risks**

The deposition of nanoparticles in heat exchanger tubes can lead to fouling and clogging, which reduces the effective heat transfer area and can significantly impair system performance. Mitigating these risks is essential for the successful implementation of nanofluids in PTCs.

v. **Environmental and Health Concerns**

The disposal of nanoparticles and their potential toxicity raises environmental and health concerns. Proper handling and disposal protocols must be established to safeguard the well-being of personnel and protect the surrounding environment. Research into eco-friendly nanoparticles and alternative materials is ongoing to address these concerns.

## 7.17 Environmental and Economic Considerations

The utilization of nanofluids in Parabolic Trough Collectors (PTCs) improves thermal performance but also introduces environmental and economic considerations that must be evaluated.

**Life Cycle Environmental Impact**

Nanofluid production, usage, and disposal have significant environmental implications at each stage. During the production phase, nanoparticle synthesis is energy-intensive, involving methods like chemical vapor deposition or milling, which contribute to greenhouse gas emissions and resource depletion. Additionally, the sourcing of raw materials, particularly for high-quality nanoparticles, can exacerbate environmental degradation. In the use phase, nanofluids enhance the thermal efficiency of PTC systems, leading to reduced energy consumption and lower carbon emissions. However, there are risks associated with nanoparticle leakage into the environment, which could threaten ecosystems. In the disposal phase, nanoparticle toxicity becomes a key concern, as improper disposal can lead to contamination of water systems, posing harm to aquatic life. To minimize these environmental risks, sustainable disposal methods or recycling processes are necessary.

**Economic Feasibility**

The economic assessment of nanofluid deployment in PTC systems involves considering both initial investment and operational costs. The initial investment is higher due to the cost of nanoparticles and necessary system modifications to accommodate the use of nanofluids. These adjustments can include changes in the design and material requirements of the system, which can increase upfront expenses. However, nanofluids can significantly lower long-term operational costs by enhancing thermal efficiency, which reduces energy consumption. Additionally, nanofluids can decrease the need for frequent maintenance and lower pumping power requirements due to their improved heat transfer capabilities. Over time, the savings from improved system performance can offset the initial investment. The return on investment (ROI) is positive when the long-term savings exceed the upfront costs, making nanofluids economically viable and justifying their use in Parabolic Trough Collector (PTC) systems for sustainable energy production.

**Advancements for Cost Reductions**

Improvements in nanoparticle manufacturing and recycling can significantly reduce costs associated with nanofluid deployment. More efficient and sustainable production methods, such as green synthesis or advanced fabrication techniques, can lower the cost of nanoparticles while simultaneously improving their quality and performance. Additionally, developing effective recycling processes for spent nanofluids can address disposal concerns, allowing for the reuse of nanoparticles. This not only mitigates environmental risks but also makes nanofluid deployment more cost-effective by reducing the need for new materials. By combining advancements in manufacturing efficiency with recycling innovations, the overall cost of using nanofluids in applications like PTC systems can be substantially reduced, enhancing both economic and environmental sustainability.

## 7.18 Study of Parabolic Trough Collector Based on Different Types of Nanofluid

Research on nanofluids in PTCs has shown significant improvements in thermal efficiency. Bellos et al. [3] found enhancements of 0.4–1.4% in thermal efficiency using Syltherm 800/Cu as a heat transfer fluid, noting a highest thermal loss increase of 7.16% for bare collector tubes. Heyhat et al. [4] demonstrated that CuO-based nanofluids, when combined with metal foam, increased friction factors significantly (50–80 times) and achieved maximum temperature differences of 16.6 °C. Fathabadi [5] reported a thermal efficiency of 76.3% using a CuO-based water thermosiphon system, outperforming other receiver types. Hatami et al. [6] identified Cu nanoparticles as the most effective for PTCs with porous tubes, enhancing heat transfer parameters. Mwesigye et al. [7] evaluated the application of single-walled carbon nanotubes (SWCNTs), achieving a 4.4% thermal efficiency increase at a 2.5% concentration. Rehan et al. [8] analyzed various nanofluids with Therminol VP-1, noting thermal parameter changes in LS-2 trough collectors. Bellos and Tzivanidis [9] tested the combined use of nanofluids and metal foams, while Valizade et al. [10] found a maximum thermal efficiency of 64.12% with SiO2/Ag–CuO nanofluids at 90 lph. Joseph et al. [11] reported enhancements in collector efficiency using CNT coatings with $Al_2O_3$ nanofluids. The studies illustrate the promising capability of nanofluids to augment the thermal efficiency of PTC systems. Figure 7.2 shows the Electron Microscope (EM) images of constructed nanofluids. Table 7.1 elaborates on the different studies based on PTC with different types of nanofluids.

The literature review demonstrated that the integration of nanofluids for heat transfer enhanced the collector's performance; however, nanoparticle agglomeration

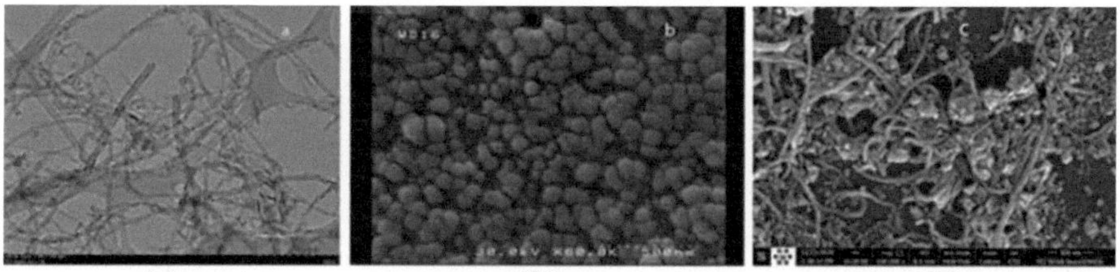

(a) (b) (c)

**Fig. 7.2** Electron Microscope (EM) images of nanofluids: **a** MWCNT/Water, showing entangled multi-walled carbon nanotube structures; **b** $Al_2O_3$/Water, illustrating spherical aluminum oxide nanoparticles; and **c** $Al_2O_3$–MWCNT/water [12]

**Table 7.1** Studies on PTC with various types of working liquids, along with modifications

| Sl. No. | Source | Type of invention | Modification | Heat transfer medium | Volume fraction (%) | Findings | Solar irradiation ($W/m^2$) |
|---|---|---|---|---|---|---|---|
| 1. | Chukami and Heyhat [13] | Numerical | Direct generation of steam | Water-based $Al_2O_3$ | (1, 1.5, and 2%) | Increased nanoparticles reduced average Nusselt numbers by 10.24% as well as 21.24% under non-uniform and uniform heat flux, respectively, showing reduced efficiency for convective heat transfer | 570–900 |
| 2. | Al-Rabeeah et al. [14] | Experimental and Numerical | – | G-$Fe_3O_4$/water | (0.01, 0.05, 0.1 and 0.2%) | G-$Fe_3O_4$ hybrid nanofluids provided thermal efficiencies ranging from 41.02 to 45.46%, surpassing water's efficiency of40.41% | 600–1200 |
| 3. | Alhamayani and Al-lehaibi [15] | Numerical | – | $Al_2O_3$–$TiO_2$ in Solar Salt, Hitec XL, and Hitec | (1.5% $Al_2O_3$ and 1.5% $TiO_2$) | Solar salt/$TiO_2$–$Al_2O_3$ demonstrated the largest average thermal efficiency of 61.8% | 1000 |
| 4. | Kaloudis et al. [16] | Numerical | – | $Al_2O_3$ in Syltherm 800 liquid oil | (0–4%) | A 10% increment in collector efficiency was found with an $Al_2O_3$ content of 4% | Monte Carlo method |
| 5. | Chakraborty and Nath [17] | Numerical | Half-star structured fin inserts | $Al_2O_3$–CuO–GO in water | (1% vol fraction) | The thermal efficiency was improved by 23.08% | Monte Carlo method |

(continued)

**Table 7.1** (continued)

| Sl. No. | Source | Type of invention | Modification | Heat transfer medium | Volume fraction (%) | Findings | Solar irradiation ($W/m^2$) |
|---|---|---|---|---|---|---|---|
| 6. | Panja et al. [18] | Numerical | Porous blocks inserts | CuO–water, $Al_2O_3$–water | (3–7%) | The largest thermal efficiency was within 77.93–79.03% | 1000 |
| 7. | Li et al. [19] | Numerical | Insertion of the vortex creator | Ionic fluid ([EMIM][$BF_4$]) | | The Nusselt number was enhanced by 77%, while the thermal efficiency improved by 2.7% | 1000 |
| 8. | Rokhforouz and Sheikholeslami [20] | Numerical | Filled with fins | $Al_2O_3$and SWCNT in RT82 | (0 and 0.025%) | The melting time reduced from 26.2 to 30.14% | 1000 |
| 9. | Yan et al. [21] | Numerical | U-structured tube | $Fe_3O_4$/CNT/$H_2O$/CMC | ($1\% < \phi < 4\%$) | At Reynolds number 5000, dual-pass pipes showed better energy efficiency than single-pass pipes, and decreasing pipe diameter improves heat transfer efficiency | Monte Carlo method |

(continued)

**Table 7.1** (continued)

| Sl. No. | Source | Type of invention | Modification | Heat transfer medium | Volume fraction (%) | Findings | Solar irradiation ($W/m^2$) |
|---|---|---|---|---|---|---|---|
| 10. | Abdulhaleem et al. [22] | Numerical | – | $Al_2O_3$–$SiO_2$/water and $Al_2O_3$–CuO/water | $Al_2O_3$ (0.1%)–CuO (0.4%) and $Al_2O_3$ (0.75%)–CuO (0.25%) | One hybrid nanofluid increased Nusselt number by 33%, while $Al_2O_3$–CuO/water showed a 47% improvement; adding turbulators boosted heat transmission up to 35.7% and cuts entropy formation by 32.8% | Uneven heat flux |
| 11. | Esmaeili et al. [23] | Numerical | Turbulator-equipped in PTC tube | $Al_2O_3$–Cu/water hybrid nanofluid | (3% alumina and 1% copper) | Turbulators elevated heat transfer rates by 35.7% and cut entropy formation by 32.8% | Uneven heat flux |
| 12. | Rawat et al. [24] | Numerical | Parabolic troughs and parallel plates were designed to effectively harness solar energy | $Al_2O_3$–Cu–CNT/water hybrid nanofluid | | Coefficient of correlation, average forecasting error rate, and MSE assessed algorithmic predictive capability | Rosseland equation for the thermal flux of radiation |

(continued)

**Table 7.1** (continued)

| Sl. No. | Source | Type of invention | Modification | Heat transfer medium | Volume fraction (%) | Findings | Solar irradiation ($W/m^2$) |
|---|---|---|---|---|---|---|---|
| 13. | Ajbar et al. [25] | Numerical | – | $Cu–Al_2O_3$/Syltherm, Cu–MgO/Syltherm, $Cu–TiO_2$/Syltherm, Ag–MgO/Syltherm, $Ag–TiO_2$/Syltherm, $Ag–Al_2O_3$/Syltherm, Ag–CuO/Syltherm, and Cu–CuO/ Syltherm, | (15 and 3%) | Hybrid nanofluids at constant volume fraction ($\phi$ = 3%) enhanced PTSC thermal efficiency similarly to the base liquid with minimal differences | 1000 |
| 14. | Tavakoli and Soufivand [26] | Numerical | Twisted tape turbulators integrated with fins | $Al_2O_3$–MWCNT/ water hybrid nanofluid | ($0\% < \phi < 3\%$) | At $\phi = 3\%$ and Re = 30,000, changing the transition from a perforated tape configuration to a finned vane turbulator design increased entropy generation by up to 40.07% | Monte Carlo radiation model |
| 15. | Tahmasebi et al. [27] | Numerical | Considering twisted tape turbulators integrated with fins | Two-phase water/ MWCNT-iron oxide nanofluid | (1%) | The magnetic hybrid nanofluid outperforms the mono nanofluid in terms of thermal performance | – |
| 16. | Yin et al. [28] | Numerical | Revolving turbulators in the PTC tube | $Cu–Fe_3O_4$/Water hybrid nanofluid | (1, 3, and 5%) | For volume fraction ($\phi$) = 1%, Prandtl number (Pr) = 0.22, and Reynolds number (Re) = 7000, a highest efficiency of 2.53% | – |

(continued)

**Table 7.1** (continued)

| Sl. No. | Source | Type of invention | Modification | Heat transfer medium | Volume fraction (%) | Findings | Solar irradiation ($W/m^2$) |
|---|---|---|---|---|---|---|---|
| 17. | Javidan et al. [29] | Numerical | Utilizing a combination of a PTC and a reception tube with fins | Hybrid nanofluid Ag–MgO Syltherm-oil-800 | (1, 3, and 5%) | When compared to competing products, the Ag–MgO Syltherm-oil-800 hybrid nanofluid generated 3% less heat waste | – |
| 18. | Vahidinia et al. [30] | Numerical | A parabolic trough collector with fins on its receiver tubes | MWCNT–$Fe_3O_4$/ Therminol VP-1 hybrid nanofluid | | The consideration of mono or hybrid nanofluids indicated no notable impact on the Levelized Cost of Energy (LCOE) across different receiver types and intake temperatures | Non-uniform |
| 19. | Esfahani et al. [31] | Numerical | Turbulator in PTC tube | SWCNT–CuO/water hybrid nanofluid | (2–6%) | PTCs with CuO–SWCNT/ water hybrid nanofluid outperform water-based liquids in thermal performance | – |
| 20. | Ghanbari et al. [32] | Numerical | Turbulator in PTC tube | MWCNT–$Fe_3O_4$-Cu/ Water magnetic hybrid nanofluid | (1.25 and 2.25%) | The magnetic hybrid nanofluid in PTC outperforms distilled water in heat transfer rate | – |

(continued)

**Table 7.1** (continued)

| Sl. No. | Source | Type of invention | Modification | Heat transfer medium | Volume fraction (%) | Findings | Solar irradiation ($W/m^2$) |
|---|---|---|---|---|---|---|---|
| 21. | Chakraborty et al. [33] | Numerical | Smooth versus Helical absorber tube | Water, $Al_2O_3$-based water | 0.09% | The maximum outlet temperature observed is 392 K for a helical tube with 95 turns at 0.15 kg/s. The heat transfer rate is improved by 71% for helical absorber tubes. Efficiency improvements: Overall efficiency: 3.5–10%; Thermal efficiency: 4–10%; Exergy efficiency: 4–5% | 700–1000 |
| 22. | Chakraborty et al. [34] | Numerical | Spherical balls and elliptical inserts in the receiver tube | $CuO–Al_2O_3$ hybrid nanofluid in distilled water | 1% | Improved thermal efficiency, Nusselt number, and thermal performance index by 13.3, 10.02, and 16%, respectively, for elliptical inserts | 300–1000 |

(continued)

**Table 7.1** (continued)

| Sl. No. | Source | Type of invention | Modification | Heat transfer medium | Volume fraction (%) | Findings | Solar irradiation ($W/m^2$) |
|---|---|---|---|---|---|---|---|
| 23. | Chakraborty et al. [35] | Numerical | Rotating receiver tubes with elliptical inserts | CuO–$Al_2O_3$ hybrid nanofluid in distilled water | 1% | Enhanced temperature at outlet, thermal efficiency, and heat transfer coefficient observed for oval inserts with a major diameter of 28 mm (in longitudinal orientation) at a receiver speed of 15 rad/s resulted in the highest enhancements in heat transfer coefficient and thermal efficiency, achieving increments of 44.21 and 29.21%, respectively | 700 |

(continued)

**Table 7.1** (continued)

| Sl. No. | Source | Type of invention | Modification | Heat transfer medium | Volume fraction (%) | Findings | Solar irradiation ($W/m^2$) |
|---|---|---|---|---|---|---|---|
| 24. | Chakraborty et al. [36] | Numerical | Smooth and helical absorber tube | Water | 0.0 | The helical absorber tube model exhibited improvements in thermal, exergy, and overall efficiencies, with increases ranging from 3.5 to 10%, 4 to 10%, and 4 to 5%, respectively<br>Maximum outlet temperature of 392 K observed for the 95-turn helical tube at 0.15 kg/s. Heat transfer rate improved by 71% compared to the smooth absorber tube model | 700 |

(continued)

**Table 7.1** (continued)

| Sl. No. | Source | Type of invention | Modification | Heat transfer medium | Volume fraction (%) | Findings | Solar irradiation ($W/m^2$) |
|---|---|---|---|---|---|---|---|
| 25. | Singh et al. [37] | Numerical | – | Mono nanofluids ($Al_2O_3$ and $TiO_2$) and hybrid nanofluids ($Al_2O_3 + TiO_2$). The base liquid is water | 1–5% | The study finds that turbulence enhances heat transfer, with hybrid nanofluids showing the highest improvement (86.88%). Energy efficiency peaks at lower temperatures and Reynolds numbers, while exergy efficiency increases with higher inlet temperature and Reynolds number. GA optimization suggests optimal conditions for thermal efficiency | 650–1000 |
| 26. | Kumar [38] | Experimental | – | CuO and $Al_2O_3$ nanofluids in water | 0.05, 0.075, and 0.1% | For CuO nanofluid at 70 L/h flow rate, efficiency increases were 28.02, 34.47, and 41.26% for 0.05, 0.075, and 0.1% concentrations, respectively. For $Al_2O_3$, efficiency increases were 14.6, 18.17, and 24.79% at the same flow rate | 950 |

(continued)

**Table 7.1** (continued)

| Sl. No. | Source | Type of invention | Modification | Heat transfer medium | Volume fraction (%) | Findings | Solar irradiation ($W/m^2$) |
|---|---|---|---|---|---|---|---|
| 27. | Mohanavel et al. [39] | Experimental | – | Ferric oxide ($Fe_3O_4$) in water | 10:100 ratio of $Fe_3O_4$ to water | The study found that using $Fe_3O_4$ nanofluid with a 10:100 mass ratio to water and a fin setup in a parabolic trough solar collector improves performance. At 0.136 kg/s flow rate, it achieved higher fluid temperature, heat generation, efficiency, and lower entropy generation | Variable |
| 28. | Gholinia et al. [40] | Numerical | DEA—dual-element annular<br>DEA-f—dual-element annular with fins<br>FEA—fin-enhanced annular<br>FEA-f—fin-enhanced annular with fins | GAGNPs/$H_2O$ (gallic acid-graphene nanoplatelets in water) | 0.025–0.1% | The study found that using turbulator models (DEA, DEA-f, FEA, FEA-f) improved heat transfer, with FEA-f showing the highest increase in Nusselt number. Increasing fin height and nanofluid concentration (0.025–0.1%) enhanced efficiency, while higher inlet temperature reduced performance | 1000 |

(continued)

**Table 7.1** (continued)

| Sl. No. | Source | Type of invention | Modification | Heat transfer medium | Volume fraction (%) | Findings | Solar irradiation ($W/m^2$) |
|---|---|---|---|---|---|---|---|
| 29. | Panja et al. [41] | Numerical | Porous strips | Copper oxide (CuO) and aluminum oxide ($Al_2O_3$) in water | 3–7% | The study found that using CuO and $Al_2O_3$ nanofluids increased the Nusselt number by up to 51.39%. Adding porous obstacles improved heat transfer by 9.86%. Maximum thermal efficiency of 62.96% was achieved with CuO nanofluid at optimal conditions | 1000 |
| 30. | Arun et al. [42] | Numerical | Dimpled tubes | $TiO_2$ nanoparticles dispersed in deionized water (DI-$H_2O$) | 0.3% | The study found that $TiO_2$ nanofluids in dimpled tubes improved the convective heat transfer coefficient by 34.25% at a 0.3% volume concentration and 2.5 kg/min flow rate. This combination enhanced thermal efficiency, showcasing nanofluids' potential for optimizing PTSC performance | 952 |

(continued)

**Table 7.1** (continued)

| Sl. No. | Source | Type of invention | Modification | Heat transfer medium | Volume fraction (%) | Findings | Solar irradiation ($W/m^2$) |
|---|---|---|---|---|---|---|---|
| 31. | Byiringiro et al. [43] | Numerical | Fined tube | MWCTN-$TiO_2$/ Syltherm800 | 1–5% | The study found that using MWCTN-$TiO_2$/ Syltherm800 hybrid nanofluid with turbulators improved Nusselt number by up to 149%, thermal efficiency by up to 5.12%, and decreased temperature gradients by up to 47.4%, significantly enhancing PTSC performance | 1000 |
| 32. | Hou et al. [44] | Numerical | Axial helical fins | Cu–Therminol-VP-1 nanofluid | 4% | The study found that replacing annular fins with axial helical fins improved thermal efficiency by 4.5%, despite increasing the friction factor by 30%. Optimal performance occurred with 10 turns of helical fins, enhancing efficiency, Nusselt number, and reducing friction factor | 1000 |

(continued)

**Table 7.1** (continued)

| Sl. No. | Source | Type of invention | Modification | Heat transfer medium | Volume fraction (%) | Findings | Solar irradiation (W/m$^2$) |
|---|---|---|---|---|---|---|---|
| 33. | Talem et al. [45] | Numerical | Receiver tube diameter variation | Dowtherm A oil-based nanofluid with Pd, Au, and NiO nanoparticles dispersed in the fluid | 0.0097 wt% of Au nanoparticles in Dowtherm A oil | The study showed that using Au-based nanofluid in a parabolic trough solar collector improved the outlet temperature by 8%, reaching 347.5 K. The optimal performance occurred at 0.0097 wt% Au nanoparticles, achieving the highest thermal energy efficiency of 50% | 650 |
| 34. | Samiezadeh et al. [46] | Numerical | | Copper (Cu) and Aluminum Oxide ($Al_2O_3$) in synthetic oil | 1–2% | Increasing the volume fraction of Cu nanoparticles by 0.02, enhances thermal efficiency by 12.2% at low Reynolds numbers but increases the friction factor by 5.2%. Cu nanoparticles are three times more effective than $Al_2O_3$. Higher permeability slightly reduces the friction factor | 800–1100 |

(continued)

**Table 7.1** (continued)

| Sl. No. | Source | Type of invention | Modification | Heat transfer medium | Volume fraction (%) | Findings | Solar irradiation (W/m$^2$) |
|---|---|---|---|---|---|---|---|
| 35. | Alnaqi et al. [47] | Numerical | Twisted tape inserts | MgO–MWCNT (80:20%) nanoparticles in thermal oil | 0–2% | The study found that using a 2% nanoparticle volume fraction of MgO–MWCNT/ thermal oil hybrid nanofluid with a 15 mm turbulator diameter optimized PTSC performance. For Re < 30,873, the LCRW model is best, while LWRC is ideal for Re > 30,873 | 725 |
| 36. | Mohammed et al. [48] | Numerical | Wavy promoters | $Fe_2O_3$–GO/Syltherm oil | 80:20% ratio of MgO to MWCNT with volume fraction 2% | Using $Fe_2O_3$–GO/Syltherm oil hybrid nanofluid with wavy promoters in the PTSC receiver tube improved thermal performance, increasing the Nusselt number by 150.4% and reducing outlet temperature by 7–31 °C. Thermal efficiency rose by 18.51%, and exergetic efficiency by 16.21% | 1000 |

(continued)

**Table 7.1** (continued)

| Sl. No. | Source | Type of invention | Modification | Heat transfer medium | Volume fraction (%) | Findings | Solar irradiation ($W/m^2$) |
|---|---|---|---|---|---|---|---|
| 37. | Ghasemi and Ranjbar [49] | Numerical | – | CuO or $Al_2O_3$ in water | 3% | The study finds that using CuO–water and $Al_2O_3$–water nanofluids in a solar parabolic trough collector enhances heat transfer compared to pure water. Increasing nanoparticle volume fraction improves the average Nusselt number, with heat transfer coefficients rising by up to 35% | 650, 750 |

(continued)

**Table 7.1** (continued)

| Sl. No. | Source | Type of invention | Modification | Heat transfer medium | Volume fraction (%) | Findings | Solar irradiation ($W/m^2$) |
|---|---|---|---|---|---|---|---|
| 38. | Kasaeian et al. [50] | Experimental and numerical | – | Carbon nanotube/oil-based nanofluids | 0.2 and 0.3% | The study found that using carbon nanotube/oil-based nanofluids (0.2 and 0.3% concentrations) as the working fluid in a solar parabolic trough collector improved thermal performance. Vacuumed copper tubes showed 11% higher efficiency compared to bare tubes | Non-uniform |
| 39. | Bellos and Tzivanidis | Numerical | – | (Cu, CuO, $Fe_2O_3$, $TiO_2$, $Al_2O_3$ and $SiO_2$) in thermal oil (Syltherm 800) | Up to 6% | The study found that Cu nanoparticles provide the highest thermal efficiency enhancement in parabolic trough collectors. Efficiency increases with lower flow rates, higher inlet temperatures, and higher nanoparticle concentrations. For typical conditions, Cu concentrations of 2, 4, and 6% enhanced efficiency by 0.31, 0.54, and 0.74%, respectively | 1000 |

within the fluid flow impeded efficiency, while the associated costs of nanofluid deployment remained substantial.

## 7.19 Future Prospects and Research Directions

The utilization of nanofluids in Parabolic Trough Collectors (PTCs) has demonstrated significant promise in improving thermal efficiency and overall system performance. However, challenges such as high production costs, long-term stability, and integration with existing systems still need to be addressed. This section outlines key prospects and research directions to optimize nanofluids for solar thermal applications, focusing on advancements in nanoparticle synthesis, hybrid nanofluids, stability and dispersion techniques, new base fluids, integration with Thermal Energy Storage (TES), and machine learning for performance prediction.

**Advancements in Nanoparticle Synthesis and Customization**

The future of nanofluids in solar thermal applications, particularly in Parabolic Trough Collectors (PTCs), is shaped by several critical research areas focused on optimizing performance, sustainability, and scalability. One of the most important aspects is the formulation of sustainable and scalable nanoparticle synthesis methods**.** As nanoparticles directly influence the thermal conductivity and performance of nanofluids, developing cost-effective, customizable nanoparticles tailored for specific operational conditions in PTC systems is crucial. Green chemistry methods for nanoparticle synthesis are gaining attention as they reduce environmental impact and minimize toxic by-products, making nanofluids more suitable for large-scale deployment.

**Hybrid Nanofluids**, which combine multiple types of nanoparticles (metallic, ceramic, and carbon-based), offer significant potential by leveraging the strengths of different substances. These hybrid fluids can optimize heat transfer properties while maintaining low viscosity and long-term stability. Identifying the optimal combinations of nanoparticles will be key to enhancing thermal performance across a wide range of temperatures without compromising fluid stability.

**Nanofluid Stability** remains a primary challenge, as nanoparticles tend to agglomerate or sediment over time, reducing performance. Research into advanced stabilization techniques, such as surface modification of nanoparticles and developing self-stabilizing nanofluids, can help improve long-term stability, preventing sedimentation and ensuring efficient operation over extended periods.

Exploring new base fluids like ionic liquids, supercritical fluids, and innovative combinations can further improve heat transfer by offering larger thermal conductivity, lower viscosity, and superior thermal stability. Integrating nanofluids with Thermal Energy Storage (TES) systems can enhance energy storage and release efficiency, especially when combined with Phase Change Materials (PCMs).

The integration of Computational Fluid Dynamics (CFD) and machine learning will accelerate the optimization of nanofluids by enabling simulations that predict performance under varying conditions. Scaling up nanofluids for commercial solar thermal plants requires addressing cost-effectiveness, real-world performance, and durability while ensuring environmental sustainability and health and safety protocols for workers handling nanomaterials.

## References


1. Tiwari AK, Kumar V, Said Z, Paliwal HK (2021) A review on the application of hybrid nanofluids for parabolic trough collector: recent progress and outlook. J Clean Product 292:126031
2. MWCNT/Water, depicting a hybrid system where $Al_2O_3$ particles interact with MWCNT networks
3. Bellos E, Tzivanidis C, Said Z (2020) A systematic parametric thermal analysis of nanofluid-based parabolic trough solar collectors. Sustain Energy Technol Assess 39:100714
4. Heyhat MM, Valizade M, Abdolahzade S, Maerefat M (2020) Thermal efficiency enhancement of direct absorption parabolic trough solar collector (DAPTSC) by using nanofluid and metal foam. Energy 192:116662
5. Fathabadi H (2020) Novel low-cost parabolic trough solar collector with TPCT heat pipe and solar tracker: performance and comparing with commercial flat-plate and evacuated tube solar collectors. Sol Energy 195:210–222
6. Hatami M, Geng J, Jing D (2018) Enhanced efficiency in concentrated parabolic solar collector (CPSC) with a porous absorber tube filled with metal nanoparticle suspension. Green Energy Environ 3(2):129–137
7. Mwesigye A, Yılmaz İH, Meyer JP (2018) Numerical analysis of the thermal and thermodynamic performance of a parabolic trough solar collector using SWCNTs-Therminol VP-1 nanofluid. Renew Energy 119:844–862
8. Rehan MA, Ali M, Sheikh NA, Khalil MS, Chaudhary GQ, Rashid T, Shehryar M (2018) Experimental performance analysis of low concentration ratio solar parabolic trough collectors with nanofluids in winter conditions. Renew Energy 118:742–751
9. Bellos E, Tzivanidis C (2018) Thermal analysis of parabolic trough collector operating with mono and hybrid nanofluid. Sustain Energy Technol Assess 26:105–115
10. Valizade M, Heyhat MM, Maerefat M (2019) Experimental comparison of optical properties of nanofluid and metal foam for using in direct absorption solar collectors. Solar Energy Mater Solar Cells 195:71–80
11. Joseph A, Sreekumar S, Kumar CS, Thomas S (2019) Optimisation of thermo-optical properties of $SiO_2$/Ag–Cu Onanofluid for direct absorption solar collectors. J Mol Liq 296:111986
12. Joseph A, Sreekumar S, Thomas S (2020) Energy and exergy analysis of $SiO_2$/Ag–CuO plasmonic nanofluid on direct absorption parabolic solar collector. Renew Energy 162:1655–1664
13. Chukami BS, Heyhat MM (2024) Numerical investigation on the effect of $Al_2O_3$-water nanofluid on direct steam generation in parabolic trough collectors. Appl Therm Eng 241:122300
14. Al-Rabeeah AY, Seres I, Farkas I (2024) Experimental and numerical investigation of parabolic trough solar collector thermal efficiency enhanced by graphene–$Fe_3O_4$/water hybrid nanofluid. Results Eng 21:101887
15. Alhamayani A, Al-lehaibi M (2024) The effect of adding hybrid nanoparticles ($Al_2O_3$–$TiO_2$) on the performance of parabolic trough solar collectors using different thermal oils and molten salts. Case Stud Therm Eng 59:104593

16. Kaloudis E, Papanicolaou E, Belessiotis V (2016) Numerical simulations of a parabolic trough solar collector with nanofluid using a two-phase model. Renew Energy 97:218–229
17. Chakraborty O, Nath S (2024) Thermal performance simulation of parabolic trough collector with ternary nanofluid flows and half-star-shaped fins inserts in different receivers. J Brazil Soc Mech Sci Eng 46:114
18. Panja SK, Das B, Mahesh V (2024) Numerical study of parabolic trough solar collector's thermo-hydraulic performance using CuO and $Al_2O_3$ nanofluids. Appl Therm Eng 248:123179
19. Li X, Yan P, Wang J, Yang T (2024) Effect of novel vortex generator on parabolic trough solar collectors using ionic liquid. Appl Therm Eng 236:121830
20. Rokhforouz M, Sheikholeslami M (2024) Numerical investigation of the process of melting and solidification within a parabolic trough solar collector using nanoparticles and fins. J Clean Product 434:140269
21. Yan SR, Golzar A, Sharifpur M, Meyer JP, Liu DH, Afrand M (2020) Effect of U-shaped absorber tube on thermal-hydraulic performance and efficiency of two-fluid parabolic solar collector containing two-phase hybrid non-Newtonian nanofluids. Int J Mech Sci 185:105832
22. Abdulhaleem SM, Abed IM, Said NM, Abed AM (2023) Numerical investigation of the effect of two-phase hybrid nanofluids on thermal-hydraulic performance and PEC of parabolic solar collectors. Eng Anal Bound Elem 157:12–20
23. Esmaeili Z, Akbarzadeh S, Rashidi S, Valipour MS (2023) Effects of hybrid nanofluids and turbulator on efficiency improvement of parabolic trough solar collectors. Eng Anal Bound Elem 148:114–125. https://doi.org/10.1016/J.ENGANABOUND.2022.12.024
24. Rawat SK, Yaseen M, Pant M, Ujarari CS, Joshi DK, Chaube S, Negi AS, Kumar M (2023) Designing soft computing algorithms to study heat transfer simulation of ternary hybrid nanofluid flow between parallel plates in a parabolic trough solar collector: case of artificial neural network and particle swarm optimization. Int Commun Heat Mass Transf 148:107011. https://doi.org/10.1016/J.ICHEATMASSTRANSFER.2023.107011
25. Ajbar W, Hernández JA, Parrales A, Torres L (2023) Thermal efficiency improvement of parabolic trough solar collector using different kinds of hybrid nanofluids. Case Stud Therm Eng 42:102759
26. Tavakoli M, Soufivand MR (2023) Investigation of entropy generation, PEC, and efficiency of parabolic solar collector containing water/$Al_2O_3$−MWCNT hybrid nanofluid in the presence of finned and perforated twisted tape turbulators using a two-phase flow scheme. Eng Anal Bound Elem 148:324–335. https://doi.org/10.1016/J.ENGANABOUND.2022.12.022
27. Tahmasebi A, Abed AM, Aghaei A, Izadi F, Malekshah EH, Aybar H (2023) Investigation into the effect of twisted tape on the thermo-hydraulic performance and entropy generation of turbulent flow of mono and hybrid magnetic nanofluids inside a parabolic solar collector absorber tube by applying two-phase analysis. Eng Anal Bound Elem 150:318–328. https://doi.org/10.1016/J.ENGANABOUND.2023.02.007
28. Yin P, Hasan YM, Bashar BS, Abdul-Zahra MM, Radhy-Al-Kubaisy MM, Majed H, Alhani I, Abood ES, Hadrawi SK, Alizadeh A, Hekmatifar M (2023) Evaluation of efficiency, thermohydraulic performance evaluation criterion, and field synergy principle improvement of the parabolic solar collector containing the hybrid nanofluid using spring turbulators. Case Stud Therm Eng 41:102571. https://doi.org/10.1016/J.CSITE.2022.102571
29. Javidan M, Gorji-Bandpy M, Al-Araji A (2023) Investigation and simulation of parabolic trough collector with the presence of hybrid nanofluid in the finned receiver tube. Theor Appl Mech Lett 13:100465. https://doi.org/10.1016/J.TAML.2023.100465
30. Vahidinia F, Khorasanizadeh H, Aghaei A (2023) Energy, exergy, economic and environmental evaluations of a finned absorber tube parabolic trough collector utilizing hybrid and mono nanofluids and comparison. Renew Energy 205:185–199. https://doi.org/10.1016/J.RENENE.2023.01.085
31. Esfahani AH, Aliehyaei M, Joshaghani AH, Najafizadeh MM (2023) Energy, exergy, economic and environmental analysis of parabolic trough collector containing hybrid nanofluid equipped with turbulator. Eng Anal Bound Elem 150:492–506. https://doi.org/10.1016/J.ENGANABOUND.2023.02.031

32. Ghanbari G, Marzban A, Yousefzadeh S (2023) Improving the thermal efficiency of parabolic trough collector equipped with combined turbulator containing two-phase magnetic hybrid nanofluid. Eng Anal Bound Elem 155:565–583. https://doi.org/10.1016/J.ENGANABOUND.2023.06.022
33. Chakraborty O, Das B, Gupta R, Debbarma S (2020) Heat transfer enhancement analysis of parabolic trough collector with straight and helical absorber tube. Therm Sci Eng Prog 20:100718
34. Chakraborty O, Das B, Gupta R (2023) Thermal performance evaluation of parabolic trough collector having different inserts and working with hybrid nanofluid. Energy Environ 35:3010–3037
35. Chakraborty O, Das B, Gupta R (2022) Performance of parabolic trough collector having rotating receiver: effect of elliptical insert and hybrid nanofluid. Int J Energy Res 46:15254–15275
36. Chakraborty O, Roy S, Das B, Gupta R (2021) Effects of helical absorber tube on the energy and exergy analysis of parabolic solar trough collector: a computational analysis. Sustain Energy Technol Assess 44:101083
37. Singh SK, Tiwari AK, Ajbar W (2025) Optimization and performance enhancement of parabolic trough collectors using hybrid nanofluids and ANN modeling. J Taiwan Instit Chem Eng 169:105984
38. Kumar A (2025) Comparative investigations on thermal performance assessment of a linear parabolic trough solar collector using various nanofluids and incident angles. Energy 317:134707
39. Mohanavel V, Logesh K, Venkatesh R, Singh PK, Hossain I, Soudagar MEM, Obaid SA, Alharbi SA, Ramachandaramurthy VK (2025) Ferric oxide nanofluid on functional properties of parabolic trough solar collector under different flow rate. Appl Therm Eng 265:125608
40. Gholinia M, Ghobadi AH, Shahcheraghi E, Armin M (2025) Thermofluids analysis of four novel anchor-shaped turbulator and eco-friendly nanofluid (GAGNPs/$H_2O$) in a parabolic trough solar collector: a CFD modeling approach. Int J Thermofluids. 25:101028
41. Panja SK, Das B, Mahesh V (2025) Performance evaluation of a novel parabolic trough solar collector with nanofluids and porous inserts. Appl Therm Eng 258:124495
42. Arun M, Barik D, Ahmad H, Alraddadi I, Shenashen MA, Khan ZA (2024) Parabolic trough solar collector technology using $TiO_2$ nanofluids with dimpled tubes. Case Stud Therm Eng 61:104840
43. Byiringiro J, Chaanaoui M, Halimi M (2024) Heat transfer enhancement of a parabolic trough solar collector using innovative receiver configurations combined with a hybrid nanofluid: CFD analysis. Renew Energy 233:121169
44. Hou G, Yadav A, Ali E, Naeem YA, Ahmed FF, Muzammil K, Falih KT, Abdulridui HA, Kadhum EH, Omran AA, Elawady A (2024) Thermal performance improvement for parabolic trough solar collectors integrated with twisted fin and nanofluid: modeling and validation. Case Stud Therm Eng 60:104763
45. Talem N, Mihoub S, Boumia L, Safa A, Navas J, Estelle P, Benayad Z (2024) Thermal performance of parabolic trough collector using oil-based metal nanofluids. Appl Therm Eng 256:124128
46. Samiezadeh S, Khodaverdian R, Doranehgard MH, Chehrmonavari H, Xiong Q (2022) CFD simulation of thermal performance of hybrid oil-Cu-$Al_2O_3$ nanofluid flowing through the porous receiver tube inside a finned parabolic trough solar collector. Sustain Energy Technol Assess 50:101888
47. Alnaqi AA, Alsarraf J, Al-Rashed AAAA (2021) Hydrothermal effects of using two twisted tape inserts in a parabolic trough solar collector filled with MgO-MWCNT/thermal oil hybrid nanofluid. Sustain Energy Technol Assess 47:101331

48. Mohammed HA, Vuthaluru HB, Liu S (2022) Thermohydraulic and thermodynamics performance of hybrid nanofluids based parabolic trough solar collector equipped with wavy promoters. Renew Energy 182:401–426
49. Ghasemi SE, Ranjbar AA (2016) Thermal performance analysis of solar parabolic trough collector using nanofluid as working fluid: a CFD modelling study. J Mol Liq 222:159–166
50. Kasaeian A, Daviran S, Azarian RD, Rashidi A (2015) Performance evaluation and nanofluid using capability study of a solar parabolic trough collector. Energy Conv Manag 89:368–375

MIX
Papier aus verantwortungsvollen Quellen
Paper from responsible sources
FSC® C105338

If you have any concerns about our products,
you can contact us on
**ProductSafety@springernature.com**

In case Publisher is established outside the EU,
the EU authorized representative is:
**Springer Nature Customer Service Center GmbH**
**Europaplatz 3, 69115 Heidelberg, Germany**

Printed by Libri Plureos GmbH
in Hamburg, Germany